CELL BIOLOGY RESEARCH PROGRESS

BILIRUBIN: CHEMISTRY, REGULATION AND DISORDER

CELL BIOLOGY RESEARCH PROGRESS

Additional books in this series can be found on Nova's website under the Series tab.

Additional E-books in this series can be found on Nova's website under the E-book tab.

HEPATOLOGY RESEARCH AND CLINICAL DEVELOPMENTS

Additional books in this series can be found on Nova's website under the Series tab.

Additional E-books in this series can be found on Nova's website under the E-book tab.

Cell Biology Research Progres

Bilirubin: Chemistry, Regulation and Disorder

Jakub F. Novotny and
Florian Sedlacek
Editors

Nova Science Publishers, Inc.
New York

Copyright © 2012 by Nova Science Publishers, Inc.

All rights reserved. No part of this book may be reproduced, stored in a retrieval system or transmitted in any form or by any means: electronic, electrostatic, magnetic, tape, mechanical photocopying, recording or otherwise without the written permission of the Publisher.

For permission to use material from this book please contact us:
Telephone 631-231-7269; Fax 631-231-8175
Web Site: http://www.novapublishers.com

NOTICE TO THE READER

The Publisher has taken reasonable care in the preparation of this book, but makes no expressed or implied warranty of any kind and assumes no responsibility for any errors or omissions. No liability is assumed for incidental or consequential damages in connection with or arising out of information contained in this book. The Publisher shall not be liable for any special, consequential, or exemplary damages resulting, in whole or in part, from the readers' use of, or reliance upon, this material. Any parts of this book based on government reports are so indicated and copyright is claimed for those parts to the extent applicable to compilations of such works.

Independent verification should be sought for any data, advice or recommendations contained in this book. In addition, no responsibility is assumed by the publisher for any injury and/or damage to persons or property arising from any methods, products, instructions, ideas or otherwise contained in this publication.

This publication is designed to provide accurate and authoritative information with regard to the subject matter covered herein. It is sold with the clear understanding that the Publisher is not engaged in rendering legal or any other professional services. If legal or any other expert assistance is required, the services of a competent person should be sought. FROM A DECLARATION OF PARTICIPANTS JOINTLY ADOPTED BY A COMMITTEE OF THE AMERICAN BAR ASSOCIATION AND A COMMITTEE OF PUBLISHERS.

Additional color graphics may be available in the e-book version of this book.

Library of Congress Cataloging-in-Publication Data

Bilirubin : chemistry, regulation, and disorder / editors, Jakub F. Novotny and Florian Sedlacek.
    p. cm.
 Includes bibliographical references and index.
 ISBN 978-1-62100-911-5 (hardcover)
1. Bilirubin. I. Novotny, Jakub F. II. Sedlacek, Florian.
 QP671.B55B555 2011
 612.1'111--dc23
                        2011039665

Published by Nova Science Publishers, Inc. † New York

Contents

Preface		vii
Chapter I	Biophysical and Technical Aspects of Phototherapy for Neonatal Hyperbilirubinemia *V. Yu. Plavskii*	1
Chapter II	Genetic Variants in Bilirubin Metabolism Pathway, Serum Bilirubin and Cardiovascular Disease *Rong Lin and Li Jin*	67
Chapter III	Molecular Simulation of Bilirubin Transport and Its pK_a Values *Rok Borštnar*	111
Chapter IV	Hyperbilirubinemia-Associated Renal Disorders *Khajohn Tiranathanagul, Asada Leelahavanichkul, Somchit Eiam-Ongand Somchai Eiam-Ong*	153
Chapter V	Serum Bilirubin and the Genetic Epidemiology of Complex Disease *Phillip E. Melton*	185
Chapter VI	Bilirubin Oxidase *Takeshi Sakurai and Kunishige Kataoka*	213
Chapter VII	Brainstem Auditory Impairment in Infants with Hyperbilirubinemia *Ze Dong Jiang*	237
Chapter VIII	Transcriptional Regulation of Human Bilirubin UDP-Glucuronosyltransferase UGT1A1 Gene and Implication of Defects in the UGT1A1 Gene Promoter *Junko Sugatani and Masao Miwa*	259
Chapter IX	Drug-Induced Cholestatic Liver Injury *Karel Urbánek, Ondřej Krystyník and Vlastimil Procházka*	279

Chapter X	Genetic Service, Counseling and Research for Hyperbilirubinemic Patients in Taiwan *Ching-Shan Huang*	**295**
Index		**305**

Preface

Bilirubin is the yellow breakdown product of normal heme catabolism. Heme is found in hemoglobin, a principle component of red blood cells. In this book, the authors present a brief review of what has already been done in the field of bilirubin research and provide some new insights into the future of the molecular simulation of bilirubin transport and studies of its chemistry. Topic discussed include biophysical and technical aspects of phototherapy for neonatal hyperbilirubinemia; genetic variants in bilirubin metabolism pathway, serum bilirubin and cardiovascular disease; molecular simulation of bilirubin transport and its pKa values and drug-induced cholestatic liver injury.

Chapter I - In the work the photophysical, photochemical and technical aspects of phototherapy for neonatal hyperbilirubinemia (jaundice) – widespread optical technology used to reduce the level of Z,Z-bilirubin IXα in the blood of newborns are studied. Therapeutic effect of light in the treatment of hyperbilirubinemia syndrome is defined by the formation of the configuration (*cis-trans*) and structural (lumirubin) photoisomers of the pigment, with a higher rate of excretion. The reasons determined the dependence of quantum yield of Z,Z-bilirubin IXα photoisomerization on the wavelength of radiation and pigment microenvironment, as well as the causes of preferential (selective) formation of one of the photoisomer (Z,E-bilirubin IXα) have been explored. Investigation of the spectral characteristics of different types of tube light sources (mercury, halogen and metal-halide lamps) used for phototherapy have been conducted. The causes of their low therapeutic efficacy and adverse side effects on the organism of the newborn have been analyzed. It is shown that both the ultraviolet and infrared components presented in the spectrum of emission of these lamps and intense visible light through photosensitized processes involving endogenous pigments (including bilirubin and its photoproduct) and medicines are able to provide a negative impact on the infant. The most promising to improve the therapeutic effectiveness of the method and to reduce the adverse side-effects is the use of quasi-monochromatic radiation sources based on the super light-emitting diodes (LED), the spectral range which corresponds to the long-wave slope of the absorption band of bilirubin for phototherapy of neonatal hyperbilirubinemia.

Chapter II - Due to the potent antioxidant property of bilirubin, increasing attention has been drawn to the potential protective effects of bilirubin against cardiovascular disease (CVD), and the polymorphisms that may play roles in influencing serum total bilirubin (TBIL) levels and CVD. This article evaluates associations between serum bilirubin, bilirubin metabolism gene polymorphisms and CVD. A series of variants in the bilirubin metabolism genes, including polymorphisms in heme oxygenase-1(*HMOX1*), uridine diphosphate glycosyltransferase 1 (*UGT1A1*) and solute carrier organic anion transporter family member 1B3 (*SLCO1B3*), were found to be associated with TBIL levels. Serum bilirubin has consistently been shown to be inversely associated with various CVD, mostly among men in different populations. However, the findings on bilirubin-related genetic polymorphisms are controversial in association with CVD. These previous association studies of bilirubin metabolism gene polymorphisms with CVD typically focus on single-locus analysis, especially on the *HMOX1* (GT)n repeat or *UGT1A1* (TA)n repeat polymorphism. In addition to the *HMOX1* (GT)n repeat and *UGT1A1* (TA)n repeat polymorphisms, other variants, which have been identified associated with TBIL levels by genome-wide association studies (GWAS) and non-GWAS studies, should be examined in the genetic association studies of CVD. And examination beyond individual variant hits, by focusing on bilirubin metabolism pathway, is also important to unleashing the true power of association studies.

Chapter III - Nowadays there is a large increase of interest in interdisciplinary sciences that have the ability to interconnect and integrate many different fields of science which have in recent years become very specific, oriented, and focused only on a certain problem without the introduction and consideration of the accompanying features, and therefore they often seem to yield unreliable results, without any applicable value. At first sight this might be true, but when we take a closer look, and in conjunction with the results of other sciences, these results can make a lot of sense, which is why we have to take them into consideration when we want to study chemistry with molecular simulations. It is also very important in the design of novel drugs, because we have to understand and to know how to design and later to synthesize a drug in a way that it will be able to reach and have favorable interaction energies with the desired target. Since there are many "borders" (membranes) a drug has to cross on its journey to the final destination, if the transport mechanism through a certain biological membrane is understood, it is then much easier to design a new drug. With molecular simulations some new insights and useful directions can be given in order to simplify the synthesis of a novel drug, consequently resulting in the lab process being much easier and more efficiently planned, and in results being much more reliable at the end. Hence we would like to present a brief review of what has already been done in the field of bilirubin research (transport, etc.) and give some new insights into the future of the molecular simulation of bilirubin transport and studies of its chemistry.

Chapter IV - Renal dysfunction in patients with extensive jaundice has been firstly described a century ago. The renal disorders after jaundice vary from tubular injury-induced electrolyte imbalances (eg. hyponatremia, hypokalemia, hypouricemia, normal gap metabolic acidosis, and Fanconi syndrome) to acute kidney injury (AKI). Hyponatremia is commonly found in more than 70% of reversible obstructive jaundice (OJ) patients. Moreover, intestinal malabsorption also causes hypocalcemia, hypophosphatemia, and hypomagnesemia. Billirubin and bile acid have been hypothesized to be the major renal toxic components in jaundice patients. From the OJ animal model studies, 2 phases of renal response have been purposed, early natriuresis (diuresis) and late renal sodium absorption phase. The natriuresis phase is the result of hepatic metabolite excretion from the kidney, instead of the liver, and/or bile acid-induced direct tubular toxicity due to detergent effect. Sulfated bile acid also neutralizes amiloride-sensitive Na^+/H^+ antiporter in the proximal tubule, leading to more distal tubular Na^+ delivery which could accelerate cortical collecting duct function and kaliuresis. Moreover, bilirubin inhibits uric acid absorption at S1 and S3 segments of the proximal tubule, causing hyperuricosuria. In the late phase of OJ (3-7 days after OJ), renal adaptation leads to more salt and water retention as measured by reduced fractional excretion of Na^+ and solute free-water whereas the urine osmolality is still persistently low. In comparison, the natriuresis phase even during sodium restriction had been reported in cholangiocarcinoma patients with severe hyperbilirubinemia (up to 40 mg/dL). Then, the liver parenchymal damage occurs at 10-20 years later and leads to hypoalbuminemia, ascites, and sodium conservation. In OJ patients, AKI develops in approximately 10-30% and contributes to more than 70% mortality. The pathophysiology of AKI after OJ, is more complicated and might consist of 1). renal macrocirculatory changes from natriuresis with negative cardiac inotropic and defective vasoactive vascular responses 2). renal microcirculatory alterations from endotoxinemia, nitric oxide, prostaglandin, or other mediators3). direct toxicity of the biliary products. Nevertheless, the pre-existing deleterious cardiovascular function and hypovolemia are the main predisposing factors to AKI. In addition, the prognosis of AKI in OJ patients seems to relate to the jaundice severity. The mortality rates of postoperative AKI in OJ are 33 and 85% in patients with serum bilirubin< 10 mg/dL and > 20 mg/dL, respectively. Thus, several modalities of extracorporeal liver support systems have been shown to reduce jaundice and biochemical parameter severity although the survival benefit has not yet been demonstrated. This article outlines the scope of various important topics and studies that contribute to the current knowledge regarding hyperbilirubinemia-associated renal disorders, including electrolyte imbalances and AKI.

Chapter V - There has been increased interest in the underlying genetic mechanisms influencing serum bilirubin due to its antioxidant properties that have been shown to be protective against the development of cardiovascular disease and certain types of neoplasms, including Hodgkin's lymphoma and endometrial cancer. Recent research has shown genetic variation between ethnic populations and these differences have important implications for pharmacogenetic development, due to the role of serum bilirubin in the removal of biochemical toxins through glucuronidation. This chapter provides a detailed review of genes previously identified that control serum bilirubin and an overview of genetic variation and epidemiology studies that demonstrate an association of these loci with complex diseases and bilirubin. These genes include uridine diphosphate glucuronosyltransferase (*UGT1A1*), heme oxygenase-1 (*HMOX1*), biliverdin reductase (*BLVR-A* and *BLVR-B*), and the solute carrier organic anion transporters (*SLCO1B1* and *SLCO1B3*). Epidemiological studies have consistently demonstrated an inverse relationship between a promoter region polymorphism in *UGT1A1* and complex diseases in European populations. However, this protective *UGT1A1* promoter association has not been demonstrated in other global populations including Africans, Asians, and American Indians, suggesting that either other variants in this gene or other genetic factors may play a role in serum bilirubin levels in these communities. Therefore, understanding this genetic variation between extant human populations has important implications given the importance of bilirubin levels in protecting against certain complex diseases and may have profound significance for drug development and therapy. Genetic epidemiology is involved in understanding the underlying genetic components that contribute to human disease. Traditionally, this field focused on the identification of Mendelian or monogenic disorders though the investigation of families or populations isolates. More recently, these studies have been expanded to examine multiple genetic factors in common complex chronic diseases such as cardiovascular disease (CVD) and cancer. This latter type of analysis has rapidly advanced due to increased sequencing technology and reduced costs allowing for large-scale studies of genetic association with common complex diseases and intermediate risk factors using large numbers of single nucleotide polymorphisms (SNPs) [1]. One of these intermediate phenotypes where there has been growing interest is in genes involved in serum bilirubin production, due to its role in glucuronidation and its antioxidant properties that have made it important potential protective risk factor for understanding the development of complex chronic diseases [2,3]. Serum bilirubin is the principal product of heme degradation and a powerful antioxidant that suppresses lipid oxidation and retards atherosclerosis formation. Historically, high levels of serum bilirubin have been viewed by clinicians as a marker of liver dysfunction [4]. While high levels of serum bilirubin are potentially toxic it is normally rendered harmless by tight binding to albumin and excretion through the liver [4]. Epidemiological research has demonstrated an inverse relationship between elevated serum bilirubin levels and the development of CVD [5-16] and some forms of cancer [17-24] due to its antioxidant properties [25]. This has led to increased interest in serum bilirubin in genetic epidemiology and pharmacogenetic studies [26-29]. This chapter reviews bilirubin genetics, its inherited disorders, genetic frequencies in different ethnic populations, and its protective components and risk factors in chronic diseases.

Chapter VI - Bilirubin oxidase is classified into multicopper oxidase together with ceruloplasmin, laccase, ascorbate oxidase etc. due to having four copper ions, a type I copper, a type II copper, and a pair of type III coopers in the active site. The former copper site functions to oxidize the substrate, and the latter two types of copper sites function to reduce the final electron acceptor, O_2 to H_2O by forming a trinuclear center. Bilirubin oxidase has been found in bacteria such as *Myrothecium verrucaria* and *Bacillus subtilis*, and has been used in the assay of bilirubin in serum and the diagnosis of jaundice because it effectively catalyzes the oxidation of bilirubin to biliverdin and further purple pigment(s). Structure, reaction mechanisms, and properties of bilirubin oxidase are reviewed in this article, followed by applications of it to clinical tests, the cathodic catalyst for biofuel cells, and formation and degradation of pigments.

Chapter VII - The neonatal auditory system is sensitive to high level of serum bilirubin, and can be damaged in neonates who suffer hyperbilirubinemia. Irrespective of etiology, elevation in the levels of unconjugated bilirubin places infants at risk for the developing bilirubin encephalopathy, including brainstem auditory impairment. Early detection of bilirubin neurotoxicity to the brain is crucial for timing treatment to reduce the risk of occurring kernicterus. Examination of functional integrity of the brainstem auditory pathway can provide important information regarding the damage of hyperbilirubinemia to the neonatal auditory brainstem and the neonatal brain in general. The brainstem auditory evoked response has been used as an important tool to study and assess bilirubin neurotoxicity to the brain, specifically the auditory system. A considerable body of research has described changes in the response in infants with hyperbilirubinemia. Most authors found some abnormalities, although others did not. Recent studies further show that neonatal hyperbilirubinemia affects brainstem auditory evoked response. The abnormalities in the response reflect functional impairment of the auditory brainstem, including impaired neural conduction and depressed electrophysiology. The degrees of these abnormalities are related to the severity of neonatal hyperbilirubinemia, though may not precisely indicate the severity. More recently, a relative new method - the maximum length sequence has been introduced to record and analyze brainstem auditory evoked response to further our understanding of functional integrity of the auditory brainstem. With this technique, recent studies demonstrated that hyperbilirubinemia does affect the functional integrity of the auditory brainstem, confirming the adverse effect of hyperbilirubinemia on the auditory brainstem. In addition, these studies revealed that neonatal hyperbilirubinemia not only damages the functional integrity of the more peripheral or caudal regions of the auditory brainstem, but also affects the more central or rostral regions. Compared with conventional response, the maximum length sequence brainstem auditory evoked response is more sensitive to bilirubin neurotoxicity to the more central regions. Therefore, this relatively new technique improves early detection of bilirubin encephalopathy, particularly for the impairment at the more central regions of the brainstem that may not be clearly shown by conventional response. An increase in BAER wave latencies and interpeak intervals and the reduction in wave amplitudes are useful indicators of bilirubin neurotoxicity to the neonatal auditory brainstem.

Chapter VIII - The human UDP-glucuronosyltransferase (UGT) 1A1 plays a critical role in the detoxification and excretion of endogenous compounds such as bilirubin, many drugs and other xenobiotics by conjugating them with glucuronic acid in the liver. Defective or reduced UGT1A1 activity causes unconjugated hyperbilirubinemia (Gilbert's syndrome and Crigler-Najjar syndrome). Based on the finding that phenobarbital treatment dramatically reduces hyperbilirubinemia in patients with inherently reduced UGT1A1 enzyme activity, we succeeded in determining the phenobarbital response enhancer module in the human UGT1A1 gene and its transcriptional factor. This chapter describes the transcriptional regulation of human UGT1A1 gene through distal and proximal promoter and nuclear receptors. A 290-bp distal enhancer module at -3499/-3210 of *UGT1A1* gene fully accounts for constitutive androstane receptor (CAR)-, pregnane X receptor (PXR)-, aryl hydrocarbon receptor (AhR)-, glucocorticoid receptor (GR)-, peroxisome proliferator-activated receptor α (PPARα)- and NF-E2-related factor 2 (Nrf2)-mediated activation of the *UGT1A1* gene. In addition, hepatocyte nuclear factor 1α (HNF1α) bound to the proximal promoter motif not only enhances the basal reporter activity of *UGT1A1*, but also influences the transcriptional regulation of *UGT1A1* by these nuclear factors. Moreover, the TA repeat polymorphism and T-3279G mutation of the 290-bp distal enhancer module affects the transcriptional activation of *UGT1A1* by the nuclear receptors. Thus, activation by specific inducers of the *UGT1A1* gene is offered an excellent clinical trial in treating patients with unconjugated hyperbilirubinemia and in preventing side effects of drug treatment such as SN-38-induced toxicity. Therefore, we screened some UGT1A1 inducers in foods and plants and found that dietary flavones were able to induce the UGT1A1 gene expression. Taken together, these results indicate that even in subjects with inherited deficiency of the *UGT1A1* gene, it would be possible to prevent drug side effects by increasing expression of the enzyme. Screening dietary compounds affecting the expression of UGT1A1 leads to protection against drug side effects, achieving more advanced therapeutic effects by smaller dosages of drugs, and reducing the medical cost.

Chapter IX - The review deals with the probable etiology, diagnostics, classification, most likely causative drugs, risk factors and disease course of drug-induced cholestasis. Cholestatic and mixed forms of drug-induced liver injury (DILI) account for nearly half of all reported cases. Medications are probably responsible for 2 – 5 % of cases of jaundice requiring hospital admission; moreover, all forms of DILI are currently the most common adverse drug reaction resulting in withdrawal of new drugs from clinical research. Cholestatic syndromes caused by drugs can be divided into acute (bland cholestasis, cholestatic hepatitis and cholangiolitis) and, less frequent, chronic (vanishing bile duct syndrome and extrahepatic biliary obstruction). The etiology seems to be mostly idiosyncratic, with a supposed genetic predisposition. Bile salt export pump (BSEP) is known to be subject to drug inhibition in susceptible patients. Besides rare mutations that have been linked to drug-induced cholestasis, the common p.V444A polymorphism of BSEP, DRB1*1501 HLA class II haplotype and homozygosity for GSTM1 null and/or GSTT1 null alleles have been identified as a potential risk factors. No specific tests are available for establishing drug etiology of cholestasis; therefore causality assessment is performed in the same way as in other adverse drug reactions. Several structured causality assessment methods for DILI have also been proposed, e.g. RUCAM, CIOMS score etc. Drugs known to cause cholestatic syndromes include various antibiotics, oral contraceptives, oral antidiabetics and numerous other drugs including herbal medicines.

Chapter X - Neonatal hyperbilirubinemia was a serious problem in Taiwan before 1980 and glucose-6-phosphate dehydrogenase (G6PD) deficiency was found being an important factor for the development of such a disease. For example, among male newborn infants suffering from neonatal hyperbilirubinemia, bilirubin concentration; number of bilirubin concentration ≥342 µM (20.0 mg/dL) and mortality rate were significantly higher in G6PD-deficient subjects than in G6PD-normal analogs. The results of further analysis showed that, in G6PD-deficient newborn infants, contact of oxidant chemical or drugs; such as naphthalene and some herbals; was the main cause of severe neonatal hyperbilirubinemia. Therefore, the Cathay General Hospital (CGH) in Taipei established a quantitative G6PD screening test for every neonate born at the CGH in 1981. The incidence of G6PD deficiency detected by this quantitative method was approximate 3%. At the CGH, the medical staffs were aware of any G6PD deficiency within 3 days of birth and health education was provided to parents before the infants' discharge from the hospital. Such a service resulted in that severity of neonatal hyperbilirubinemia caused by G6PD deficiency had improved. The Department of Health, Taiwan (DOH) has decided G6PD being one item of nationwide neonatal screening tests since 1987. At present, dry blood spots of neonates are sent to three neonatal screening institutions, which were certified by the DOH, and the turn around time of G6PD report is restrict less than 10 days of birth in order to health education being given as earlier as possible. In Taiwan, study of molecular biology for G6PD was started in 1992. Up to now, it is known that at least 15 different types of single-point mutations responsible for G6PD deficiency among Taiwanese. Among the 15 types, the single-point mutation at nucleotide (nt) 1376 (G>T, Arg459Leu) is predominant, accounts for about 50% of G6PD deficiency in Taiwanese. Further studies of DNA-based conformation for G6PD deficiency offered useful medical information for Taiwanese. Although the number of Taiwanese neonates suffering from severe hyperbilirubinemia was reduced after nationwide-neonatal G6PD screening test being performed, incidence of neonatal hyperbilirubinemia was found to be 7.8% and 15.6% in G6PD-normal and G6PD-deficienct neonates, respectively, with 14.6% of the neonates with hyperbilirubinemia having peak bilirubin levels ≥342 µM. It was concerned that other genetic issues may be associated with hyperbilirubinemia in Taiwanese. Serum bilirubin levels are dependent on both bilirubin production and elimination. The relationship between G6PD deficiency and hyperbilirubinemia may be attributable to that life span of G6PD-deficienct erythrocytes is shorter than G6PD-normal erythrocytes and thus more amount of bilirubin is produced by G6PD-deficienct subjects. Uridine-diphospho-glucuronosyl transferase (UGT) 1A1 had been known being the sole enzyme responsible for glucuronidation for bilirubin. After glucuronidation, bilirubin is more water-soluble and feasible for elimination. However, the variation status of UGT gene was never studied for Taiwanese until 2000. The first article concerning UGT 1A1 gene in Taiwanese indicated that among the four variant sites found within the coding region, 211 G>A (Gly71Arg) was the predominate one, 1091 C>T (Pro364Leu) was a novel variation and 686 C>A (Pro229Gln) was associated with A(TA)$_6$TAA>A(TA)$_7$TAA at nt -53 in the promoter area. Thereafter, further studies revealed that variation in UGT1A1 gene was an important risk factor for the development of hyperbilirubinemia in Taiwanese. In conclusion, G6PD deficiency and variation of UGT1A1 gene are the main genetic issues associated with hyperbilirubinemia in Taiwanese. However, in Taiwanese, mutation type of G6PD and variation status of UGT1A1 gene are different from those in other ethnic groups. Understandings of such genetic problems are useful for health counseling for our citizens.

In: Bilirubin: Chemistry, Regulation and Disorder
Editors: J. F. Novotny and F. Sedlacek
ISBN: 978-1-62100-911-5
© 2012 Nova Science Publishers, Inc.

Chapter I

Biophysical and Technical Aspects of Phototherapy for Neonatal Hyperbilirubinemia

V. Yu. Plavskii

B. I. Stepanov Institute of Physics,
National Academy of Sciences of Belarus, Nezavisimosti Ave.,
Minsk, Belarus

Abstract

In the work the photophysical, photochemical and technical aspects of phototherapy for neonatal hyperbilirubinemia (jaundice) – widespread optical technology used to reduce the level of Z,Z-bilirubin IXα in the blood of newborns are studied. Therapeutic effect of light in the treatment of hyperbilirubinemia syndrome is defined by the formation of the configuration (*cis-trans*) and structural (lumirubin) photoisomers of the pigment, with a higher rate of excretion. The reasons determined the dependence of quantum yield of Z,Z-bilirubin IXα photoisomerization on the wavelength of radiation and pigment microenvironment, as well as the causes of preferential (selective) formation of one of the photoisomer (Z,E-bilirubin IXα) have been explored. Investigation of the spectral characteristics of different types of tube light sources (mercury, halogen and metal-halide lamps) used for phototherapy have been conducted. The causes of their low therapeutic efficacy and adverse side effects on the organism of the newborn have been analyzed. It is shown that both the ultraviolet and infrared components presented in the spectrum of emission of these lamps and intense visible light through photosensitized processes involving endogenous pigments (including bilirubin and its photoproduct) and medicines are able to provide a negative impact on the infant.

The most promising to improve the therapeutic effectiveness of the method and to reduce the adverse side-effects is the use of quasi-monochromatic radiation sources based on the super light-emitting diodes (LED), the spectral range which corresponds to the long-wave slope of the absorption band of bilirubin for phototherapy of neonatal hyperbilirubinemia.

Introduction:
Phototherapy an Alternative to Operating Methods of Treatment for Neonatal Jaundice

The phototherapy of hyperbilirubinemia (jaundice) of newborns is one of the clearest examples of the effective use of optical technologies in medicine. According to the data of various authors, the jaundice syndrome is observed in 50–60% of full-term newborns [1–3] and about 80% of premature infants [4] and is most pronounced by day 3–4 of their life. The indicated noninfectious disease is caused by the excess accumulation in the blood, as well as in subcutaneouous fat cells, of bile pigment (a product of the exchange of hemoglobin) – Z,Z-bilirubin IXα – which imparts a characteristic golden yellow color to the skin. For most newborns who manifest the attributes of hyperbilirubinemia, as the operation of the system for excreting bilirubin improves and the biochemical systems of the organism normalize, the jaundice disappears in 1–2 weeks without causing any harm to the child. However, in 6–10% of the infants [1, 5] the jaundice proceeds to a serious form as a consequence of a high bilirubin level and requires intensive therapy. If emergency measures are not taken, the presence of a high concentration of the indicated toxic pigment in the infant's organism can have an effect on his physical and neuropsychic development, as well as being a direct cause of death.

The main methods of treatment aimed at reducing the level of bilirubin in the blood of newborns is phototherapy and exchange transfusion [6, 7]. The standard treatment for neonatal jaundice is phototherapy. And the tactic of therapeutic interventions is usually aimed at achieving a favorable clinical outcome with the help of phototherapy, without resorting to the methods of interventional therapy. Exchange transfusion was the first successful therapy for severe neonatal jaundice [6, 7]. This technique rapidly eliminates bilirubin from the circulation. Circulating antibodies that target the erythrocytes are also removed. Exchange transfusion is especially beneficial in infants who have ongoing hemolysis from any cause. One or two central catheters are placed, and small aliquots of blood are removed from the infant and replaced with similar aliquots of red cells from a donor, mixed with plasma. This procedure is repeated until twice the blood volume has been replaced. During the procedure, serum electrolytes and bilirubin should be measured periodically. The amount of bilirubin removed from the circulation varies according to both the amount of bilirubin stored in tissues that reenters the circulation and the rate of hemolysis. In some cases, the procedure needs to be repeated to lower the serum bilirubin concentration sufficiently.

Typically, invasive methods are only applicable if using phototherapy fails to reduce the concentration of bilirubin in the blood of newborn babies to a safe level [6, 7]. It should be noted that the operations of exchange transfusion of large amounts of blood (to an infant from an adult donor) are heavily beared by the babies, and in some cases exacerbate intoxication of child's body. Mortality of children under this procedure ranges from 0.3% [6] to 5.2% [8]. Thus, one of the main objectives of phototherapy of neonatal hyperbilirubinemia of newborns is the reduction to a safe level of bilirubin in the blood, without resorting to the methods of interventional therapy, potentially causing organism infection and infant mortality.

The principal and most widely used method of treating hyperbilirubinemia of newborns is phototherapy [6–22] consisting of the total action on the child's body surface of light with a power density of $P = 0.3$–2 mW/cm^2, whose spectral composition corresponds to the long-wavelength absorption band of bilirubin ($\lambda = 400$–530 nm). The method began to be widely used in medical practice in the 1970s [23–26], and, according to the data of the American Academy of Pediatrics, more than a million infants throughout the world have by now been treated [6]. It is assumed [27–33] that the determining role in lowering the bilirubin level in the organism of newborns who undergo phototherapy is mainly played by the processes of photoisomerization of the pigment – the formation of its configurational isomers (*Z,E*-bilirubin IXα, *E,Z*-bilirubin IXα, and *E,E*-bilirubin IXα) and structural isomers (*Z*-lumirubin and *E*-lumirubin) (see Figure 1). The indicated isomers, and above all lumirubin, being more hydrophilic compounds than native *Z,Z*-bilirubin IXα, are characterized by an elevated excretion rate [27–33]. Besides the process of isomerization of *Z,Z*-bilirubin IXα when exposed to light, oxidation reactions also occur (Figure 2), accompanied by accumulation of colorless products based on monopyrrole and dipyrrole moieties and also biliverdin, formed as a result of oxidation of the methylene bridge and characterized by an extended absorption band with a maximum at about $\lambda = 660$–750 nm (depending on the type of solvent) [34, 35].

According to [34–37], one of the mechanisms for photodegradation of bilirubin involves participation of singlet oxygen in this process (self-sensitized oxidation).Publications evidence that using phototherapy for treatment of neonatal hyperbilirubinemia often leads to a marked therapeutic effect [6–22]. However, phototherapy devices based on vacuum-tube light sources in some cases, particularly in the treatment of hemolytic disease of the newborn, do not provide a reduction of bilirubin levels to safe concentrations. Thus, according to [4] 6% of 672 infants, that had been treated with phototherapy, required surgical intervention.

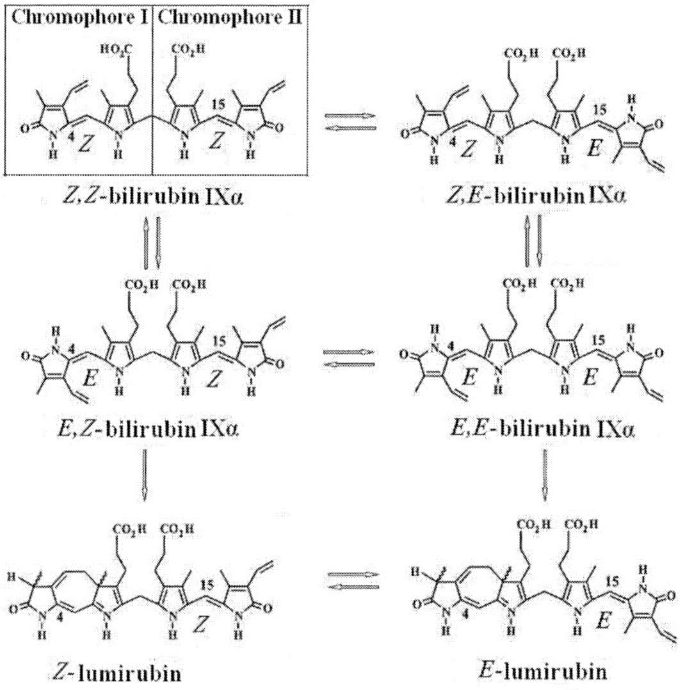

Figure 1. Scheme for configurational (*cis-trans*) and structural photoisomerization of *Z,Z*-bilirubin IXα.

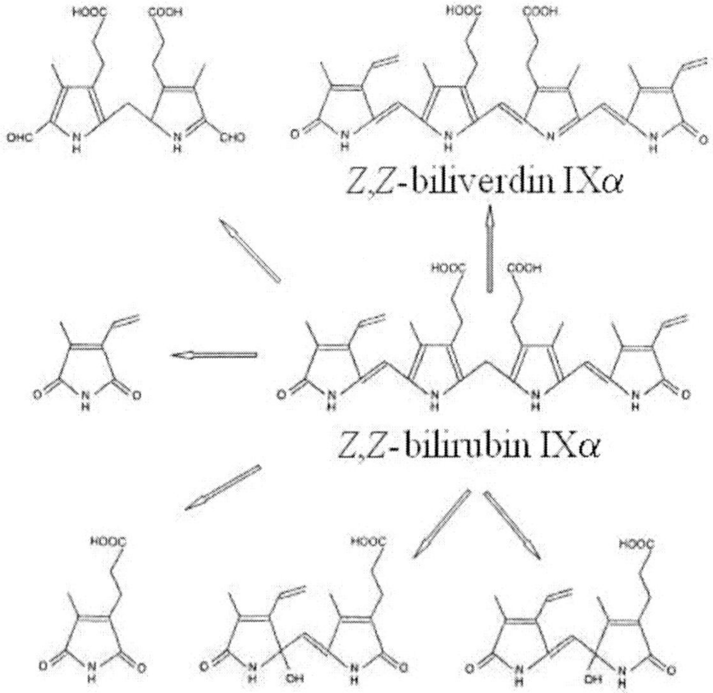

Figure 2. Structural formulas for photodegradation products of Z,Z-bilirubin IXα.

For infants with hemolytic disease, the use of modern phototherapy does not prevent the need of exchange transfusion in 17% of cases [10].

Thus, infants hyperbilirubinemiya attracts the attention of specialists, not only because of its high incidence among newborns, but also because the hyperproduction of bilirubin may be dangerous to a child, causing central nervous system damage and other equally dangerous complications and consequences.

For this reason the development of new high technologies and equipment for phototherapy of neonatal hyperbilirubinemia children is an urgent task.

However, it should be noted that progress in this direction can not be achieved without a detailed study of the mechanism of photophysical and photochemical processes that underlie the above method of phototherapy. One of the indicators pointing to the lack of knowledge in processes of bilirubin photoconversion by its excitement is a great variety of phototherapy devices used to treat hyperbilirubinemia, characterized by the type of light source spectral range of radiation power density of luminous flux, but in fact they can hardly be characterized with good results in phototherapy.

1. Materials and Methods

Chemicals and Solutions. Z,Z-bilirubin IXα (Fluka, Germany) purified and crystallized as described by McDonagh and Assisi [38]. The spectral luminescence and polarization studies of Z,Z-bilirubin IXα were carried out using chloroform (Fluka) and isobutyl alcohol (Aldrich) as the solvents and also in buffered aqueous solutions within the complex with HSA (Sigma, USA, fraction V, degree of purification about 99%) and BSA (Sigma, degree of purification at least 96%).

Because of the low solubility of bilirubin in water with physiological values of pH (about 7 nM, [39]), it was predissolved in 0.05 M NaOH. Then 8–10 μL of the indicated bilirubin solution was introduced into 10 mL of a 0.1-M phosphate buffer solution of serum albumin (HSA or BSA) with a concentration of $C_{HSA} = C_{BSA} = 4$ μM with pH 7.4. Control measurements showed that adding micro amounts of base in the indicated ratio of the volumes to be mixed has practically no effect on the pH of the buffer. The final concentration of bilirubin reached $C_{Br} = 2$ μM. When choosing the concentrations of protein and pigment ($C_{HSA}/C_{Br} = 2$), we started from the fact that the HSA molecule has two different bilirubin binding sites, with binding constants (association constants) $K_{as} = 5.5 \cdot 10^7$ M^{-1} and $K_{as} = 4.4 \cdot 10^6$ M^{-1} [40], while the BSA molecule has one site, with $K_{as} = 5.5 \cdot 10^6$ M^{-1} [41].

With the concentrations of the components indicated above, the bilirubin molecules are localized at only one coupling center (characterized by a higher association constant). Moreover, with $C_{Br} = 2$ μM, the optical density A of the solution at the maximum of the long-wavelength absorption band of bilirubin in standard cells with a layer thickness of $l = 10$ mm equals about $A \approx 0.1$. This eliminated the possibility of distorting the fluorescence spectra of the solutions because of the "internal-filter" effect (screening of the radiation at the excitation and fluorescence wavelengths).

Extinction Coefficients. When determining the concentrations of the compounds of interest, we started from the following molar extinction coefficients at the maximum of their long-wavelength band: $\varepsilon_{280} = 3.55 \cdot 10^4$ M^{-1} cm^{-1} for HSA [42], $\varepsilon_{280} = 4.33 \cdot 10^4$ M^{-1} cm^{-1} for BSA [42], $\varepsilon_{453} = 5.85 \cdot 10^4$ M$^{-1} \cdot$cm^{-1} for bilirubin in chloroform (according to the manufacturer's data), while bilirubin as part of a complex with HSA or BSA has, respectively, $\varepsilon_{458} = 4.70$ M^{-1} cm^{-1} [43] and $\varepsilon_{462} = 5.80 \cdot 10^4$ M^{-1} cm^{-1} [44].

Absorption Spectra. The absorption spectra were recorded on a Specord M40 UV–VIS spectrometer (Carl Zeiss, Jena, Germany).

Fluorescence Spectra and Fluorescence Polarization Spectra. The corrected fluorescence spectra and fluorescence-excitation spectra, as well as the fluorescence-excitation polarization spectra and the fluorescence polarization spectra, were measured on an SDL-2 high-sensitivity automatic spectrofluorimeter [45], consisting of a fast excitation MDR-12 monochromator and an MDR-23 recording monochromator. The excitation and recording optical axes made an angle of 90°. A correction factor was automatically introduced when the polarization spectra were recorded to allow for the difference of the fluorimeter photosensitivity for polarized radiation with vertical and horizontal orientation of the electric vector, the value of which varies as the fluorescence-recording wavelength varies [46].

When recording the polarization spectra, the scanning step was 2 nm. The degree of polarization of the fluorescence was determined from the relation $p = (I_{vv} - kI_{vh})/(I_{vv} + kI_{vh})$, where $k = I_{vh}/I_{vh}$ [sic] is a correction factor characterizing the difference between the fluorimeter sensitivity for polarized radiation with vertical (v) and horizontal (h) orientations of the electric vector; I_{vv}, I_{vh}, I_{vh}, I_{hh} [sic] are the components of the radiation intensity for different orientations of the electric vector for linearly polarized light, where the first subscript describes the orientation of the electric vector for the exciting radiation and the second subscript describes the orientation of the electric vector for the fluorescence radiation.

When carrying out the polarization measurements, we assumed that the factor k varies as the fluorescence detection wavelength varies.

The spectral slit width of the excitation and fluorescence monochromators was 4.2 nm. Rather narrow slits were chosen for the excitation monochromator especially because of the low photochemical stability of the pigment, which for intense excitation may lead to the appearance of photoisomerization products in solutions of the pigment with spectral luminescence characteristics close to those of Z,Z-bilirubin IXα [47–50]. In order to eliminate possible artifacts due to accumulation of photoproducts, each solution in the cuvet was used to record only one spectrum and all the manipulations with the solutions were carried out under low-light conditions with dim red illumination..

Temperature of the Solution. In studying the spectral characteristics of buffered bilirubin-HSA, bilirubin-BSA solutions and also bilirubin solutions in chloroform, we used a thermostatted cuvet compartment maintaining the temperature at $T = 20°C \pm 0.2°C$; the bilirubin solutions in isobutanol were studied at $T = -100°C \pm 2°C$. For the low-temperature studies, we used a special cryostat providing the required temperature conditions by varying liquid nitrogen pumping rate. The temperature of the solution was monitored using a calibrated thermocouple.

Time-Resolved Fluorescence Techniques. The fluorescence intensity decay kinetics for bilirubin solutions in isobutanol at $T = -100°C \pm 2°C$ were recorded on a PRA-3000 (Canada) pulsed spectrofluorimeter operating in single photon counting mode. The excitation wavelength (λ_{ex} = 460 nm) corresponded to the maximum of the bilirubin absorption spectrum in this solvent.

Sensitizing Effect of Z,Z-Bilirubin IXα and its Photoproducts on Enzyme. In this paper we also investigated the sensitizing effect of Z,Z-bilirubin IXα and its photoproducts on the enzyme lactate dehydrogenase (LDH) in model buffer solutions containing LDH and the HSA-bound pigment over a broad range of bilirubin/albumin concentration ratios, and also for different wavelengths of the radiation used. We used the bilirubin complex with albumin as the photosensitizer rather than free bilirubin because in the blood, bilirubin is mainly found in the bound state. Furthermore, in the free state in the neutral pH range, it is extremely unstable and rapidly undergoes chemical reactions. We needed to vary the pigment/protein concentration ratio and the wavelengths of the radiation used because the amount by which the bilirubin concentration exceeds the binding capacity of albumin determines the severity of the hyperbilirubinemia, and phototherapy is considerably more effective when using radiation corresponding to the long-wavelength edge of the bilirubin absorption band than when the spectral range of the radiation is located in the region of the maximum of the bilirubin absorption spectrum [18, 51–56].

The enzyme molecules were selected as a test compound for testing the photosensitizing effect of bilirubin and its photoproducts because of the determining role of this type of macromolecule in regulation of metabolic processes in the body and their ability to act as the most sensitive target for photodynamic reactions occurring both according to a radical mechanism and with participation of singlet oxygen.

The contribution of the above-indicated reactions to the degradative effect of light under conditions of photosensitization by bilirubin (or its photoproducts) was estimated using as an example LDH, which does not contain prosthetic groups and does not absorb in the visible region of the spectrum, and therefore a change in its enzyme activity on exposure to low-intensity radiation in the visible region of the spectrum can be initiated only by photosensitized processes.

The final concentration of bilirubin was brought to $C_{BR} = 4$ µM; the LDH concentration in solution at the time of exposure was $C_{LDH} = 8 \cdot 10^{-9}$ M. The bilirubin/HSA concentration ratio was varied by changing the initial concentration of albumin; in this case, the pigment concentration remained unchanged. The concentration of the LDH were determined taking into account the following molar extinction coefficient at the maximum of their long-wavelength band: $\varepsilon_{280} = 1.96 \cdot 10^5$ M^{-1}·cm^{-1} [42].

Enzyme Activity. The relative enzyme activity of LDH (Serva, Germany) before and after irradiation in the presence of bilirubin and HSA was characterized by the rate of the biochemical reaction catalyzed by it: conversion of pyruvic acid to lactic acid.

The method for determining the enzyme activity of LDH is based on measurement of the rate of decrease in the optical density of one of the components as a result of its consumption during the reaction, and is described in detail in [57, 58].

The dose curves are plotted on a semilog scale, since this allows us to clearly identify any deviation of the experimental results from an exponential dependence.

Photoirradiation Procedures. As the radiation sources, we used both a Westinghouse F20T12 special blue fluorescent lamp, which is widely used for phototherapy of hyperbilirubinemia in newborn infants, and the following lasers: an LG-70 helium-cadmium laser ($\lambda = 441.6$ nm) and an LGN-503 argon laser (with wavelength selection from 457.9 nm to 514.5 nm) made by Polyaron NPO, L'vov (Ukraine).

The solutions were irradiated in spectral cuvets with an $l = 5$ mm base. The optical density at the maximum of the bilirubin absorption spectrum in this case was no greater than $D = 0.1$, which allowed us to avoid shielding the radiation because of the "internal filter" effect. In order to increase the uniformity of the radiation exposure and saturation with oxygen, the solutions were periodically stirred with an automatic pipet.

Emission Spectrum of Lamps and LEDs. The emission spectrum of the fluorescent lamps was recorded on a KSVU-23 spectral system (LOMO, St. Petersburg, Russia).

Measurement of Optical Power. The optical radiation power was monitored by an IMO-3S power meter for measuring the average power and energy of laser radiation (relative intrinsic error $\delta \leq 6\%$).

2. Spectral Fluorescence and Polarization Characteristics of Z,Z-Bilirubin IXα

Investigation of the spectral luminescence characteristics and the characteristic features of deactivation of the electronic excitation energy in the molecule of the yellow pigment Z,Z-bilirubin IXα is not only of purely academic interest but also of considerable practical importance for optimization of phototherapy for hyperbilirubinemia (jaundice) in newborn infants.

However, despite the fact that the photonics of bilirubin under *in vitro* conditions have been studied in detail using steady-state [47, 59–70], picosecond [48–50], and subpicosecond [71] spectroscopy, some fundamentally important characteristics of the Z,Z-bilirubin IXα photoisomerization process have not been properly explained so far. One unresolved question involves clarification of the reasons for the experimental dependence of the quantum yield for the formation of photoproducts on the wavelength of the acting radiation when varied within the long-wavelength absorption band of bilirubin [43, 72–83]. Thus while when solutions of the bilirubin-HSA complex are exposed to radiation with $\lambda = 457.9$ nm (argon laser), the quantum yield for the formation of Z,E-bilirubin IXα is $\varphi_{ZE} = 0.109$ and the quantum yield for the formation of lumirubin is $\varphi_{Lr} = 0.72 \cdot 10^{-3}$, going to $\lambda = 514.5$ nm is accompanied by a two-fold decrease in the quantum yield of Z,E-bilirubin IXα ($\varphi_{ZE} = 0.054$) and an increase in the yield of lumirubin by a factor of 2.5 ($\varphi_{Lr} = 1.80 \cdot 10^{-3}$) [78].

We believe that some information about the reasons for the unusual dependence of the photoisomerization quantum yield on the wavelength of the acting radiation can be obtained from spectral luminescence and polarization studies of the characteristics of Z,Z-bilirubin IXα, and that is the goal of this work. Considering that in the blood vessel, bilirubin molecules are not found in the free state but rather in a complex with the major blood transport protein (HSA), it is of special interest to study the spectral luminescence and polarization characteristics of bilirubin bound to the indicated protein. This is especially so because bilirubin photoisomerization processes in the newborn's body occur precisely in the state of such an equilibrium complex, and binding with HSA has a considerably effect on the course of bilirubin photoisomerization [44, 84–88].

Absorption, Spectral Fluorescence and Polarization Characteristics of Z,Z-Bilirubin IXα in Chloroform at T = 20 °C. Figure 3 shows the absorption and fluorescence spectra of Z,Z-bilirubin IXα in chloroform at room temperature. We see that the long-wavelength absorption band of bilirubin (curve 1) is located in the 350–520 nm region, $\lambda_{max} = 453$ nm; the fluorescence spectrum (curves 2–4) is located in the 480–650 nm region, $\lambda_{max} = 518–521$ nm. And when the excitation wavelength is varied within the indicated absorption band, the position of the fluorescence spectrum and accordingly also the position of its maximum vary by $\Delta\lambda_{max} \sim 3$ nm (bathochromic shift). Thus for $\lambda_{ex} = 400$ nm, 453 nm, 485 nm, the maximum of the fluorescence spectrum is located at $\lambda_{max} = 518$ nm, 519 nm, 521 nm respectively.

A typical feature of the fluorescence of bilirubin solutions at room temperature is the extremely low quantum yield (φ_{fl}) and the very short lifetime of the excited singlet state of the pigment [47–50].

According to data in [47] for Z,Z-bilirubin IXα in chloroform at $T = 22°C$, $\varphi_{fl} < 2 \cdot 10^{-4}$, while according to [49] the fluorescence intensity decay kinetics are described by an exponential with $\tau = 14 \pm 2$ picoseconds.

Even shorter fluorescence lifetimes are given by the authors of [71]. According to their data, the luminescence decay kinetics for bilirubin in chloroform are polyexponential, and are described best by four components with $\tau_1 = 0.134$ psec, $\alpha_1 = 0.51$; $\tau_2 = 0.580$ psec, $\alpha_2 = 0.37$; $\tau_3 = 2200$ psec, $\alpha_3 = 0.09$; $\tau_4 = 9.400$ psec, $\alpha_4 = 0.03$.

Considering the extremely short fluorescence lifetime, we can expect that when it is excited by polarized radiation, rotational depolarization processes cannot develop and polarization spectra can be recorded even in nonviscous solutions in chloroform. In fact, as follows from Figure 3 (curve 5), when fluorescence is excited at the maximum of the absorption spectrum of bilirubin ($\lambda_{ex} = 453$ nm), the emission polarization spectrum is recorded. In the region corresponding to the maximum of the fluorescence spectrum, the degree of polarization is $p = 0.24 \pm 0.03$.

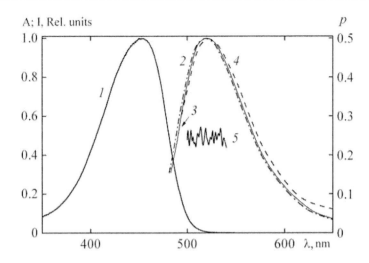

Figure 3. Absorption spectrum (1), fluorescence spectra (2–4), emission polarization spectrum (5) of bilirubin solutions in chloroform at $T = 20°C$, $\lambda_{ex} = 400$ (2), 453 (3, 5) and 485 (4) nm.

Unfortunately, the very low fluorescence quantum yield and the low photochemical stability of bilirubin in chloroform do not allow us to reliably study its polarization spectrum over the entire range of fluorescence and long-wavelength absorption of bilirubin. We know [47, 48, 61, 70] that a substantial increase in fluorescence quantum yield is achieved for bilirubin when the temperature of the solution is lowered. In particular, it was shown earlier [70] that when using isobutyl alcohol as the solvent and cooling the bilirubin solution down to $T = -100°C$, $\varphi_{fl} = 0.10$. Such a high fluorescence quantum yield allows us to study in detail the spectral polarization characteristics of bilirubin. Measurements of the fluorescence intensity decay kinetics for bilirubin showed that under these conditions, the fluorescence lifetime is in the nanosecond range ($\tau = 1.1 \pm 0.1$ nsec) [70].

Spectral Fluorescence and Polarization Characteristics of Z,Z-Bilirubin IXα in Isobutyl Alcohol at T = –100 °C. Figure. 4 shows the fluorescence excitation spectra, the fluorescence spectra, and also the fluorescence emission and absorption polarization spectra for solutions of Z,Z-bilirubin IXα in isobutyl alcohol at $T = -100°C$. When bilirubin is excited in the region of the maximum of its absorption spectrum ($\lambda_{ex} = 460$ nm), we observe intense fluorescence in the 500–650 nm region with maximum at $\lambda_{max} = 535$ nm (curve 5). Varying the excitation wavelength over a 100 nm range ($\lambda_{ex} = 400$ nm, 460 nm, 500 nm) is accompanied by a bathochromic shift of the maximum of the fluorescence band over an ~2.5 nm range ($\lambda_{max} = 533$ nm, 535 nm, 535.5 nm, curves 4–6). The position of the maximum in the fluorescence excitation spectrum also varies within a 2.5-3.0 nm range (curves 1–3) when the detection wavelength is varied within an 80 nm range ($\lambda_{em} = 500$ nm, 535 nm, 580 nm).

Important information about the pathways for deactivation of the electronic excitation energy in the Z,Z-bilirubin IXα molecule is obtained from the fluorescence excitation polarization spectra (Figure 4, curve 7) and fluorescence emission polarization spectra (curves 8–10).

When recording the polarization spectrum at the maximum of the fluorescence band ($\lambda_{em} = 535$ nm, curve 7), we observe a strong dependence of the degree of polarization on the excitation wavelength. When fluorescence is excited in the region $\lambda_{ex} > 490$ nm, the degree of polarization of the fluorescence ($p = 0.47$) is close to the limiting value ($p = 0.50$); but when the excitation wavelength is decreased, the degree of polarization monotonically decreases down to $p = 0.12$ for $\lambda_{ex} = 415$ nm.

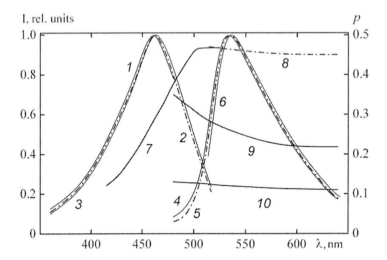

Figure 4. Fluorescence excitation spectra (1–3), fluorescence spectra (4–6), fluorescence excitation polarization spectrum (7), and fluorescence emission polarization spectrum (8–10) of a bilirubin solution in isobutanol at $T = -100°C$; $\lambda_{em} = 500$ nm (1), 535 nm (2, 7), and 580 nm (3); $\lambda_{ex} = 400$ (4, 10), 460 (5, 9), and 500 (6, 8) nm.

The behavior of the emission polarization spectrum is also interesting. When fluorescence is excited on the long-wavelength slope of the bilirubin absorption band ($\lambda_{ex} = 500$ nm, curve 8), the degree of polarization of the fluorescence is close to the limiting value over the entire emission spectrum, slightly decreasing from 0.47 to 0.45 as the detection wavelength increases. When fluorescence is excited to high vibrational sublevels of the $S1$ state of bilirubin (short-wavelength slope of the absorption band, $\lambda_{ex} = 400$ nm, curve 10), the degree of polarization is low ($p = 0.11$–0.12) over the entire range of the fluorescence spectrum. We see different behavior in the emission polarization spectrum when fluorescence is excited in the region of the maximum of the bilirubin absorption spectrum (curve 9). In this case, there is a pronounced dependence of the degree of polarization of the fluorescence on the detection wavelength (the degree of polarization drops by 0.12 for a shift from the short-wavelength slope of the fluorescence spectrum toward longer wavelengths).

As noted, the HSA molecule can bind at least two bilirubin molecules with high affinity (association constant $K_{as} = 5.5 \cdot 10^7$ M^{-1} for the first binding site and $K_{as} = 4.4 \cdot 10^6$ M^{-1} for the second binding site [40]). And while formation of the complex with the second HSA binding site at room temperature has practically no effect on the bilirubin fluorescence intensity ($\varphi_{fl} < 2 \cdot 10^{-5}$ [47]), for binding with the first site the fluorescence quantum yield increases up to $\varphi_{fl} = (1-3) \cdot 10^{-3}$ [47, 48, 59]. We also know [47, 50, 71] that the fluorescence lifetime for Z,Z-bilirubin IXα at the stronger HSA binding site corresponds to the picosecond range. Thus according to [50], for the indicated site the fluorescence intensity decay kinetics for bilirubin at room temperature are described by $\tau = 64$ picoseconds. Shorter fluorescence lifetimes ($\tau = 18 \pm 3$ psec) were obtained in [47] when using a streak camera. In this connection, we should mention the data in [71], according to which the luminescence decay kinetics for HSA-bound bilirubin are polyexponential and best described by five components with $\tau_1 = 0.077$ psec, $\alpha_1 = 0.39$; $\tau_2 = 0.60$ psec, $\alpha_2 = 0.28$; $\tau_3 = 3.65$ psec, $\alpha_3 = 0.17$; $\tau_4 = 20$ psec, $\alpha_4 = 0.12$; $\tau_5 = 95$ psec, $\alpha_5 = 0.04$.

Spectral Fluorescence and Polarization Characteristics of Z,Z-Bilirubin IXα Bound to the First HSA Binding Site at T = 20 °C. Figure 5 shows the spectral luminescence and polarization characteristics of bilirubin bound to the first HSA binding site. We see that when fluorescence is excited at $\lambda_{ex} = 400$ nm, 460 nm, and 500 nm, the maxima in the fluorescence of bilirubin within the complex are located at $\lambda_{max} = 529$ nm, 532 nm, and 536 nm respectively (see curves 4–6). Conversely, for detection of fluorescence at $\lambda_{em} = 500$ nm, 535 nm, and 580 nm, the maxima in the excitation spectrum change in the series $\lambda_{max} = 452$ nm, 457 nm, and 462 nm (curves 1–3). Consequently, varying the excitation wavelength within the long-wavelength absorption band of bilirubin (a range of ~100 nm) or varying the detection wavelength within the fluorescence band (a range of ~ 80 nm) is accompanied by a pronounced shift (by ~8–10 nm) of the position of the maxima in the fluorescence or fluorescence excitation spectrum respectively. The indicated bathochromic shift of the spectra is more pronounced than the spectral shifts for bilirubin in the case of organic solvents (chloroform and isobutyl alcohol).

The fluorescence excitation polarization spectra (Figure 5, curve 7) and emission polarization spectra (curve 8) for bilirubin within the HSA complex at $T = 20°C$ have characteristics similar to those for solutions of the pigment in isobutanol at $T = -100°C$ (see Figure 4). We should especially point out the presence of a very pronounced dependence of the degree of polarization of the fluorescence on the excitationn wavelength when it is varied within the long-wavelength absorption band (curve 7).

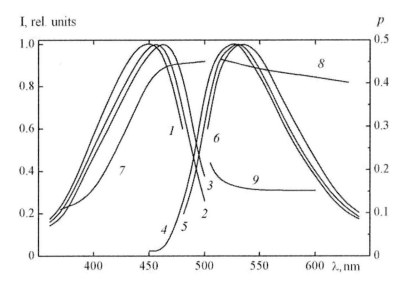

Figure 5. Fluorescence excitation spectra (1–3), fluorescence spectra (4–6), fluorescence excitation polarization spectra (7), and fluorescence emission polarization spectra (8, 9) of the bilirubin–HSA complex at $T = 20°C$, $\lambda_{em} = 500$ (1), 535 (2, 7), and 580 (3) nm; $\lambda_{ex} = 400$ (4, 9), 460 (5), and 500 nm (6, 8).

As in the case of solutions in isobutanol, for the bilirubin–HSA complex we observe a decrease in the degree of polarization of the fluorescence as the excitation wavelength is shifted toward the shorter wavelength portion of the absorption spectrum. Furthermore, for long-wavelength excitation of bilirubin (λ_{ex} = 500 nm, curve 8), the degree of polarization of the fluorescence (p = 0.46) when detected on the short-wavelength slope of the fluorescence band (λ_{em} = 500 nm) is close to the limiting value. As the fluorescence detection wavelength increases (λ_{em} = 630 nm), the degree of polarization decreases (p = 0.4).

Discussion of Research Results of Spectral Fluorescence and Polarization Characteristics of Z,Z-Bilirubin IXα. A distinguishing feature of the fluorescence of Z,Z-bilirubin IXα solutions at room temperature is its extremely low quantum yield: $\varphi_{fl} < 2 \cdot 10^{-4}$ for solutions of the pigment in chloroform, and $\varphi_{fl} = (1-3) \cdot 10^{-3}$ for its complex with HSA in buffered aqueous medium [47, 59–64]. And the low value of φ_{fl} is not due to the high efficiency of intersystem crossing $S_1 \to T_1$: the quantum yield in the triplet state for solutions of bilirubin in chloroform is $\varphi_t < 0.005$ [89], while for the BR–HSA complex $\varphi_t < 0.01$ [47]. Hence neither emission of radiation as fluorescence nor intersystem crossing $S_1 \to T_1$ are major deactivation pathways for the electronic excitation energy of the singlet state of the bilirubin molecule at room temperature. Consequently, electronic relaxation of singlet-excited bilirubin under these conditions may occur mainly by internal conversion and/or photochemical reactions. However, the quantum yields for bilirubin self-sensitized photooxidation $\varphi_{PO} < 0.01$ [47] and structural photoisomerization (formation of lumirubin) $\varphi_{Lr} < 0.002$ [79, 87] are also low. Geometric (*cis-trans*) photoisomerization of bilirubin is more likely: the overall quantum yield for the formation of Z,E and E,Z isomers is $\varphi_{CT} < 0.11$ [43, 78]. From the given values of the quantum yields, it follows that $\varphi_{fl} + \varphi_t + \varphi_{PO} + \varphi_{Lr} + \varphi_{CT} \approx 0.13 \ll 1.0$, i.e., the major pathway for deactivation of the electronic excitation energy of the singlet state of the bilirubin molecule at room temperature is internal conversion. The pronounced dependence of the fluorescence intensity on the temperature and viscosity of the solution allowed the authors of [47, 48] to conclude that the major reason for the weak fluorescence of bilirubin solutions at room temperature is the rotational mobility of the outer pyrrole rings of the molecule.

The studies presented in this paper on the spectral luminescence and polarization characteristics of Z,Z-bilirubin IXα under different microenvironment conditions (organic solvents: chloroform and isobutyl alcohol) and also within a complex with the major blood protein transport protein HSA in buffered aqueous solution and for different temperature conditions (T = 20°C for the solutions in chloroform and the complexes with HSA in the buffer; T = –100°C for the solutions in isobutanol) allow us to add to current ideas about the photonics of bilirubin.

We have observed a pronounced dependence of the position of the fluorescence spectrum for Z,Z-bilirubin IXα on its excitation wavelength (Figure 3, curves 2-4; Figure 4 and Figure 5, curves 4–6) and a dependence of the position of the excitation spectrum on the fluorescence detection wavelength (see Figure 4 and Figure 5, curves 1–3). In this case, the bathochromic shift of the fluorescence (fluorescence excitation) spectrum significantly increases on going from isotropic organic solvents ($\Delta\lambda_{max}$ = 2.5–3.0 nm) to the bilirubim–HSA complex in buffered aqueous solution ($\Delta\lambda_{max}$ = 8–10 nm). The purity of the Z,Z-bilirubin IXα preparation used allows us to say that the dependence of the position of the fluorescence spectrum on wavelength is not due to the presence of impurities in the solution with similar spectral characteristics.

A distinguishing feature of the fluorescence polarization spectra of Z,Z-bilirubin IXα is the rather high degree of polarization of the fluorescence when it is excited both at room temperature in nonviscous bilirubin solutions (in chloroform and in the HSA complex) and at $T = -100°C$ in high viscosity solutions in isobutanol. The degree of polarization of the fluorescence in isobutanol and in the protein complex ($p = 0.46–0.47$ for $\lambda_{em} = 500$ nm) is close to the limiting value ($p = 0.50$) when it is excited on the long-wavelength slope of the bilirubin absorption band (Figure 4 and Figure 5, curves 7, 8) and sharply drops as we go to the short-wavelength slope of the bilirubin absorption band (Figure 4, curve 10). Also note the variability of the degree of polarization of the fluorescence over the emission spectrum: an increase in the fluorescence detection wavelength is generally accompanied by some decrease in the degree of polarization (Figure 4, curves 8 and 9; Figure5, curve 8). When fluorescence is excited to high vibrational sublevels of the S_1 state of bilirubin (short-wavelength slope of the absorption band with $\lambda_{ex} = 400$ nm, curve 10), the degree of polarization has low values ($p = 0.11–0.12$) over the entire range of the fluorescence spectrum, and is practically independent of the detection wavelength.

Certainly the rather high degree of polarization of the fluorescence of Z,Z-bilirubin IXα at room temperature is due especially to the extremely short fluorescence lifetime of the pigment under the indicated conditions. According to the data in [71], obtained using (for excitation of the fluorescence of Z,Z-bilirubin IXα femtosecond laser technology, the fluorescence decay kinetics for bilirubin in chloroform and in the HSA complex are polyexponential and best approximated by four to five components. In this case, the lifetime of the most intense component is no longer than $\tau = 0.134$ psec, while the "longest-lived" component is described by $\tau = 95$ psec. Of course, for such a short fluorescence lifetime, rotational depolarization processes cannot develop, and when the bilirubin molecule is excited by polarized radiation, the emission continues to be predominantly polarized even in nonviscous solutions. Moreover, some role in the increase in the degree of polarization of the fluorescence we detected (Figure 5) when bilirubin binds to HSA at room temperature is probably played by (along with the short lifetime of the excited state) the decrease in the rotational mobility of the pigment, due to the rigid microenvironment of the HSA protein framework. One of the signs of a decrease in the rotational mobility of bilirubin under the given conditions is the sharp increase in the quantum yield for its fluorescence when it binds to HSA [47, 48, 59]. On going to low temperature conditions (Figure.4), the determining contribution to the presence of a high degree of polarization of the fluorescence for its long-wavelength excitation probably comes from the constrained rotational mobility of bilirubin due to the high viscosity of the solution, since at $T = -100°C$ in isobutanol, the fluorescence decay time for bilirubin sharply increases (compared with room temperature): $\tau = 1.1 \pm 0.1$ nsec. Also note the results in [48], according to which the fluorescence lifetime for bilirubin in the HSA complex at liquid nitrogen temperature ($T = -196°C$) and when using a mixture of water and ethylene glycol (50%/50%) as the solvent also corresponds to the nanosecond range ($\tau = 2.8$ nsec).

In order to analyze the reasons for the observed dependence of the position of the fluorescence (fluorescence excitation) spectrum on the excitation (detection) wavelength, the variability of the degree of polarization of the fluorescence over the emission spectrum, and also its sharp decrease when the excitation wavelength is shifted to the short-wavelength portion of the absorption spectrum, let us turn to data on the chemical structure and the spatial structure of the bilirubin molecule.

Structurally, Z,Z- bilirubin IXα can be considered (Figure 6) as a tetrapyrrole molecule consisting of two rather similar but not completely symmetric moieties which each include two pyrrole rings and which are linked by a methylene bridge (CH$_2$ group). And due to the methylene group breaking up the general conjugation chain, the molecule of the indicated pigment should be considered as a combination of two chromophores (I and II): a bichromophore [47, 90].

The spatial conformation of the bilirubin molecule in an isotropic solvent is a folded structure, similar to a ridge tile of a roof, in which the two pyrrole moieties of one half of the molecule are located in a plane making a dihedral angle of $\Theta \sim 98–104°$ with the plane of the other two pyrrole rings [91, 92]. The distance between the dipyrrole moieties of the molecule is 5–6 Å. The reason for the formation of a closed, bent conformation of the bilirubin molecule is formation of six strong intramolecular hydrogen bonds formed by the –COOH group of one half of the tetrapyrrole molecule with the –NH and –C=O groups of the other half of the tetrapyrrole molecule [91].

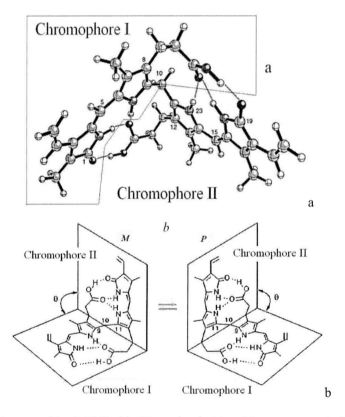

Figure 6. Spatial structure of the Z,Z-bilirubin IXα molecule (a) and the arrangement of its chromophores in space (b).

In order to determine the pathways for deactivation of the electronic excitation energy in the bilirubin molecule, it is important to establish whether there are differences between the absorption spectra of the chromophores making up the molecule, taking into account the fact that their structures are different. The attempt made in [40] to determine (calculate) the energy of the electron transitions of the first excited singlet state ($S_0 \to S_1$) of the moieties of the bilirubin molecule (in vacuo) by quantum chemical methods using the Hyperchem program did not provide unambiguous results. Thus in optimization of the geometry for the moieties of the molecule by the MINDO/3, MM, and AM1 methods, the S_1 level of chromophore I is located 400–700 cm^{-1} above the level for chromophore II, while the PM3 and MNDO methods yield the opposite result (–(100–300) cm^{-1}).

For comparison, we note that according to [94], the maximum of the absorption spectrum in chloroform for Z,Z-bilirubin IIIα (see Figure 7), which is a symmetric bichromophore based on chromophore II, is located in the region λ_{max} = 455–458 nm, while the absorption spectrum for Z,Z-bilirubin XIIIα (a symmetric bichromophore based on chromophore I) is located in the region λ_{max} = 449–453 nm. Consequently, the S_1 level of Z,Z-bilirubin XIIIα is located at an energy 250–300 cm^{-1} above the analogous level in Z,Z-bilirubin IIIα. As expected (also see Figure 2), the maximum of the absorption spectrum for Z,Z-bilirubin IXα occupies an intermediate position (λ_{max} = 453–455 nm) [94].

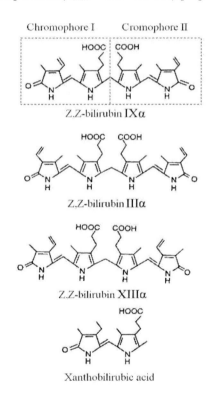

Figure 7. Structural formulas for Z,Z- bilirubin IXα, Z,Z- bilirubin IIIα, Z,Z- bilirubin XIIIα and xanthobilirubinic acid.

Thus in an isotropic solvent (chloroform, isobutanol), the energy differences for the electronic transitions of chromophores I and II in Z,Z-bilirubin IXα most likely are insignificant and the S_1 level of chromophore I is located slightly above the level for chromophore II (i.e., the absorption band for chromophore I is shifted toward shorter wavelengths relative to chromophore II). The question concerning the relationship between the absorption spectra of chromophores I and II in HSA-bound Z,Z-bilirubin IXα is not clear.

Modeling studies [95] indicate that **Z,Z**-bilirubin IXα may be able to fit within the binding site in sub-domain IB HSA in a partially extended conformation in which the interplanar angle between the dipyrrinone groups is increased from 100°. To date, the best substantiated model is the one in [40], according to which the bilirubin molecule is completely incorporated into the structure of the HSA macromolecule while retaining its folded conformation. In [40, 95] it is assumed that each of the dipyrrole moieties of bilirubin occupies a grooveshaped section in the HSA structure and is characterized by a different microenvironment created by different sections of the polypeptide chain. Furthermore, the bilirubin molecule is a monoanion in buffer solution in the region of physiological pH values, while within the HSA complex it is a dianion [96]. This also can lead to a change in the spectral characteristics of the chromophores, i.e., spectral differences between the chromophores can be enhanced within the HSA complex as a result of the heterogeneity of their microenvironments and the "charge" effect [67].

Consequently, due to differences in the spectral characteristics of chromophores I and II as a result of the asymmetry of their structure and/or the heterogeneity of the microenvironments of the chromophores in the protein matrix, a change in the fluorescence excitation wavelength leads to a change in their contributions to the total absorption of bilirubin. Assuming that for the same reason chromophores I and II are characterized by different fluorescence spectra and that each of them contributes to the fluorescence spectrum of bilirubin to some degree (see below), it becomes obvious that a change in the excitation wavelength should lead to a change in the position of the fluorescence spectrum (see Figures 3–5). Considering (see Figure 1) that depending on where the absorbed photon is localized (on chromophore I or chromophore II), different photoisomers are formed (formation of *E,Z*-bilirubin IXα and lumirubin is associated with structural rearrangement and for lumirubin is followed by vinyl cyclization in chromophore I, while structural rearrangement of chromophore II leads to formation of *Z,E*-bilirubin IXα), a change in the wavelength of the acting radiation leads to a change in the quantum yield (calculated based on the total absorption of bilirubin) for photoproduct formation and accordingly a change in their quantitative ratio.

We should note that besides a change in the contributions of chromophores I and II to the overall absorption and fluorescence of bilirubin as the excitation wavelength changes, there may be other reasons for the bathochromic shift of the fluorescence spectrum. These reasons may include the conformational heterogeneity of the pigment, due to the presence in solution of its two mirror-symmetric helical forms: *P* and *M* enantiomers [96] (Figure 6). And while in an isotropic organic solvent the concentrations of left-handed and right-handed helical forms are the same, according to [96], in the HSA complex some excess of the *P* enantiomer is observed (greater by ~17%).

Under our experimental conditions (extremely short fluorescence lifetime at room temperature and high viscosity of the solution in isobutanol at $T = -100°C$), some contribution to the change in the position of the fluorescence spectrum with a change in excitation wavelength may come from molecules which did not completely vibrationally relax after their excitation ("hot fluorescence") [67]. This effect appears when the fluorescence rate is comparable with the vibrational relaxation rate, i.e., the absence of an equilibrium between the environment and the molecules in the excited state, combined with inhomogeneous orientational broadening of the levels [97 – 99], can also contribute to the bathochromic shift of bilirubin fluorescence. And the bathochromic shift of the fluorescence spectrum as a result of inhomogeneous broadening should be observed only for excitation at the long-wavelength edge in a polar solvent [97 – 99].

Thus reasons for the dependence of the position of the fluorescence (fluorescence excitation) spectra of Z,Z-bilirubin IXα on the excitation (detection) wavelength may be: structural and spectral differences between the chromophores making up the bilirubin molecule; conformational heterogeneity of the pigment in solution; contribution to the fluorescence from molecules which have not completed vibrational relaxation in the S_1 state ("hot fluorescence"); inhomogeneous orientational broadening of the levels; heterogeneity of the microenvironment of the chromophores in the protein matrix [67]. Among the listed factors, probably the first one is predominant in an isotropic organic solvent while the last one predominates for the bilirubin–HSA complex [67].

For analysis of the characteristic features of interchromophore interaction in the Z,Z-bilirubin IXα molecule, let us turn to the fluorescence absorption polarization spectra (Figure 4 and Figure 5, curve 7) and the fluorescence emission polarization spectra (Figure 4 and 5, curves 8–10). Especially note the pronounced dependence of the degree of polarization of the fluorescence on the excitation wavelength: a high (close to the limiting value) degree of polarization when excited on the long-wavelength slope of the bilirubin absorption band, and a relatively low degree of polarization on going to excitation on the short-wavelength slope of the absorption band of the pigment. Such a shape for the excitation polarization spectrum is typical for interchromophore interactions accompanied by intramolecular or intermolecular singlet-singlet transfer of electronic excitation energy [100, 101]. From the data presented, it follows that when fluorescence is excited to high vibrational sublevels of the S_1 state (the short-wavelength portion of the spectrum), despite the anomalously short fluorescence lifetime we observe efficient intramolecular interchromophore energy transfer. The presence of intramolecular transfer rather than intermolecular transfer is indicated by the fact that bilirubin is a monomer at a concentration of $C_{Br} = 2$ μM in the solvents used, while the intermolecular distances in solution are too large for inductive resonance energy transfer. Moreover, replacing the cuvet with optical path length $l = 10$ mm with a cuvet with $l = 2$ mm, with a corresponding five-fold increase in the pigment concentration, had practically no effect on its polarization spectra. When the excitation is shifted to the long-wavelength ("red") edge of the bilirubin absorption band, there are increasingly fewer centers which could be acceptors of electronic excitation energy, and transfer becomes first less likely and then ($\lambda_{ex} \geq 480$ nm) even impossible: the degree of polarization of the fluorescence is close to the limiting value (the "red-edge effect" or the Weber effect [100]).

Some information about interchromophore interactions in the bilirubin molecule is also provided by analysis of the emission polarization spectra (Figure 4, curves 8–10; Figure 5, curve 8). In this connection, the pronounced dependence of the fluorescence emission polarization spectrum of bilirubin when excited in the region of the maximum in the absorption spectrum ($\lambda_{ex} = 460$ nm, Figure 4, curve 9) is most likely due to the changing contribution over the fluorescence band of bilirubin from chromophores directly absorbing photons (relatively high degree of polarization for shortwavelength detection) and chromophores receiving electronic excitation energy as a result of intramolecular transfer (the long-wavelength portion of the emission spectrum, characterized by a relatively low value of p). The presence of depolarized fluorescence over the entire spectrum when it is excited on the short-wavelength slope (λ_{ex} = 400 nm, Figure 4, curve 10) of the bilirubin absorption band allows us to say that in this case, we are observing efficient interchromophore energy transfer, and the contribution to the fluorescence from chromophores directly absorbing radiation is insignificant. In turn, for excitation on the long-wavelength edge (λ_{ex} = 500 nm, Figure 4 and Figure 5, curves 8), a contribution to the fluorescence comes only from chromophores not participating in the process of intramolecular transfer of electronic excitation energy.

We should note that data are not available in the literature on the fluorescence decay time for Z,Z-bilirubin IXα (τ) as a function of excitation and detection wavelength. We can only expect *a priori* that τ is variable over the emission spectrum, considering its possible shortening in the presence of a nonradiative energy transfer process. In turn, the dependence $\tau = f(\lambda_{em})$ can also affect the fluorescence polarization spectrum of bilirubin [67].

Analysis of the experimental data (Figure 4 and Figure 5) shows that for short-wavelength excitation, rapid intramolecular interchromophore energy transfer delocalizes the excitation between the two absorbing chromophores, with higher probability populating the levels with lower energy. A decrease in the excitation energy (a shift toward longer wavelengths) reduces the probability of transfer, since the barrier between exciton states becomes significant. In this case, the excitation remains localized on the chromophore which absorbed the photon. This is why a change in the wavelength of the acting radiation (when it is varied within the long-wavelength absorption band of Z,Z-bilirubin IXα leads to a change in both the ratio between the photoisomerization products of bilirubin and the quantum yield for their formation (calculated based on the overall absorption by the pigment).

Conclusion of the Study of Spectral Fluorescence and Polarization Characteristics of Z,Z-Bilirubin IXα. We have studied the spectral fluorescence and polarization characteristics of Z,Z-bilirubin IXα, at room temperature in chloroform and in aqueous buffer medium, within an equilibrium complex with human serum albumin (HSA), and also under low temperature conditions (T = –100°C) in isobutyl alcohol. We have observed a bathochromic shift of the fluorescence spectra, which is most pronounced for the bilirubin-albumin complex. The following are considered as possible reasons for the observed dependence of the position of the fluorescence (fluorescence excitation) spectra on the excitation (detection) wavelength: structural and spectral differences between the chromophores making up the bilirubin molecule; conformational heterogeneity of the pigment in solution; a contribution to the fluorescence from molecules which have not completed the vibrational relaxation process; inhomogeneous orientational broadening of the levels; heterogeneity of the microenvironment of the chromophores in the protein matrix. We show that polarized fluorescence of bilirubin occurs at room temperature, due to the anomalously short fluorescence lifetime τ (picosecond or subpicosecond ranges). Despite such a short τ, the absorption and emission polarization spectra suggest the presence of intramolecular nonradiative singlet-singlet energy transfer when bilirubin is excited to high vibrational sublevels of the S_1 state (degree of polarization p = 0.11–0.12). When fluorescence is excited on the long-wavelength slope of the absorption band, no transfer occurs: the degree of polarization (p = 0.46–0.47) is close to the limiting value (p = 0.50).

Studies of the spectral luminescence and polarization characteristics of Z,Z-bilirubin IXα have shown that the major reason for the dependence of the quantum yield for the formation of configurational photoisomers (Z,E-bilirubin IXα, E,Z-bilirubin IXα, E,E-bilirubin IXα) and structural (lumirubin) photoisomers on the wavelength of the acting radiation is the bichromophore nature of the absorption (and emission) of light by the bilirubin molecules and differences between the chromophores making up the bilirubin molecule. When bilirubin is excited to high vibrational sublevels of the S_1 state, we observe efficient intramolecular singlet-singlet energy transfer. The process of interchromophore transfer of the electronic excitation energy under conditions of an anomalously short fluorescence lifetime for bilirubin (picosecond or subpicosecond range) promotes the presence of exciton interactions between the chromophores. When the excitation is shifted toward the long-wavelength edge of the bilirubin absorption band, there are increasingly fewer centers which can be acceptors of electronic excitation energy, and transfer becomes first less likely and then even impossible.

3. Photophysical Processes that Determine the Photoisomerization Selectivity of Z,Z-Bilirubin IXα in Complexes with Albumins

This section of the paper discusses the photophysical processes that determine the regularities of Z,Z-bilirubin IXα photoisomerization – the mechanism that forms the basis of the widely used optical technology for treating hyperbilirubinemia (jaundice) of newborns. Based on a study of the spectrofluorescence and polarization characteristics of complexes of Z,Z-bilirubin IXα with human serum albumin (HSA) and bovine serum albumin (BSA), it is established that one cause of the predominant formation of Z,E-bilirubin IXα among the *cis-trans* photoisomerization products of Z,Z-bilirubin IXα is the bichromophore character of the structure of the pigment, the nonidentity of the chromophores that form it, and the presence of interchromophore intramolecular energy transfer. The highest efficiency of nonradiative transfer is found for the bilirubin–HSA complex, and this correlates with the literature data concerning the higher selectivity of the bilirubin photoisomerization process in the indicated complex by comparison with BSA. The factor that promotes the process of intermolecular energy transport when the fluorescence lifetime of bilirubin is anomalously short the picosecond or subpicosecond range) is the presence of excitonic interactions between the chromophores.

Despite the obvious practical success in developing a method of phototherapy and its fairly high effectiveness, some fundamental questions of the photoisomerization process of Z,Z-bilirubin IXα remain unanswered. The unsolved problems include the causes of the pronounced selectivity of the *cis-trans* photoisomerization process of Z,Z-bilirubin IXα, associated with the main transport protein of blood – human serum albumin (HSA). The selectivity consists of the predominant formation among the products of one of the configurational photoisomers of the pigment – Z,E-bilirubin IXα [43, 44, 73–84]. Thus, in the state of dynamic equilibrium established between the isomeric forms of the pigment when bilirubin–HSA complexes are irradiated under oxygen-free conditions, the concentration of Z,E-bilirubin IXα (C_{ZE}) exceeds the concentration of E,Z-bilirubin IXα (C_{EZ}) by about a factor of 20 (λ = 436 nm) [44] or even by a factor of 100 (λ = 450–460 nm) [84, 86]. At the same time, for complexes of Z,Z-bilirubin IXα with albumins of other specific origin, such as from a bull (bovine serum albumin or BSA), a horse, or a rat, the selectivity is significantly lower ($C_{ZE}/C_{EZ} \approx 1.6-2.4$, [44]) or completely absent ($C_{ZE}/C_{EZ} \approx 1$ [47, 102]). Recalling that the E,Z-isomer is an intermediate in the lumirubin-formation stage (see Figure 1) and plays a key role in lowering the bilirubin level in the organism of a newborn undergoing phototherapy [27–34], the investigation of the causes of such selectivity is not only of purely academic interest but can also provide information for optimizing the methods of phototherapy of hyperbilirubinemia in newborn children.

In our opinion, definite data for determining what causes the selectivity of the isomerization process and its unusual dependence on the type of transfer protein can be obtained by investigating the spectroluminescence and polarization characteristics of complexes of Z,Z-bilirubin IXα with HSA and BSA. BSA is chosen as an object of comparison because, for complexes of Z,Z-bilirubin IXα with the indicated protein (along with HSA), it provides the most information on the structure and certain spectral characteristics [47–50, 59, 65, 71].

Absorption Spectra, Spectral Fluorescence and Polarization Characteristics at T = 20 °C of Z,Z-Bilirubin IXα Bound to BSA. The absorption spectra of a buffer solution of Z,Z-bilirubin IXα (pH 7.4, $C_{Br} = 2$ μM) in complexes with HSA ($C_{HSA} = 4$ μM) and BSA ($C_{BSA} = 4$ μM) are shown in Figure 2 (curves 1 and 2, respectively).

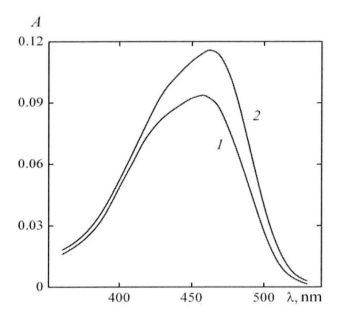

Figure 8. Absorption spectra of buffer solutions of Z,Z-bilirubin IXα in a complex with HSA (1) and BSA (2).

As follows from the figure, the long-wavelength absorption spectra of the indicated pigment–protein complexes lie in the blue–green region of the spectrum (λ = 380–520 nm), with maxima at $\lambda_{max} = 458$ nm for bilirubin–HSA and $\lambda_{max} = 462$ nm for bilirubin–BSA. In this case, a higher molar extinction coefficient at the band maximum is noted for the complex of the pigment with BSA.

The fluorescence spectra, the fluorescence-excitation spectra, the fluorescence-excitation polarization spectrum, and the fluorescence polarization spectrum for the bilirubin–BSA complex are shown in Figure 9. As follows from the figure, variation of the excitation wavelength within the long-wavelength absorption band of Z,Z-bilirubin IXα is accompanied by displacement of the fluorescence band toward longer wavelengths.

Thus, with fluorescence excitation at wavelengths λ_{exc} = 400, 460, and 500 nm, the maxima of the fluorescence spectra of bilirubin as part of a complex with BSA are located at λ_{max} = 530, 534, and 541 nm (curves 4–6). Conversely, as the fluorescence-recording wavelength varies over λ_{em} = 500, 535, and 580 nm, the corresponding maxima in the excitation spectrum for the bilirubin–BSA complex are at λ_{max} = 458, 465, and 471 nm (Figure 9, curves 1–3). Thus, variation of the excitation wavelength within the longwavelength absorption band of bilirubin (about 100 nm) or of the recording wavelength within the fluorescence band (about 80 nm) is accompanied by a pronounced displacement of the absorption of the maxima of the fluorescence spectra (by 11 nm) and of the fluorescence-excitation spectra (by 13 nm) of the pigment embedded in the protein matrix.

Important information on the deactivation paths of the electronic excitation energy of the Z,Z bilirubin IXα molecule associated with albumins of various specific origin is obtained from the fluorescence-excitation polarization spectra (see Figure 9, curves 7) and the fluorescence polarization spectra (Figure 9, curves 8 and 9).

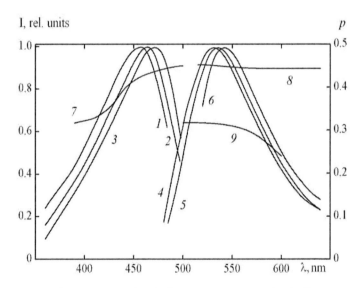

Figure 9. Fluorescence-excitation spectra (1–3), fluorescence spectra (4–6), fluorescence polarization spectra with respect to excitation (7) and emission (8, 9) of buffer solutions of Z,Z bilirubin IXα in a complex with BSA. λ_{em} = 500 (1), 535 (2, 7), 580 nm (3), λ_{exc} = 400 (4, 9), 460 (5), and 500 nm (6, 8).

A Comparison of the Spectral-Polarization Characteristics at T = 20 °C of Z,Z-Bilirubin IXα Bound to BSA and HSA. As can be seen from Figure 5 (curve 7), a strong dependence of the degree of polarization of the fluorescence on the excitation wavelength is observed for the bilirubin–HSA complex (polarization spectrum with respect to excitation, λ_{em} = 535 nm). Thus, if the degree of fluorescence polarization p = 0.45–0.46 is close to the limiting value (p = 0.50) when fluorescence is excited in the region λ_{exc} > 480 nm, it is observed to sharply decrease to $p = 0.135$ at $\lambda_{exc} = 390$ nm as the excitation wavelength is reduced. Depolarization of the fluorescence is also observed for the bilirubin–BSA complex as the excitation wavelength is displaced on the short-wavelength slope of the absorption band of the pigment (Figure 9, curve 7). However, unlike the bilirubin–HSA solution, the degree of fluorescence polarization of the pigment embedded in the structure of the bovine albumin is significantly greater and equals $p = 0.32$ at $\lambda_{exc} = 390$ nm. Here, as for the bilirubin–HSA complex, the degree of fluorescence polarization is close to the limiting value in the case of long-wavelength excitation, $\lambda_{exc} > 480$ nm.

There is also interest in the behavior of the polarization spectrum with respect to emission. As follows from Figure 5 (curves 8 and 9), in the case of long-wavelength excitation of the bilirubin fluorescence ($\lambda_{exc} = 500$ nm, curve 8), the degree of polarization ($p = 0.46$) on the short-wavelength slope of the fluorescence band ($\lambda_{em} = 515$ nm) is close to the limiting value, regardless of the type of transfer protein (HSA or BSA). If the recording wavelength is increased ($\lambda_{em} = 630$ nm), a certain decrease of the degree of the fluorescence polarization is observed ($p = 0.40$) for the bilirubin–HSA complex (Figure 5, curve 8), whereas it shows virtually no decrease for the bilirubin–BSA solution ($p=0.45$, Figure 9, curve 8).

When fluorescence is excited to high vibrational sublevels of the S_1 state of bilirubin (the short-wavelength slope of the absorption band $\lambda_{exc} = 400$ nm, Figures 5 and 9, curves 9), the degree of polarization has lower values over the entire range of the fluorescence spectrum. A more pronounced dependence of the degree of the fluorescence polarization on the recording wavelength is observed in this case (the degree of polarization drops off as one goes from the short-wavelength to the long-wavelength slope of the fluorescence spectrum). It should also be pointed out that, with short-wavelength excitation ($\lambda_{exc} = 400$ nm), the degree of fluorescence polarization of bilirubin coupled with BSA (Figure 9, curve 9) is somewhat higher than the corresponding value (for the same recording wavelength) for the bilirubin–HSA complex (Figure 5, curve 9).

Analysis of the Photophysical Processes that Determine the Photoisomerization Selectivity of Z,Z-Bilirubin IXα in Complexes with Albumins. An outstanding feature of the fluorescence of solutions of Z,Z-bilirubin IXα at room temperature is its extremely low quantum yield (φ_{fl}): for aqueous solutions, $\varphi_{fl} = 2 \cdot 10^{-5}$ [47], which increases by an order of magnitude when the pigment is incorporated in the protein structure of HSA or BSA. According to the data [47, 48, 59], $\varphi_{fl} = (1-3) \cdot 10^{-3}$ for the complex of the pigment with HSA in an aqueous buffer medium, and, according to [31], $\varphi_{fl} = 8.5 \cdot 10^{-4}$ for the complex with BSA. In this case, an increase of the fluorescence quantum yield is observed only when the bilirubin is incorporated at the strong-coupling center of the HSA, and the formation of complexes with the weaker-coupling centers has virtually no effect on the fluorescence intensity. It should be pointed out that the low quantum yield of the fluorescence of bilirubin is not caused by high intercombination conversion efficiency $S_1 \to T_1$: the quantum yield in the triplet state for the bilirubin–HSA complex is $\varphi_t < 0.01$ [47]. That is, neither radiation emission in the form of fluorescence nor intercombination conversion $S_1 \to T_1$ is the main deactivation path of the electron-excitation energy of the singlet state of the bilirubin molecule at room temperature. It is customarily assumed [47, 48] that the main cause of the weak fluorescence of bilirubin solutions under the indicated conditions is internal conversion – high rotational mobility of the external pyrrole rings of the molecule. In this connection, the increased quantum yield of the bilirubin fluorescence when it is incorporated in the protein matrix of HSA or BSA can be a consequence of a certain limitation of the degrees of freedom [36].

The studies presented in this paper show that there is a pronounced dependence of the position of the fluorescence spectrum of Z,Z-bilirubin IXα as part of a complex with albumin (HSA or BSA) on the excitation wavelength (see Figures 5 and 9, curves 4–6) and of the position of the excitation spectrum on the fluorescence-recording wavelength (see Figure 5 and 9, curves 1–3). In this case, the shift of the fluorescence spectrum toward longer wavelengths is more pronounced for the bilirubin–BSA complex: A change of the excitation wavelength from λ_{exc} = 400 nm to λ_{exc} = 500 nm is accompanied by a displacement of the position of the maximum of the fluorescence spectrum by 11 nm. Under similar conditions, the fluorescence spectrum of bilirubin coupled with HSA is displaced by 7 nm. The high degree of purification of the Z,Z-bilirubin IXα sample used here, as well as the results of the studies carried out earlier [67] using isotropic organic solvents, make it possible to assert that the observed wavelength dependence of the position of the fluorescence spectrum is not caused by the presence of impurities with similar spectral characteristics in the solution.

A characteristic feature of the fluorescence polarization spectra of Z,Z-bilirubin IXα incorporated in the structure of the HSA or BSA protein globule is a fairly high (close to the limiting value) degree of fluorescence polarization (p = 0.45−0.46) when it is excited on the long-wavelength slope of the absorption band of bilirubin and recorded on the shortwavelength slope of the fluorescence band (curves 7 and 8 in Figures 5 and 9). However, when the excitation wavelength is displaced toward the shorter wavelength part of the absorption spectrum of bilirubin, pronounced depolarization of the fluorescence of the pigment coupled with HSA is observed (at λ_{exc} = 390 nm and λ_{em} = 535 nm, p = 0.135). The degree of fluorescence polarization is not so substantially reduced for the bilirubin–BSA complex under similar conditions (p = 0.32). The degree of fluorescence polarization also varies over the emission spectrum: An increase of the fluorescence recording wavelength is usually accompanied by some reduction of the degree of polarization (curves 8 and 9 in Figures 5 and 9).

There is no doubt that the fairly high degree of fluorescence polarization of Z,Z-bilirubin IXα as part of complexes with HSA and BSA at room temperature under longwavelength excitation is mainly caused by the extremely short fluorescence lifetime of the pigment, by the absence of intra- and intermolecular energy transfer, and also by the fact that the pigment is rigidly fixed in the protein matrix [67]. Thus, according to [50], the quenching time of the fluorescence intensity of bilirubin at the strongest-coupling center of HSA is characterized by τ = 64 psec (λ_{exc} = 476 nm, λ_{em} = 550 nm, pH 7.4). Shorter fluorescence lifetimes (τ = 18±3 psec) are obtained in [47] using a streak camera. In this connection, a recent paper should also be pointed out [71] according to which the luminescence-quenching kinetics of bilirubin coupled with HSA (λ_{exc} = 454 nm, λ_{em} = 525 nm, pH 8.0) is polyexponential and is best described by five components with τ_1 = 0.077 psec, α_1 = 0.39; τ_2 = 0.60 psec, α_2 = 0.28; τ_3 =3.65 psec, α_3 = 0.17; τ_4 = 20 psec, α_4 = 0.12; and τ_5 = 95 psec, α_5 = 0.04. In our opinion, the fact that the results given above on the determination of τ do not agree with our results is caused not only by the different time resolution of the technique that was used, but also by the fact that the fluorescence was excited and recorded in a different range. In the presence of nonradiative singlet–singlet energy transfer, τ can be shortened and can substantially depend on the wavelength at which the fluorescence is excited and recorded. As far as the fluorescence-quenching time of Z,Z-bilirubin IXα coupled with BSA is concerned, according to [50], $\tau =33$ psec ps (λ_{exc} = 476 nm, λ_{em} = 550 nm, pH 7.4). Rotational depolarization processes naturally cannot develop in such a short fluorescence lifetime of bilirubin coupled with HSA and BSA, and, when the bilirubin molecules are excited by polarized radiation, the emission also remains polarized even in inviscid solutions. At the same time, the decrease of the rotational mobility of the pigment caused by the rigid microenvironment of the protein skeleton of the albumin plays a definite role in the high degree of polarization (close to the limiting value) of the bilirubin fluorescence. The fact that the pigment is rigidly fixed is also indicated not only by the tenfold increase of its fluorescence intensity when it is coupled with HSA and BSA but also by the higher degree of fluorescence polarization in comparison with that for bilirubin in chloroform, for which p = 0.24±0.03 with λ_{exc} = 453 nm, λ_{em} = 550 nm [67], while the fluorescence-intensity-damping kinetics is characterized by an exponential with τ = 14±2 psec [50].

To analyze the causes of the observed dependence of the position of the fluorescence (fluorescence-excitation) spectrum on the excitation (recording) wavelength, the nonconstancy of the degree of fluorescence polarization over the emission spectrum, and also its sharp falloff as the excitation wavelength is displaced toward the shorter-wavelength part of the absorption spectrum, we turn to data on the structure and stereoisomerism of the bilirubin molecule.

According to the spectral data, the bilirubin molecule maintains the folded conformation as part of a complex with both HSA [96, 103, 104] and with BSA [105]. In this case, bilirubin in a complex with HSA is deeply and completely incorporated in the structure of the macromolecule, so that it has virtually no contact with the aqueous environment [40, 50, 65].

In the complex with BSA, the pigment is accessible to the solvent: Its spectral characteristics are sensitive to a change of the pH of the medium [50, 65]. Moreover, whereas some excess of the *P* conformer is observed in the complex of bilirubin with HSA [96], BSA preferentially bonds the *M* conformation [103]. However, the complexes of bilirubin with both HSA and BSA are characterized by nonidentical microenvironments of each of the chromophores, created by the different sections of the polypeptide chain. It is also well known [40, 95] that the chromophores in the HSA molecule occupy positions with different semidomains of the protein macromolecule.

To analyze the regularities of the interchromophore interaction in the Z,Z-bilirubin IXα incorporated in the structure of the HSA or BSA protein globule, we turn to the fluorescence-excitation polarization spectrum (curves 7 in Figure 5 and Figure 9) and the fluorescence polarization spectra (curves 8 and 9 in Figure 5 and Figure 9). Most importantly, the form of the fluorescence-excitation polarization spectrum is typical of interactions accompanied by intra- or intermolecular singlet–singlet energy transfer of electron excitation [67–70]. Unlike the bilirubin–HSA complex (curve 7 in Figure 5), fluorescence depolarization for bilirubin coupled with BSA is much less pronounced when the excitation wavelength is displaced toward the shorter-wavelength part of the absorption spectrum (curve 7 in Figure 9). That is, under the indicated excitation conditions, the intramolecular interchromophore energy transfer is not as efficient for the bilirubin–BSA complex as for bilirubin coupled with HSA.

As the excitation is displaced toward the long-wavelength ("red") edge of the absorption band of bilirubin as part of a complex with HSA or BSA, there are fewer and fewer centers that can be acceptors of electron-excitation energy, and energy transfer becomes first improbable and then ($\lambda_{exc} \geq 480$ nm) virtually impossible: the degree of fluorescence polarization is close to the limiting value.

In this connection, the pronounced dependence of the polarization spectrum of the bilirubin fluorescence when it is excited in the region of the maximum of the absorption spectrum ($\lambda_{exc} = 460$ nm, curve 9 in Figure 5 and Figure 9) is most likely caused by the varying contribution over the pigment's fluorescence band of chromophores that directly absorb photons (a relatively high degree of polarization with short-wavelength recording) and of chromophores that obtain electron-excitation energy as a result of intramolecular transport (the longwavelength part of the emission spectrum, characterized by a lower value of p).

The fluorescence depolarization of Z,Z-bilirubin IXα recorded in this paper (more pronounced for its complex with HSA and less for that with BSA) for excitation on the shortwavelength slope of the absorption band is evidence that the interchromophore nonradiative energy-transfer processes are characterized by high rate, since they compete with the radiative deactivation processes of the electron-excitation energy of the chromophores that enter into the composition of Z,Z-bilirubin IXα.

An analysis of the literature data makes it possible to conclude that the factor that promotes this process is the presence of dipole–dipole resonance interactions between the chromophores – Davydov exciton coupling [106–108], which is evidence that its components are not completely independent in the bichromophore. A number of factors indicate that excitonic interactions are present between the chromophores of bilirubin. Most importantly, splitting of the band (sign inversion) in the circular dichroism spectrum is observed for complexes of the pigment with HSA [96, 104] and BSA [105], whereas an unsplit band is recorded for monochromophore analogs of bilirubin (the "semibilirubin" constituted by xanthobilirubic acid) [40, 109, 110]. In this case, two molecules of xanthobilirubic acid for which there is no splitting occupy the same site on the HSA molecule as on the bilirubin molecule [40].

Another confirmation of the presence of excitonic interactions between the chromophores is the significant broadening of the absorption spectrum of bilirubin and its displacement toward the long-wavelength region relative to the monochromophore analog. The cause of the broadening of the resultant spectrum of the bichromophore according to the theory of excitonic interactions is the splitting of the S_1 level of each chromophore into two (Davydov exciton splitting): one of them is energetically below (E_1) and the other above (E_2) the electronic level of the corresponding chromophore (see Figure 10).

The literature data show that the maximum of the absorption spectrum of xanthobilirubic acid in a complex with HSA is located at $\lambda_{max} = 425$ nm, while the spectrum is characterized by a half-width of $\Delta v = 3980$ cm^{-1} [40], whereas $\lambda_{max} = 458$ nm and $\Delta v = 4740$ cm^{-1} for a complex of Z,Z bilirubin IXα with HSA, and $\lambda_{max} = 462$ nm and $\Delta v = 4520$ cm^{-1} for that with BSA. In accordance with the theory of excitonic interactions [106–108], depending on the angle Θ between the chromophores, the resulting absorption spectrum of the bichromophore can be displaced either toward longer wavelengths or toward shorter wavelengths with respect to the absorption spectra of the component chromophores. The direction of the displacement is determined by the energy of the resonance dipole-dipole interaction V_{dd}: a long-wavelength shift must be observed when $V_{dd} < 0$, and a short-wavelength shift when $V_{dd} > 0$.

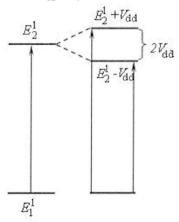

Figure 10. Schematic energy diagram of excition (Davydov) splitting.

The value of V_{dd} strongly depends on the dipole moments μ_1 and μ_2 of each of the chromophores, the distance R between the dipoles, and their orientation [106–108]: $V_{dd} = [\mu_1 \mu_2 (\cos \Theta - 3 \cos \beta_1 \cos \beta_2)]/4\pi\varepsilon_0 R^3$, where ε_0 is the permittivity, Θ is the angle between the dipoles, and β_1 and β_2 are the angles between each dipole and the radius-vector that connects the centers of the dipoles. In this case, the value of the excitonic splitting of the energy is $E_2 - E_1 = 2V_{dd}$.

As already pointed out, for bilirubin in chloroform, $\Theta \approx 98° - 104°$ [96], whereas, for the pigment–protein complexes studied here, the angle between the dipyrrole fragments is not determined. From the possible versions ($\Theta \approx 30°$, $\beta_1 \approx \beta_2 \approx 10° - 20°$; $\Theta \approx 90° - 100°$, $\beta_1 \approx \beta_2 \approx 60° - 70°$; $\Theta \approx 140°$, $\beta_1 \approx \beta_2 \approx 130° - 170°$) of the location of the chromophores proposed for such complexes in [96] on the basis of a study of their optical activity, it follows that $V_{dd} < 0$ in all cases. In accordance with theory, the resultant absorption spectrum of the bichromophore must be displaced toward longer wavelengths with respect to the monochrophore analog, and this is confirmed by the data given above for bilirubin and xanthobilirubic acid.

Thus, there is a basis for assuming that the factor that promotes excitonic excitation-energy transfer between the chromophores under conditions of an anomalously short fluorescence lifetime of bilirubin (the picosecond or subpicosecond range) is the presence of excitonic interactions between the chromophores. As a consequence of the small interchromophore distance (5–6 Å), the electric field vibrations of one chromophore (when it is excited) can be perceived by the other, causing excitation transfer. In this case, the electronic excitation becomes delocalized between the chromophores, creating a molecular excited state (an exciton).

An analysis of the experimental data (Figures 5 and 9) shows that, with short-wavelength excitation, rapid intramolecular interchromophore energy transfer delocalizes the excitation between the two absorbing chromophores, populating the levels at lower energy with higher probability. Reducing the excitation energy (displacement toward longer wavelengths) reduces the transfer probability, since the barrier between the excitonic states becomes significant. In this case, the excitation becomes localized on the chromophore that has absorbed a photon.

Because of this, the excited singlet state of Z,Z-bilirubin IXα splits into two energy levels and may be represented simplistically as a double minimum potential well (see Figure 11) from which decay, via twisted intermediates, can lead to ground state photoisomers (Z,E-bilirubin IXα or E,Z bilirubin IXα) or native Z,Z-bilirubin IXα.

The structural regrouping of chromophore I then implies the formation of E,Z-bilirubin IXα, and the structural reconstruction of chromophore II causes the formation of Z,E-bilirubin IXα (see Figure 1).Since, for bilirubin–HSA complexes, when the wavelength of the exciting radiation is displaced toward the longwavelength slope of the absorption, a reduction is recorded in the quantum yield of the Z,E-bilirubin IXα and an increase of the quantum yield of lumirubin [43], it could be assumed that the absorption spectrum of chromophore I is displaced toward longer wavelengths with respect to the spectrum of chromophore II. However, as follows from the data [44, 84, 103], when bilirubin–HSA complexes are irradiated within the long-wavelength absorption band of the pigment, the concentration of Z,E-bilirubin IXα always exceeds that the E,Z-isomer.

The data given above make it possible to conclude that, for Z,Z bilirubin IXα coupled with HSA, the absorption spectrum of chromophore II is displaced toward the longer wavelengths with respect to chromophore I. In this case, when bilirubin is excited to high vibrational sublevels of the S_1 state, the excited level of chromophore II is populated both by direct electronic excitation and by energy migration from the S_1 state of chromophore I. As the wavelength of the active radiation is displaced toward the longwavelength slope of the absorption band of bilirubin, the interchromophore energy-transfer probability is reduced (see Figures 5 and 9). Accordingly, the quantum yield of the formation of Z,E bilirubin IXα must also be reduced, and this is confirmed by the data [43].

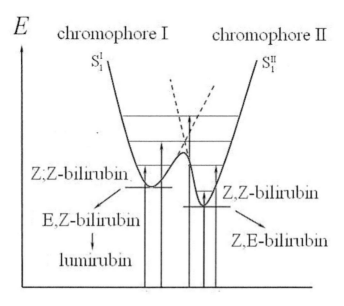

Figure 11. Scheme of Z,Z-bilirubin IXα photoisomerization and the relative position of levels of singlet excited state of chromophore I and II.

In this case, as a consequence of the more intense absorption of chromophore II (by comparison with chromophore I), with long-wavelength edge excitation, the relative number of molecules of Z,E bilirubin IXα (formed as a result of cis-trans photoisomerization of chromophore II) exceeds that for E,Z bilirubin IXα. With short- wavelength excitation, the absorption intensity of chromophore I is greater than that of chromophore II. However, as a consequence of the very efficient intramolecular "efflux" of electronic-excitation energy to chromophore I, the predominant formation of the pigment Z,E bilirubin IXα is also observed among the photoproducts.

The efficiency of interchromophore energy transfer is not so pronounced for the bilirubin–BSA complex, and this causes the low selectivity of the photoisomerization process for the indicated complex. It can be assumed that the high efficiency of the interchromophore energy transfer in bilirubin coupled with HSA and its lower value for the complex with BSA are caused by the differences of the conformations of the pigment in the indicated complexes (differences in the angles between the dipyrrole fragments and in the interchromophore distances). The available information [39, 40, 50, 65, 96, 103, 104] concerning the structure of the complexes of bilirubin with HSA and BSA is evidence in favor of the given assumption. It should be pointed out in this case that the results given above do not at all exclude a possible role of steric factors capable of having an effect on the degree of selectivity of the process of configurational cis–trans photoisomerization of Z,Z bilirubin IXα in its complexes with HSA and BSA.

Conclusion on the Results Study of the Photophysical Processes that Determine the Photoisomerization Selectivity of Z,Z-Bilirubin IXα in Complexes with Albumins. For the first time experimental evidence was obtained, that the molecule of Z,Z-bilirubin IXα, which excessive accumulation in the body of a newborn leads to the development of jaundice syndrome, is a compound of two interacting chromophores (bichromophore) the contribution of each into total absorption and emission of light pigment, as well as its geometric (cis-trans) and structural photoisomerization is determined by the excitation wavelength and the microenvironment.

The reasons of the ratio alteration between photoproducts (Z,E-bilirubin IXα, E,Z-bilirubin IXα, E,Z-lumirubin) of Z,Z-bilirubin IXα, related with blood albumin, by change of the wavelength of radiation within the long-wavelength absorption of tetrapyrrole are found: (i) spectral non-identity of chromophores, forming a molecule of pigment, caused by the asymmetry of their structure and heterogeneity of the environment in a protein matrix, (ii) the presence of interchromophore intramolecular energy transfer, probability of which depends strongly on the wavelength.

It's shown that the dependence of the quantum yield of lumirubin on the wavelength of influencing laser radiation is determined by: (i) changing contribution into the entire absorption of light by bilirubin chromophore I, responsible for forming of lumirubin, (ii) change of the efficiency of intramolecular "outflow" of the energy of electronic excitation from this chromophore to chromophore II, unable to photocyclization due to structural features.

The results of our studies showed that one cause of the predominant formation of Z,E-bilirubin IXα among the photoisomerization products of Z,Z-bilirubin IXα when it is irradiated as part of complexes with transport proteins HSA and BSA is the presence of interchromophore intramolecular energy transfer. The efficiency of such transfer substantially depends on the type of transfer protein that determines the conformation of the coupled pigment and is more pronounced for its complex with HSA and less for that with BSA.

The factor that promotes interchromophore energy transfer under conditions of anomalously short fluorescence lifetime of bilirubin (the picosecond or subpicosecond range) is the presence of excitonic interactions between the chromophores.

4. Sensitizing Effect of Z,Z-Bilirubin IXα and its Photoproducts on Enzymes in Model Solutions

As noted that the determining role in reducing the bilirubin level in the blood of newborns during phototherapy is played by processes of photoisomerization of the pigment: formation of its configurational isomers (Z,E-bilirubin IXα, E,Z-bilirubin IXα, E,E-bilirubin IXα) and structural isomers (Z-lumirubin and E-lumirubin) (see Figure 1). Besides the process of isomerization of Z,Z-bilirubin IXα when exposed to light, oxidation reactions also occur (Figure 2), accompanied by accumulation of colorless products based on monopyrrole and dipyrrole moieties. According to [35–37], one of the mechanisms for photodegradation of bilirubin involves participation of singlet oxygen in this process (self-sensitized oxidation). In fact, according to a number of indirect estimates, the triplet level of bilirubin has an energy of ~150 kJ/cm² [35], which makes it possible to generate singlet oxygen since for the latter, the energy of the S_1 level is $^1\Delta_g$ = 94 kJ/cm². In this case, however, we cannot rule out the possibility that the singlet oxygen generated by the triplet excited Z,Z-bilirubin IXα will cause photodynamic damage to other essential molecular structures and organelles of the cell. Thus in [111], it is demonstrated that photodegradation of amino acids of the major transport protein in blood, human serum albumin (HSA), can occur when its complexes with bilirubin are exposed to radiation in the visible region of the spectrum. The authors of [112] note bilirubin-sensitized damage to blood plasma lipoproteins. We discuss the possibility of photodynamic damage folic acid photosensitization of bilirubin [113]. They also discussed the need to reduce the dose load on the baby during phototherapy, because of the observed possibility of damage to nuclear DNA [114, 115], breakdown of the permeability of the cell membranes, and death of bilirubin-stained human cells in culture [116–120] when they are irradiated.

All this indicates the need to study the sensitizing effect of bilirubin on very important biological structures to eliminate (reduce) possible unfavorable side effects of exposure to light, especially considering that the total phototherapy time in treatment of severe forms of hyperbilirubinemia is sometimes more than 100 h, with a total dose of up to 200 J/cm² [1–6].

The aim of this work was to study the sensitizing effect of Z,Z-bilirubin IXα and its photoproducts on the enzyme lactate dehydrogenase (LDH) in model buffer solutions containing LDH and the HSA-bound pigment over a broad range of bilirubin/albumin concentration ratios, and also for different wavelengths of the radiation used [121]. We used the bilirubin complex with albumin as the photosensitizer rather than free bilirubin because in the blood, bilirubin is mainly found in the bound state.

The enzyme molecules were selected as a test compound for testing the photosensitizing effect of bilirubin and its photoproducts because of the determining role of this type of macromolecule in regulation of metabolic processes in the body and their ability to act as the most sensitive target for photodynamic reactions occurring both according to a radical mechanism and with participation of singlet oxygen. The contribution of the above-indicated reactions to the degradative effect of light under conditions of photosensitization by bilirubin (or its photoproducts) was estimated using as an example LDH, which does not contain prosthetic groups and does not absorb in the visible region of the spectrum, and therefore a change in its enzyme activity on exposure to low-intensity radiation in the visible region of the spectrum can be initiated only by photosensitized processes.

Spectral Characteristics of Irradiated Components and the Radiation Sources. The electronic absorption spectra of HSA and LDH are located in the UV region of the spectrum, and are characterized by a maximum at $\lambda = 280$ nm, which is mainly determined by the tryptophan, tyrosine, and phenylalanine amino acid residues within the indicated protein molecules [57].

The absorption spectrum of bilirubin bound to HSA in a pH 7.4 buffer solution is shown in Figure 12 (curve 1); the maximum is located at $\lambda = 458$ nm. In the emission spectrum of the Westinghouse F20T12 fluorescent lamp (curve 2), besides the main fluorescence band with maximum at $\lambda_{em} = 447$ nm, there is a series of mercury lines (313, 365, 405, 436, 546, 577 nm) located in the near UV and visible ranges.

We emphasize that usually phototherapy for hyperbilirubinemia is carried out in an neonatal incubator, the upper wall of which is made of polymethylmethacrylate and is a unique protective shield for the emission of the mercury lines with $\lambda \leq 313$ nm, capable of causing photodegradation of protein molecules [57]. For this reason, the test solutions were exposed to light from the lamp through a polymethylmethacrylate backing (thickness 6 mm), the transmission spectrum of which is given in Figure 12 (curve 3).

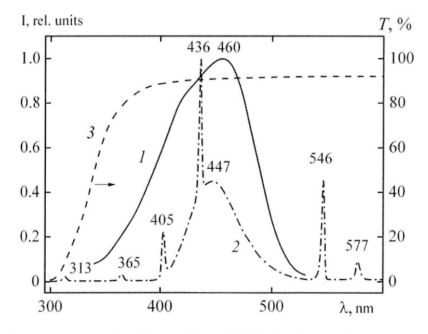

Figure 12. Absorption spectrum of a buffered solution of Z,Z-bilirubin IXα bound to HSA (1), emission spectrum of a Westinghouse F20T12 fluorescent lamp (2), light transmission spectrum of polymethylmethacrylate of thickness 6 mm (3).

The intensity of the lamp emission, at the level where the cuvet with the solution was located, was ≈1 mW/cm^2, which is a typical value for phototherapy procedures carried out with the aim of treating the hyperbilirubinemia syndrome in newborns.The emission of the laser sources corresponds to the maximum of the bilirubin absorption spectrum (λ = 457.9 nm), and also its short-wavelength (λ = 441.6 nm) and long-wavelength (λ = 514.5 nm) edges.Special studies showed that neither the lamp emission nor the laser emission in the intensity range used ($P \leq 40$ mW/cm^2) in the absence of bilirubin have an effect on the enzyme activity of LDH.

Kinetics of Sensitized Photo-Inactivation of the Enzyme. Prolonged irradiation of buffer solutions containing bilirubin–HSA (C_{Br} = 4 µM, C_{HSA} = 2 µM) and LDH (C_{LDH} = 8·10^{-9} M) leads to, in addition to photodegradation of bilirubin, a decrease in the catalytic activity of the enzyme. Figure 13 shows (on a semilog scale) the kinetics of the decrease in optical density in percent relative to the control ((A/A_0)·100%) at the maximum of the absorption spectrum for Z,Z-bilirubin IXα, and the enzyme activity of LDH (γ) in percent relative to the control when irradiated by the lamp and by the lasers. In the initial stage of the dose dependence (curves 1, 3, 5) we observe rapid bleaching of the bilirubin, then the rate of this process falls off (the dose curve deviates from a straight line).

The decrease in the LDH enzyme activity exhibits different behavior (curves 2, 4, 6). Thus over a significant dose interval, there is no photodegradation of the enzyme (the enzyme activity of LDH is no different from the unirradiated control samples). Then we note some decrease in the enzyme activity, and the rate of the LDH photo-inactivation process increases as the irradiation dose increases and consequently as the Z,Z-bilirubin IXα concentration decreases.

This type of LDH photoinactivation kinetics allows us to conclude that it is not bilirubin that acts as the enzyme photosensitizer but rather its photoproducts, the concentration of which increases according to the degree of radiation exposure for the solution.

There are no fundamental differences in the behavior of sensitized photo-inactivation of LDH when exposed to emission of either the fluorescent lamp or the lasers (see Figure 13).

In this case, the use of laser radiation corresponding to different sections of the absorption spectrum for bilirubin and its photoproducts allows us to study the sensitized effect on LDH under conditions when (at the beginning of exposure) we make sure that identical numbers of photons are absorbed by the bilirubin (N = 1.66·10^{15} cm^{-2}·sec^{-1}) at the different wavelengths. Taking into account the difference in optical density of bilirubin at the laser emission wavelengths, the indicated condition is satisfied by varying the intensity of the radiation used: P = 4.0 mW/cm^2 for λ = 441.6 nm, P = 3.7 mW/cm^2 for λ = 457.9 nm, and P = 35.4 mW/cm^2 for λ = 514.5 nm (for unchanged initial concentration of the components to be irradiated).

Note that for irradiation corresponding to identical initial absorption of photons of different wavelengths by bilirubin, the fastest initial rate of decrease in optical density at the maximum of bilirubin absorption is observed for exposure to radiation with λ = 514.5 nm (curve 5), and the slowest initial rate is observed at λ = 441.6 nm (curve 3).

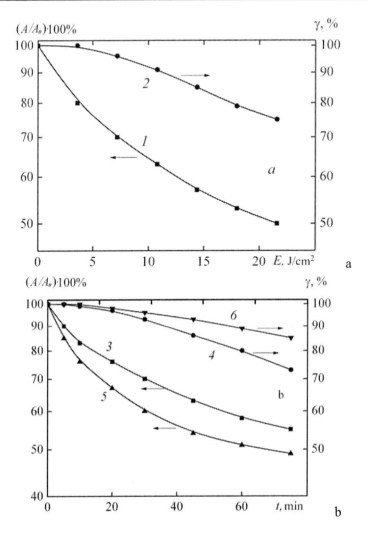

Figure 13. Kinetics of the decrease in optical density at the maximum of the absorption spectrum for Z,Z-bilirubin IXα (1, 3, 5) and photo-inactivation of LDH (2, 4, 6) when they are exposed in the presence of HSA to a fluorescent lamp (1, 2), a helium-cadmium laser with λ = 441.6 nm (3, 4), or an argon laser with λ = 514.5 nm (5, 6). $C_{Br}/C_{HSA} = 2/1$. E is the irradiation dose, t is the exposure time.

Our studies also showed that for identical exposure time and the above-indicated conditions, the greatest inactivation of the enzyme is observed for exposure to radiation with λ = 441.6 nm (Figure 13, curve 4), and the least inactivation is observed at λ = 514.5 nm (curve 6); the sensitizing activity with respect to LDH of radiation with λ = 457.9 nm is intermediate between these values (the curves are not shown to avoid clutter).

We established that sensitized damage to LDH occurs with participation of singlet oxygen ($^1O_2$), evidence for which comes from the decrease in the photo-inactivation effect when known $^1O_2$ quenching agents are added to the mixture to be irradiated: sodium azide (Figure 14, curve 2) or reduced nicotinamide adenine dinucleotide (NADH, curve 3), and also the increase in the degree of LDH photo-inactivation when the buffer based on H_2O (curve 1) is replaced by a buffer based on D_2O (curve 5). As we know, the D_2O test is an indicator of participation of $^1O_2$ in a photochemical reaction, since the lifetime of singlet oxygen and the probability of its chemical reaction with a biosubstrate is an order of magnitude higher in heavy water than in H_2O. Note that both D_2O and NaN_3 also affect photolysis of bilirubin, but the rate of bleaching of the solution is increased (D_2O) or decreased (NaN_3) by no more than 3%–5%. Therefore it is most likely that the increase in the degradative effect on LDH on going to D_2O and the decrease in the degree of photo-inactivation of the enzyme on addition of NaN_3 are not due to a change in the concentration of the sensitizer photoproducts, but rather are evidence for participation of $^1O_2$ in the studied process. At the same time, the higher protective effect of NADH (characterized by a $^1O_2$ quenching rate constant $k_q = 7.9 \cdot 10^7$ $M^{-1} \cdot sec^{-1}$ and being an electron and proton donor) compared with NaN_3 ($k_q = 4.4 \cdot 10^8$ $M^{-1} \cdot sec^{-1}$) does not rule out participation of radical processes in the studied enzyme photo-inactivation effects. Participation of hydroperoxides in photodegradation of LDH is also possible, since addition of catalase (an enzyme with a high H_2O_2 cleavage rate) to the irradiated solution has a pronounced protective effect with respect to LDH (curve 4).

As already noted, for an initial concentration ratio $C_{Br}/C_{HSA} = 2/1$, it is not bilirubin that acts as a sensitizer with respect to the enzyme present in solution but rather the photoproducts formed during its irradiation. The situation changes when the bilirubin concentration significantly exceeds the concentration of the carrier protein. In this case, bilirubin is not only capable of being incorporated into two strong binding sites on the HSA molecule, but also can occupy 10–14 sites on the surface of the protein globule, characterized by weak binding with the macromolecule [122]. As an example, we show the dose dependence of LDH photo-inactivation for $C_{Br} = 4$ µM, $C_{HSA} = 1$ µM (Figure 14, curve 6).

As we see, for excess bilirubin (compared with HSA), we observe an increase in the LDH photo-inactivation rate, where the dose dependence is close to exponential. It is most likely that for the condition $C_{Br}/C_{HSA} > 2$, bilirubin can act as a photosensitizing agent with respect to LDH.

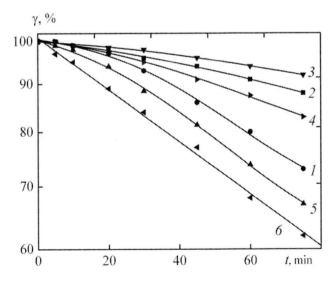

Figure 14. Kinetics of photo-inactivation of LDH in the presence of the bilirubin-albumin complex on exposure to radiation from a helium-cadmium laser with $\lambda = 441.6$ nm: 1, 6) no additives; 2) in the presence of $5 \cdot 10^{-3}$ M NaN_3; 3) in the presence of $1 \cdot 10^{-4}$ M NADH; 4) in the presence of $1 \cdot 10^{-8}$ M catalase; 5) solution of bilirubin–HSA–LDH in D_2O; $C_{Br} = 4$ µM; $C_{HSA} = 2$ µM (1–5) and 1 µM (6).

Spectral Characteristics of Photoproducts of Z,Z-Bilirubin IXα. In order to determine the contribution of bilirubin photoproducts to photo-inactivation of the enzyme, let us focus on the spectral characteristics of the pigment photoisomerization and photooxidation products. Note that both configurational and structural photoisomers of bilirubin are very unstable compounds [36]. Generally the absorption spectra of the isomeric forms of the pigment can be recorded only in the main band (with maximum in the region $\lambda = 450-460$ nm) using a polychromator mounted at the exit from the chromatographic column after separation of the components. Therefore the absorption and spectral luminescence characteristics of the isomers have not been studied much, and data from different authors are quite contradictory [29, 75, 77, 87, 123–125]. Binding to albumin has some stabilizing effect on the isomers [126], but in this case a large set of bilirubin photoproducts are present in solution, and so information on the spectral characteristics of the isomeric forms of bilirubin within an HSA complex is fragmentary [75, 77].

Figure 15 shows the typical changes induced in the absorption spectra of the bilirubin-albumin complex ($C_{BR}/C_{HSA} = 2/1$) when exposed for $t = 75$ min to the emission from a helium-cadmium laser with $\lambda = 441.6$ nm, $P = 4$ mW/cm^2 and the emission from an argon laser with $\lambda = 514.5$ nm, $P = 35.4$ mW/cm^2, under conditions when the initial numbers of photons absorbed by bilirubin are identical for different wavelengths. Irradiation of the solutions is accompanied by a decrease in the optical density at the maximum of the pigment absorption band, an increase in the optical density on the short-wavelength edge of the bilirubin absorption spectrum, and also formation of a new absorption band in the region $\lambda = 530-700$ nm. Despite the identical number of photons initially absorbed by the bilirubin on exposure to light with $\lambda = 441.6$ nm (curve 2) and $\lambda = 514.5$ nm (curve 3), radiation in the green region causes more pronounced changes at the maximum of the absorption band even in the UV part of the spectrum. The increase in optical density in the red region is more pronounced for exposure to radiation at $\lambda = 441.6$ nm.

There is a basis for assuming that for a dose load causing sensitized damage to LDH, the decrease in optical density at the maximum of the bilirubin absorption band is due mainly to formation of structural photoisomers (Z-lumirubin and E-lumirubin), while the contribution of the *cis/trans* photoisomers (Z,E-bilirubin IXα, E,Z-bilirubin IXα, and E,E-bilirubin IXα) and the photooxidation products of the pigment is less pronounced.

Thus removing molecular oxygen from the solution before it is irradiated leads to a decrease in the initial rate of bleaching of the solution (monitored at $\lambda = 460$ nm) by no more than 10%. This indicates that the processes of self-sensitized photooxidation of the pigment, characterized by a very low quantum yield $\varphi_{PO} < 0.01$, are not of determining importance in photoconversion of bilirubin.

We may expect a significant contribution to the decrease in optical density at the maximum of the absorption band for the pigment Z,E-bilirubin IXα, the formation of which is characterized by the highest quantum yield (among the *cis/trans* photoisomerization products of bilirubin) ($\varphi_{Z,E} < 0.11$). However, according to [75], the molar extinction coefficient of Z,E-bilirubin IXα bound to HSA at the maximum of the absorption band ($\lambda = 475$ nm) is somewhat higher than the corresponding values for Z,E-bilirubin IXα at its maximum for $\lambda = 460$ nm. Therefore formation of the Z,E-isomer should not lead to a decrease in optical density in the blue region of the spectrum. At the same time, for the free form of the stereoisomers (in the absence of albumin), different data are available in the literature about the ratio of the molar extinction coefficients for Z,E-bilirubin IXα and Z,Z-bilirubin IXα ($\varepsilon_{Z,E}/\varepsilon_{Z,Z}$).

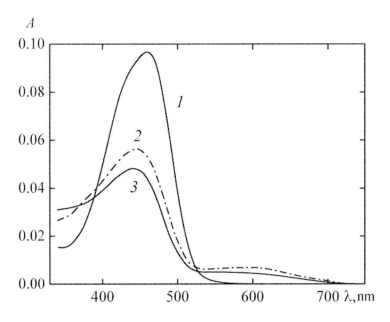

Figure 15. Absorption spectra of Z,Z-bilirubin IXα bound to HSA, before exposure (1) and with exposure to radiation from a helium-cadmium laser with λ = 441.6 nm, P = 4 mW/cm^2 (2); an argon laser with λ = 514.5 nm, P = 35.4 mW/cm^2 (3) for 75 min; C_{BR}/C_{HSA} = 2/1.

Thus according to the data in [123], $\varepsilon_{Z,E}/\varepsilon_{Z,Z}$ = 0.80; according to the data in [87], $\varepsilon_{Z,E}/\varepsilon_{Z,Z}$ = 0.70. The other geometric photoisomers of bilirubin (E,Z-bilirubin IXα and E,E-bilirubin IXα) most likely contribute insignificantly to the decrease in the optical density of the solution in the blue region of the spectrum when it is irradiated. This is indicated by the fact that the quantum yield for the formation of the E,Z-isomer (depending on the wavelength of the radiation used) is 10–80 times lower than the corresponding $\varphi_{Z,E}$ [75, 86, 87], while the molar extinction coefficient is close to that for Z,Z-bilirubin IXα: $\varepsilon_{E,Z}/\varepsilon_{Z,Z}$ = 0.98 [77, 123].

It is important to emphasize that due to the reversibility of the *cis/trans* photoisomerization reaction of bilirubin, an equilibrium is rapidly established between the Z,Z, Z,E, E,Z, and E,E stereoisomers. The time needed to establish the steady-state concentration of the isomers is determined entirely by the power of the radiation used, and in the intensity range used (P = 3.7–35.4 mW/cm^2) is no longer than 1 min from the time exposure begins [88, 124]. In contrast to stereoisomerization processes, the structural photoisomerization reactions accompanied by lumirubin formation are irreversible (see Figure 1). Therefore the lumirubin concentration increases according to the degree of exposure of the solution (for ~30–120 min [87, 88, 124]), while the concentration of the *cis/trans* isomers gradually decreases because of loss of Z,Z-bilirubin IXα (due to photooxidation and structural photoisomerization processes that occur) and establishment of an equilibrium between the stereoisomers at a lower concentration level. Typically for a concentration ratio C_{Br}/C_{HSA} = 1/2 (i.e., when the bilirubin is localized at the strongest binding site of the protein) and a dose load ensuring approximately a two-fold decrease in optical density at the maximum of the pigment absorption band, the lumirubin concentration in solution is greater than the concentration of Z,Z-bilirubin IXα and Z,E-bilirubin IXα [127]. And as the bilirubin occupies the second binding site on the albumin (as the ratio C_{Br}/C_{HSA} increases to 2/1), the yield of lumirubin increases, while the yield of Z,E-bilirubin IXα decreases by a factor of about one-and-a-half [88, 127].

Thus for the concentration ratio used, C_{Br}/C_{HSA} = 2.0, and the exposure time inducing a decrease in the optical density in the blue region of the spectrum by a factor of 1.5–2.0, the major contribution to bleaching of the solution comes from the lumirubin, characterized in the λ = 450 nm region by a molar extinction coefficient ≈2.0–2.7 times lower than for Z,Z-bilirubin IXα [87, 123].

One more typical feature of the spectrally manifested photochemical processes in the bilirubin molecules is the short-wavelength shift of the absorption band as the radiation dose increases. In our opinion, this is due to the shorterwavelength (compared with native bilirubin) position of the absorption spectrum of lumirubin, which is accumulating in the solution while it is irradiated. Evidence in favor of the indicated hypothesis comes from the data in [29, 73, 87, 124], according to which the maximum in the absorption spectrum of lumirubin in the free state (in the absence of protein) is shifted by 10-12 nm toward shorter wavelengths compared with Z,Z-bilirubin IXα in the same solvent. (Data on the spectral characteristics of lumirubin bound to HSA could not be observed.) As already noted, the absorption spectrum of the Z,E isomer bound to HSA is characterized by a maximum at λ = 475 nm [75], while the maximum of the absorption spectrum for E,Z-bilirubin IXα within the HSA complex practically coincides with that for Z,Z-bilirubin IXα (λ = 460 nm) [77].

The most likely reason for the increase in the absorption in the UV region of the spectrum as the irradiation dose increases for the HSA–bilirubin complexes is formation of bilirubin photooxidation products (based on monopyrrole and dipyrrole moieties, characterized by a maximum at λ = 320 nm [57]), and also lumirubin and its photoproducts [112]. According to [112], one of the absorption bands of lumirubin is characterized by a maximum at λ = 337 nm, while its photoproducts absorb at 350–370 nm. In the λ = 380 nm, there is also an absorption band from biliverdin, which is practically nonfluorescent and does not have a sensitizing effect [35, 36]. In addition to the indicated compounds, some contribution to the observed increase in absorption in the UV region can also come from the products of bilirubin-sensitized degradation of tryptophan amino acid residues of HSA (formylkynurenine, kynurenine), absorbing in the λ = 360 nm region and capable of generating singlet oxygen and having a photodynamic effect [57].

Another unclear question concerns the photoproducts forming after absorption in the region λ = 530–700 nm as the radiation exposure dose increases. Note that when using a broadband lamp source, the increase in the absorption in the given range of the spectrum is not so pronounced [36]. This has resulted in some authors [36] doubting the presence of a photoproduct with absorption maximum in the range λ = 530–700 nm. In this connection, we should note that one of the maxima in the absorption spectrum of lumirubin, according to data in [18], is located in the λ = 590 nm region, while the maximum for biliverdin is located in the range 660–750 nm [35, 36]. However, according to the results of chromatography studies, the increase in optical density in the 530–700 nm region does not correlate either with the increase in the lumirubin concentration or with the increase in the concentration of biliverdin, detected in trace amounts [127]. Furthermore, the increase in absorption in the given region of the spectrum is also absorbed after the lumirubin concentration goes to the steady-state level, which made it possible to hypothesize [127] a connection between the indicated increase and lumirubin photopolymerization processes.

Some information about the absorption spectra of bilirubin photoproducts when bilirubin is irradiated within an HSA complex was obtained from the fluorescence excitation spectra (Figure 16). As we know [46, 68], the fluorescence of Z,Z-bilirubin IXα bound to HSA is characterized by very low quantum yield ($\varphi_{fl} = (1-3) \cdot 10^{-3}$) with maximum at $\lambda = 530$ nm. For $\lambda_{em} = 530$ nm, for the unirradiated solution we record an excitation spectrum with maximum at $\lambda = 457$ nm (curve 1), which practically corresponds to the absorption band of the intact pigment. For the irradiated solution, the excitation spectrum is located at $\lambda = 460$ nm (curve 2) and is significantly broadened, which is probably explained by the contribution to the fluorescence of the lumirubin and cis/trans isomerization products of bilirubin present in solution, which have close but somewhat different absorption and fluorescence spectra. Furthermore for the irradiated solution of bilirubin, at $\lambda_{em} = 680$ nm, corresponding to the fluorescence band of lumirubin [18], we record the excitation spectrum with maxima at $\lambda = 335, 457,$ and 590 nm (curve 3), and for $\lambda_{em} = 720$ nm we record an excitation spectrum with maxima at $\lambda = 348$ nm and $610-620$ nm (curve 4), which according to [18] belongs to the photolysis products of lumirubin. Note that due to the multicomponent composition of the irradiated solution, the position of the indicated maxima in the fluorescence excitation spectrum (Figure 16) is shifted by $\Delta\lambda = \pm 10$ nm, depending on the fluorescence detection wavelength and the radiation exposure dose for the solution, and hence the ratio of the concentrations of the components.

Analysis Sensitizing Action of Z,Z-Bilirubin IXα and its Photoproducts on Enzymes in Model Solutions. The data presented suggest that prolonged irradiation ($t = 30-75$ min) of model solutions containing Z,Z-bilirubin IXα bound to HSA can lead to photo-inactivation of the enzyme protein (LDH) present in low concentrations in the irradiated solution. Either Z,Z-bilirubin IXα itself or its photoproducts act as the enzyme photosensitizing agent.

The contribution of the indicated photosensitizers to photo-inactivation of LDH is determined by the initial ratio of the concentrations, C_{Br}/C_{HSA}. Thus for $C_{Br}/C_{HSA} \leq 2/1$, the bilirubin molecules are localized at strong HSA binding sites and the sensitizing effect of the pigment is directed at amino acid residues of the carrier protein [111]. In this case, the photoconversion products of bilirubin act as LDH photosensitizers, which is confirmed by the enzyme photo-inactivation kinetics (Figure 13, curves 2, 4, 6): over a significant dose interval, there is no photodegradation, then we observe some decrease in the enzyme activity, and the LDH photo-inactivation rate increases as the irradiation dose increases and consequently the concentration of Z,Z-bilirubin IXα decreases (curves 1, 3, 5) and the photoproduct concentration increases.

In our opinion, for the indicated initial ratio of the bilirubin/HSA concentrations, the determining role in sensitized damage to LDH is played by lumirubin: the product of structural photoisomerization of the pigment which is quite soluble in water [73], while the contribution of the products of stereoisomerization of Z,Z-bilirubin IXα is insignificant.

This is indicated by the fact that the concentration of configurational photoisomers of bilirubin (Z,E, E,Z, and E,E isomers) goes to the steady-state level (due to the dynamic equilibrium between the stereoisomers) after ≈ 1 min from the time radiation exposure begins in the intensity level used [88, 124].

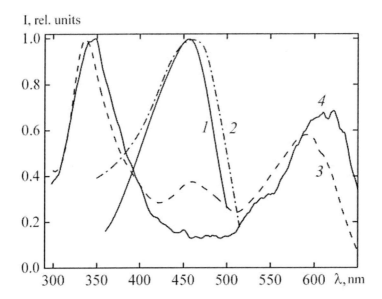

Figure 16. Typical fluorescence excitation spectra of Z,Z-bilirubin IXα bound to HSA (1) and its photoproducts (2–4) for λ_{em} = 530 nm (1, 2), 680 nm (3), and 720 nm (4).

Therefore in the case of the sensitizing effect of the *cis/trans* photoisomers on the dose curve, we should expect a decrease in the enzyme activity of LDH practically immediately after radiation exposure begins. Formation of lumirubin is a rather slow process, and its accumulation in solution occurs over a prolonged time period (t = 30–120 min [87, 88, 124]), depending on the intensity of the radiation used. Furthermore, the binding constant between lumirubin and HSA, according to estimates in [129], is two orders of magnitude lower than the corresponding value for Z,Z-bilirubin IXα; and in contrast to bilirubin, lumirubin is located on the surface of the HSA protein globule. Thus if the photochemically active intermediates formed on absorption of light by bilirubin, having practically no contact with the aqueous environment at the strong binding sites of the macromolecules [46, 68], primarily interact with the amino acid residues of HSA (as follows from [111]), then lumirubin can sensitize the process of degradation of both HSA and LDH.

Another characteristic feature of photosensitized damage to LDH is the more pronounced effect of radiation in the blue region of the spectrum (λ = 441.6 nm) compared with radiation in the green region (λ = 514.5 nm). This conclusion follows from the results shown in Figure 13 (curves 4 and 6), where for an equal number of photons absorbed by bilirubin at the initial instant of time, the intensity of the radiation with λ = 514.5 nm was greater than that for λ = 441.6 nm by a factor of ≈8.6. In this case, sensitized photo-inactivation of the enzyme occurs with participation of singlet oxygen and hydrogen peroxide, as follows from Figure 14.

In contrast to LDH, photodegradation of bilirubin under the given irradiation conditions occurs more rapidly when exposed to light with λ = 514.5 nm (Figure 15, curve 3) compared with λ = 441.6 nm (curve 2) or λ = 457.9 nm. In this case, for doses of the radiation used ensuring a one-and-a-half to two-fold decrease in optical density at λ = 460 nm, the determining contribution to the indicated transformation of the spectrum comes from formation of lumirubin, the concentration of which is greater than the concentration of the other isomers, characterized in addition by the lowest molar extinction coefficient ($\varepsilon_{Z,Z}/\varepsilon_{Lr}$ = 2.0–2.7) [87, 123].

Accordingly, the observed difference between the bilirubin bleaching rates (monitored at the maximum of its absorption spectrum) for exposure to radiation with λ = 441.6 nm and 514.5 nm (Figure 13, curves 3 and 5) is due to the higher (by a factor of ≈2) quantum yield for lumirubin formation at λ = 514.5 nm [73, 75, 77, 130]. The reasons for the dependence of the quantum yield for the formation of the photoisomerization products on the wavelength of the radiation used are the bichromophore nature of the pigment structure, the fact that the chromophores are structurally and spectrally different, and also the presence of intramolecular energy transfer (due to exciton interactions between chromophores), the efficiency of which changes when the wavelength of the radiation is varied within the limits of the long-wavelength absorption band of bilirubin [46, 68].

As already noted, despite the higher lumirubin content in solution for exposure to λ = 514.5 nm, its sensitizing effect (Figure 13, curve 6) is less pronounced than when using light with λ = 441.6 nm (curve 4). Probably the difference between the photodynamic effects is explained by the higher molar extinction of lumirubin at λ = 441.6 nm than at λ = 514.5 nm [77, 87, 123]. The negative effect of light on the biochemical activity of LDH is even more enhanced (Figure 13, curve 2) on going to a broadband radiation source: a fluorescent lamp, half the intensity of which is concentrated in the violet region of the spectrum (Figure 12). The indicated transmission band of the lamp also overlaps the intense absorption band of lumirubin, the maximum of which overlaps the intense absorption band of lumirubin, the maximum of which is located in the UV region at λ = 335 nm (Figure 16, curve 3). It is logical to assume that its presence for lumirubin is due to a chromophore containing (in contrast to bilirubin and its stereoisomers) a sevenmembered ring (see Figure 1), since according to [130], the short-wavelength maximum of the absorption spectrum of bilirubin is located at λ < 250 nm.

Our studies have also shown that for a significant excess of Z,Z-bilirubin IXα compared with albumin (C_{Br}/C_{HSA} = 4/1), the bilirubin can act as the LDH photosensitizing agent, which is confirmed by the absence of any lag period on the dose curve for photo-inactivation of the enzyme (Figure 14, curve 6). Note that in this case, the LDH inactivation kinetics are close to exponential. Probably the appearance of a photosensitizing effect of bilirubin on the enzyme for C_{BR}/C_{HSA} > 2/1 is promoted by some of the pigment molecules being located on the surface of the HSA protein globule [122]. There are grounds for assuming that such a situation may be realized in real clinical practice for babies with a high bilirubin level, especially considering that due to a change in the conformation of albumin for some types of hyperbilirubinemia, its binding capacity may be decreased many-fold [131]. For this reason, despite the closeness of the albumin concentration (370 μM) to the bilirubin concentration (170–350 μM) in the blood of newborn infants with the hyperbilirubinemia syndrome, some amount of unbound bilirubin is detected [6], capable of acting as a photosensitizer for macromolecules functioning as enzymes. However, we have to say that in blood there are also a number of components, including low molecular weight components, capable of acting as quenching agents and acceptors of active forms of oxygen, which may significantly inhibit photo-inactivation of enzymes.

A decrease in potential side effects during phototherapy of hyperbilirubinemia, due to the sensitizing effect from lumirubin, can be promoted by alternating phototherapeutic procedures and pauses between exposures (while maintaining the overall energy load). For successful realization of this approach, the pause between phototherapy sessions must be sufficient for diffusion of lumirubin into the interior of the tissue and its subsequent excretion.

Conclusion on the Results Study of the Sensitizing Action of Z,Z-Bilirubin IXα and its Photoproducts on Enzymes in Model Solutions. In model systems, we have studied side effects which may be induced by light during phototherapy of hyperbilirubinemia (jaundice) in newborn infants, with the aim of reducing the Z,Z-bilirubin IXα level. We have shown that the sensitizing effect of Z,Z-bilirubin IXα, localized at strong binding sites of the human serum albumin (HSA) macromolecule, is primarily directed at the amino acid residues of the carrier protein and does not involve the molecules of the enzyme (lactate dehydrogenase (LDH)) present in the buffer solution. The detected photodynamic damage to LDH is due to sensitization by bilirubin photoisomers, characterized by lower HSA association constants and located (in contrast to native Z,Z-bilirubin IXα) on the surface of the HSA protein globule. Based on study of the spectral characteristics of the photoproducts of Z,Z-bilirubin IXα and comparison of their accumulation kinetics in solution and the enzyme photo-inactivation kinetics, we concluded that the determining role in sensitized damage to LDH is played by lumirubin. The photosensitization effect depends on the wavelength of the radiation used for photoconversion of bilirubin. When (at the beginning of exposure) we make sure that identical numbers of photons are absorbed by the pigment in the different spectral ranges, the side effect is minimal for radiation corresponding to the long-wavelength edge of the bilirubin absorption band. Under conditions when identical numbers of photons are absorbed at wavelengths corresponding to the maximum of the absorption band for Z,Z-bilirubin IXα (λ = 457.9 nm) and also its short-wavelength (λ = 441.6 nm) and long-wavelength (λ = 514.5 nm) edges, minimal side effects are observed on exposure to radiation in the green region of the spectrum. The characteristic features of the enzyme photo-inactivation kinetics are determined by the bilirubin/HSA concentration ratio.

We have shown that for a bilirubin/HSA concentration ratio >2 (when some of the pigment molecules are sorbed on the surface of the protein globule), the bilirubin can act as a photosensitizing agent for the enzyme present in solution.

During phototherapy of hyperbilirubinemia in newborns, reduction of the unfavorable side effects can be promoted by using light-emitting diode sources and also by alternating phototherapy sessions and pauses between light exposures.

5. Spectral and Energy Characteristics of Radiation Sources Used for Phototherapy of Neonatal Hyperbilirubinemia

Characteristics of Phototherapy Treatments. Modern phototherapy methods of neonatal hyperbilirubinemia mean the effect of light blue or blue-green spectrum area corresponding to the absorption band of bilirubin on the body surface of a newborn (excepting eyes and genitals). The traditional location of sources of radiation exposure over a child is considered to effect on the 30-40% of the body surface. The therapeutic effect depends on the power density, radiation dose, and its spectral range [16, 17, 20–22]. Also significant impact on the effectiveness of treatment has the area of child's body surface exposed to light exposure, the degree of baby prematurity (gestational age), the bilirubin level and the presence of underlying disease [16, 17, 20–22, 132].

The International Electrotechnical Commission recommends to provide illumination of the horizontal area on the surface of the mattress size 60 x 30 cm while phototherapy carrying out [6]. However, most practical recommendations are based on the need to ensure the required light intensity ~700 cm^2 (35 cm x 20 cm) for term infants and ~450 cm^2 (28 cm x 16 cm) for premature infants [17, 20]. Typically, in this case the radiation source is located at a distance of 30-50 cm from the surface of a newborn [16, 17, 20–22]. However, last years phototherapy devices which provide the effect on the entire body surface due to the location of the sources both over and under a child have been started to manufacture (in the last case, the impact is carried out through a transparent mattress) [16, 17, 20, 21].

The phototherapy duration is defined by medical conditions .Therefore it varies greatly depending on the disease severity and the effectiveness of reducing the level of bilirubin. Typically, it ranges from 12 to 24 hours [11–13] per day (with breaks for feeding). Total duration of light exposure constitutes 72–96 [6, 11] or even 200 hours [16] and the total dose – up to 200 -900 J/cm^2 [16].

The Spectral Range of Radiation. According to the recommendations (1974) of Committee on Phototherapy in the Newborn Infant – National Research Council, coordinated by the National Academy of Sciences of the USA, the radiation spectrum has to correspond to the maximum of the bilirubin absorption spectrum (450–460 nm) and to be located in the area of λ = 425–475 nm for optimal therapeutic effect. 30 years later (2004) American Academy of Pediatrics recommended to use this spectral range, allowing it to wide for λ = 400–480 nm, and for λ = 430–490 nm during carrying out intensive therapy [6]. However, there is the evidence that pronounced therapeutic effect has the light in the spectral range 480–530 nm [53–55].

The Intensity and the Dose of Radiation Exposure. The parameters which characterize the light intensity used for phototherapy of neonatal jaundice require separate discussion. The main problem is that the spectral range of radiation of most phototherapy devices is much broader then the absorption spectrum of bilirubin and varies greatly for different types of lamps. Moreover, some of the sources are characterized by the fact that the maximum of emission spectrum is outside of the absorption band of bilirubin. Consequently, the integrated power density (P, mW/cm^2) of luminous flux, measured at the child body, does not adequately characterize the sources in terms of their effectiveness for bilirubin photoconversion. Therefore, along with the power density, the power spectral irradiance (E, µW/cm^2/nm) is often used as the characteristic parameter [6, 17, 20–22].

But measuring instruments, which control integrated power density of radiation or energy spectral irradiance in medical institutions, are not so perfect, and their testimony is so contradictory (differing by more than 2-fold [134]). Therefore it is necessary to ascertain the absence of standard equipment for the preliminary assessment of the therapeutic efficacy of light sources of different types.

Despite the contradictory of existing views about the optimal dosage of radiation, in the literature certain criteria for minimum and maximum allowable values of light intensity have been worked out, as well as the recommendations of its most preferred range. Fundamental in this plan is the work [9], in which the author has shown that while phototherapy the dependence of daily reduce of the bilirubin level in children blood on the intensity of blue light is a curve with saturation: the increase of power density for more than $P \sim 2,0$ mW/cm^2 (energy irradiance above $E \sim 40$ μW/cm^2/nm) does not lead to the increase of phototherapeutic effect. The minimum light intensity ($P \sim 0.05$ mW/cm^2) which may cause bilirubin photoconversion of newborns body were also have been estimated [9]. However, according to [135], the minimum level of energy irradiance by blue light, which is able to reduce level of bilirubin in the newborn blood significantly, is $E = 4$ μW/cm^2/nm ($P \sim 0.2$ mW/cm^2). According to the recommendations of the American Academy of Pediatrics [6], a standard phototherapy assumes influence by light with irradiance $E = 8–10$ μW/cm^2/nm ($P \sim 0.4–0.5$ mW/cm^2). And while intensive therapy – up to $E = 30$ μW/cm^2/nm ($P \sim 1.5$ mW/cm^2). Unlike this data British Department of Health recommends $P = 1.0$ mW/cm^2 as a minimum level of power density for phototherapy of neonatal jaundice [20]. It is necessary to note that there are supporters of higher dosages (power density is 5–7 mW/cm^2).

Modes of Light Exposure. Considering the duration of phototherapy, which is directed to reduce the bilirubin level in newborn infant's blood, in some studies [136–138] was proposed to replace continuous phototherapy with faltering light influence (alternation of phototherapy procedures and intervals between exposures).

Offer essence is that during phototherapy there is a significant discoloration of bilirubin in skin. Therefore a certain time interval (which may be dark) for the diffusion of the photoproducts of bilirubin into the depth of tissue and for the arriving into the skin of intact of bilirubin from the blood vessels is required. Indeed, it was shown [136] that alternating of phototherapy and dark period (duration ~1 h) does not reduce the rate of bilirubin excretion and does not increase the duration of treatment. However, there is other data [137, 138], which show the decrease in therapeutic effect while the transition to an intermittent model. Therefore, at high concentrations of bilirubin it is recommended to carry out continuous phototherapy [6].

Types of the Lamp Sources Used for Phototherapy. The analysis of literary sources and promotional leaflets of the phototherapy equipment manufacturers shows that fluorescent bulbs, metal halide discharge lamps and halogen lamps are used as the radiation sources for hyperbilirubinemia treatment [17, 20–22, 139]. Spectral and energy characteristics of the most common types of lamps used for treatment of neonatal jaundice, are listed in Table 1 and on Figure 17.

For comparison, absorption spectrum of bilirubin in complex with human serum albumin is shown here also. It should be noted that the lamps characteristics are the average to some extent. Because of their operation leads to decreasing of energy parameters, and to changes in the emission spectrum. For this reason, the data of various authors on the spectral energy characteristics of the lamps differs.

From all of these types of light sources the emitters on the basis of fluorescent lamps are the most widely used in clinical practice. The hard radiation of UV lamp transforms into the visible range radiation by applying of the fluorophore on the inner surface of the bulb.

The first fluorescent lamps used for phototherapy radiated in the violet ($\lambda_{max} = 419$ nm, half-width of the main band is $\Delta\lambda = 33$ nm) and blue ($\lambda_{max} = 447$ nm, $\Delta\lambda = 52$ nm) spectral regions. A major shortcoming of this type of lamps is a rapid degradation of the fluorophore. Thus after 200 hours of service the radiation intensity of blue lights lamps dropped by 20%

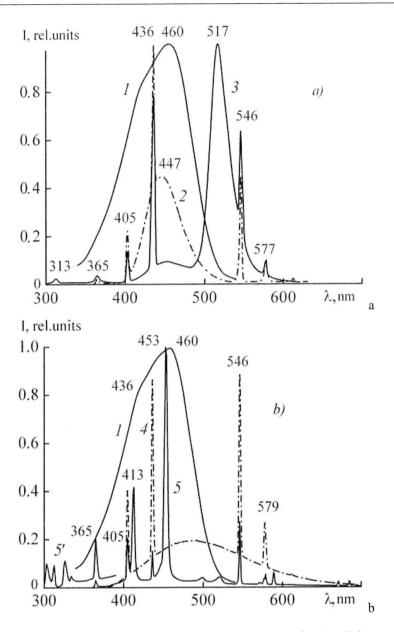

Figure 17. Normalized absorption spectra of bilirubin, emission spectra of modern light sources of various types used for phototherapy of neonatal hyperbilirubinemia, and optical transmission spectrum of polymethylmethacrylate: 1 – absorption spectra of bilirubin, 2–8 – emission spectra of lamps, respectively: 2 – Medela blue lamp, Medela Inc., Switzerland, 3 – BG160 green lamp, National, Japan, 4 – Vickers 80 030, Air-Shields-Vickers Inc., USA, 5 – Heraeus Dräger PT800, WC Heraeus GmbH, Germany, 6 – T-12 Vita-Lite, Duro-Test Corp., Canada, 7 – Ohmeda BiliBlanket, Ohmeda Medical Inc., USA, 8 – Hill-Rom Micro-Lite, Hill-Rom Air-Shields Inc ., USA, 9 – optical transmission spectrum of PMMA thickness of 6 mm.

Table 1. Spectral and energy characteristics of optical radiation sources used for phototherapy of neonatal hyperbilirubinemia (λ_{max} – position of the maximum in the emission spectra; $\Delta\lambda$ – value of the half-width of the emission spectra; λ_{Hg} – position of mercury lines in the emission spectra lamps; $I_{400-480}$ – intensity of radiation in the spectral range 400–480 nm; $I_{300-400}$ – intensity of radiation in the spectral range 300–400 nm; t – resource; n.a. – not available)

Device	Type and number of sources	λ_{max}, nm	$\Delta\lambda$, nm	λ_{Hg}, nm	$I_{400-480}$, mW/cm^2	$I_{300-400}$, mW/cm^2	t, h
Philips TK20W/03T	7 fluorescent violet tubes	419	33	405, 436, 546, 577	1.5	0.002	2000
Westinghouse F20T12/BB	7 fluorescent special blue tubes	446	34	405, 436, 546, 577	2.9	0.002	2000
Philips TL20W/52	7 fluorescent special blue tubes	452	55	405, 436, 546, 577	3.5	0.002	2000
Draeger Phototherapy 4000	6 fluorescent blue tubes	447	52	365, 405, 436, 546, 577	4.1	0.008	1000
Mediprema Cradle	16 fluorescent blue tubes	452	51	365, 405, 436, 546, 577	5.3	0.005	2000
Ampliflux 2 Hoods	12 fluorescent blue tubes	452	51	365, 405, 436, 546, 577	7.4	0.006	2000
Medela phototherapy lamp	4 fluorescent blue tubes	447	52	365, 405, 436, 546, 577, 612	4.7	0.016	1500
Medela BiliBed	1 fluorescent blue tube	447	52	405, 436, 546, 577	2.6 - 7.8	0.0004	1500
Bili-Compact	10 fluorescent blue lamps	450	n.a.	405, 436, 546	3.0	n.a.	1000
Amelux	6 fluorescent blue tubes	440	n.a.	405, 436, 546	4.3	n.a.	2000
Osram turquoise L18W/L131-UVS	4 fluorescent blue-green tubes	490	65	405, 436, 546, 577	2.7	n.a.	n.a.

Device	Type and number of sources	λ_{max}, nm	$\Delta\lambda$, nm	λ_{Hg}, nm	$I_{400-480}$, mW/cm^2	$I_{300-400}$, mW/cm^2	t, h
BiliCrystal System IV 2	(4+8) fluorescent blue-green tubes	446	n.a.	436, 546	10.0	n.a.	3000
Sylvania F20T12/G	7 fluorescent green tubes	528	38	365, 405, 436, 546, 577	1.2	n.a.	1000
BG160 National	4 fluorescent green tubes	517	29	365, 405, 436, 546, 577	2.4	n.a.	n.a.
Sylvania F20T12/DA	7 fluorescent daylight tubes	504	135	365, 405, 436, 546, 577	1.7	n.a.	1000

Table 1. (Continued)

Device	Type and number of sources	λ_{max}, nm	$\Delta\lambda$, nm	λ_{Hg}, nm	$I_{400-480}$, mW/cm^2	$I_{300-400}$, mW/cm^2	t, h
Vickers Medical 80/0155	4 fluorescent daylight tubes	485	144	365, 405, 436, 546, 577	1.2	0.002	1000
Philips TL18W/54	7 fluorescent daylight tubes	450, 580	n.a.	365, 405, 436, 546, 577	1.3	n.a.	1000
Draeger Phototherapy 800	1 metal halide discharge bulb	453	5	303, 313, 334, 365, 405, 413, 436, 496, 521, 546, 579, 589, 690	1.7-7.5	0.019	1000
Hill-Rom Micro-Lite Phototherapy System	3 quartz halogen bulbs	592	210	-	1.2-2.5	0.0024	1000
Ohmeda BiliBlanket	1 quartz halogen bulb	533	93	-	4.8	0.0248	800
Giraffe Spot PT Lite	1 quartz halogen bulb	430, 550	n.a.	-	2.5	n.a.	2500
Natus neoBLUE LED Phototherapy system	852 blue LEDs	458	30	-	1.0-2.2	0	20000
Bilitron 3006	5 blue LEDs	450	30	-	2	0	20000
Malysh	18 blue and cyan LEDs	470, 505	60	-	2.0-5.0	0	20000

As for as the purple light is concerned, its major shortcoming is the low therapeutic efficacy (the low speed of bilirubin photoconversion). It is caused by pronounced shielding effect of hemoglobin radiation at wavelengths corresponding to the maximum of the emission spectrum. It is also believed that the use of violet region of the spectrum is potentially carcinogenic, and it is necessary to refuse using of such lamps while the treatment of neonatal jaundice.

Specifically designed for the treatment of neonatal jaundice special blue fluorescent lamps with λ_{max} = 452 nm, $\Delta\lambda$ = 55 nm [140] replaced purple and blue bands. Currently these lamps are the most widespread type of light sources for phototherapy of hyperbilirubinemia. The American Academy of Pediatrics recommends them for practical use [6]. Emission spectrum of such lamp in the fundamental band fits the absorption spectrum of bilirubin in the best way. In addition, the intensity of the special fluorescent lamp is 2–3 times higher than the intensity of emission of other types of fluorescent lamps [22].

In some countries, cheaper fluorescent lights are widely used. They are characterized by broadband radiation in the visible region [140]. The instability of radiation power is a characteristic disadvantage of this type. According to [16] after 1 hour of their work the optical power density decreases for 25 %, and after 2 000 hours of work, it is not more than 44 % from the initial level.

The are reports of successful use of green (λ_{max} = 504 nm, $\Delta\lambda$ = 135 nm) and blue-green (λ_{max} = 490 nm, $\Delta\lambda$ = 65 nm) fluorescent lamps [53–55]. The maxima of the emission spectrum are located on the long-wavelength slope of the absorption band of bilirubin (Figure 17).

A characteristic feature of all types of fluorescent lamps used for phototherapy, is the presence in the emission spectrum along with the fluorescent component of the mercury lines with λ_{max} = 313, 365, 405, 436, 546, 577, 612 nm (Figure 17, Table 1.). The presence of ultraviolet light in the emission spectrum of these lamps acutely raises the question of possible adverse side effects [140], since long exposure to UV radiation is potentially carcinogenic due to genetic effects caused by them [142, 143]. Let us note, however, that during the light exposure through the wall of the incubator UV- component with λ = 313 nm is strongly screened by polymethylmethacrylate (Figure 17).

As for broadband light sources based on halogen lamps (metal halide discharge lamp, quartz halogen bulb), as follows from Figure17, the spectrum of their emission (curves 7 and 8) only partially overlaps with the absorption spectrum of bilirubin. In addition, the intensity of light in the center of the spot formed by such a source is significantly higher than the figures on the periphery [16]. Another disadvantage of this type of lamp is the temperature factor: halogen spotlights emit a significant amount of heat during operation and must comply with certain precautions. For these reasons, emitters based on halogen lamps for phototherapy of hyperbilirubinemia were not that widespread. Some hopes for the use of halogen sources have been associated with the development of phototherapy units in the U.S. (Ohmeda Biliblanket Phototherapy System, 1990 and the Wallaby Phototherapy System, 1989) with fiber-optic light delivery system of radiation to newborn and special light-carrying blankets, light of which is directed mainly on the surface of the baby's body [139]. Typically, these systems provide a sufficiently high intensity of radiation ($P\sim$10-35 $\mu W/cm^2/nm$) [139], but due to lack of body surface area that is lighted, they are less efficient than traditional methods of phototherapy using daylight fluorescent lamps or special blue fluorescent light.

Variety of existing phototherapeutic systems has initiated numerous studies aimed at determining the most efficient source for the treatment of hyperbilirubinemia of newborn children. In our opinion, the obtained results are hardly comparable and not always correct, because, as a rule, these studies were conducted on a purely pragmatic purpose: to find out which phototherapy devices are most effective (among of those possessed by an investigator) in reduction of bilirubin levels. In this regard, very different light intensity, provided by various types of sources, may be one of the reasons for contradictory results and gives no answer about the most optimal spectral range for light therapy. However, most researchers still tend to the conclusion about higher therapeutic efficiency of the special blue lamps [6, 9, 11, 13, 140].

Summarizing the results given by different authors, we cna conclude that, despite the existence of various sources, it has not been definitely decided yet which of them should be used for light therapy. Analysis of optical, energy, medical and technical characteristics of the radiation sources based on halogen and fluorescent (purple, blue, special blue, green, turquoise, daylight) lamps, as well as the results of clinical observation show that the current method of phototherapy for neonatal hyperbilirubinemia has some drawbacks, the key one is its low efficiency. There is a background to assume that one of the reasons for poor therapeutic efficiency of tube lights is low entire intensity of radiation in the spectral range corresponding to the absorption spectrum of bilirubin. More than half of the fluorescence intensity of the green lights, falls on the range $\lambda > 530$ nm, which is inefficient for bilirubin photoconversion, and more than half of the emission intensity of the violet, blue and special blue lamps accounts for the range $\lambda = 400-450$ nm, in which bilirubin is shielded by hemoglobin. All above points to the need for further improvement of devicws for phototherapy of hyperbilirubinemia of neonatal children. In our opinion, a great variety of phototherapy devices used to treat hyperbilirubinemia, differ by the type of lamp light source, spectral range of radiation, power density of luminous flux, but in practice all of them are rather ineffective for phototherapy. It is an indicator of insufficient study of the processes of bilirubin photoconversion underlying the specified method of light therapy.

6. Side Adverse Effects Caused by Long-Term Influence of Tube Lights

In addition to the low therapeutic efficacy of tube lights, another drawback of this method of phototherapy is the presence of side adverse effects. The issue of the possible complications of phototherapy has been discussing since the very beginning of its application [23, 24]. In fact, the widespread use of light to treat neonatal jaundice began only after the clinical results that have demonstrated the absence of neurotoxic effects on the infant during phototherapy procedures [25, 26]. Studies carried out in subsequent years, showed no growth retardation or abnormality of mental or physical development of children treated with phototherapy. However, with the accumulation of data of the results of phototherapy of neonatal hyperbilirubinemia some reports appeared that long-term exposure of light may have an adverse effect on the baby's body. As already noted, in some cases the duration of phototherapy exceeded 200 hours at a total dose of 900 J/cm^2 due to its low efficiency [16]. Naturally, with such a high energy load it can not be excluded that the processes of bilirubin photodegradation are not accompanied by photochemical reactions involving both bilirubin and its photoproducts, and other endogenous compounds that can act as photosensitizer are not involved in the photochemical processes.

Some of the side adverse effects caused by light exposure, appear during or after termination of phototherapy, while others are recorded in the follow-up studies after several years or decades. Among the clearly identified and described in the literature evidence of adverse effects recorded during or immediately after cessation of phototherapy there are:

- Skin erythema. Caused by long-term lamps exposure in the spectrum which contains mercury lines of the nearest UV range (λ_{max} = 313, 365 nm) [6, 141];
- Potential carcinogenic effects due to the presence in the emission spectrum of UV lamps that can cause mutagenic changes while long-term exposure [142, 143];
- Appearance of bright pigmentation (the syndrome of "bronze baby"). Caused by a violation of the photoinduced porphyrin metabolism [144] and requires a cessation of phototherapy;
- Breach of the metabolic processes (including reducing the amount of calcium, tryptophan [145], riboflavin [146]) in the body, hormonal metabolism [147–149], negative effects on the immune system [150];
- Photodestruction of hemoglobin leading to the increase of bilirubin concentration [151];
- Growth retardation at the time of phototherapy. At follow-up, it's shown [6] that the physical development of children treated with phototherapy on the hyperbilirubinemia syndrome is not different from control group;
- General overheating of the body, and as a consequence – a violation of thermoregulation and fluid loss [152]. These processes are due to a significant contribution to the emission spectrum of halogen lamps of infrared component.

It should be noted that in some cases the side effects that arise in the course of phototherapy, could become a contraindication for its further implementation (strong erythema, "bronze baby" syndrome). In some cases, theese side effects, primarily associated with decreased concentrations of blood components (riboflavin, tryptophan), or fluid loss, offset by their introduction into the body, and do not require termination of phototherapy [6].

However, in recent years, evidence appeared that showed the presence of side effects, manifested in about 10 years after the phototherapy of neonatal jaundice. Thus, according to [153–155], children subjected to intensive neonatal phototherapy with fluorescent bulbs have a higher risk of malignant melanoma of the skin. One example of this is the number of moles (nevi) among 8-9 years children treated by phototherapy in the neonatal period, that is 2 times higher then the corresponding figure for children who did not receive phototherapy. It should be noted here that the development of melanoma (in the absence of hyperbilirubinemia) is identified with prolonged exposure to ultraviolet radiation.

It is important to note that, despite of the availability of reliable data on side adverse effects that manifest as a result of phototherapy for hyperbilirubinemia; their causes are not fully understood. There is no doubt that a number of side effects (erythema of the skin, the appearance of nevi) are caused by presence of ultraviolet component in the emission spectrum of fluorescent and metal halide discharge lamps (Table 1). Apparently, the main role in the overheating of the baby and fluid loss during phototherapy using a halogen lamp is an infrared component in the spectrum of radiation. It can be expected that a number of destructive processes in the infant's body under the influence of light (photodegradation hemoglobin, riboflavin) can not be excluded in full range during phototherapy, because of their absorption spectra overlap strongly with the absorption spectrum of bilirubin. An important step aimed at eliminating these effects can be reduction of phototherapy duration (energy load on the baby) by increasing its efficiency.

Separate discussion is required about possible role in the implementation of the adverse effects on the infant of photosensitized reactions involving bilirubin and products of its photolysis. Based on the study of spectral characteristics of the photoisomerization and photooxidation products of Z,Z-bilirubin IXα, comparing the kinetics of their accumulation in the solution to the kinetics of photoinactivation of the enzyme it was concluded [121] that lumirubin makes the decisive contribution to the sensitized photoinactivation of enzyme molecules. Photosensitization effect depends on the wavelength of radiation used for bilirubin photoconversion. By providing (to the beginning of exposure) same number of photons absorbed by the pigment of different spectral range, a side effect is minimal for the radiation, corresponding to the long-wavelength slope of the absorption band of bilirubin, while maximum is under the influence of short-wave radiation (violet-blue region of the spectrum) [121].

It is noteworthy that lumirubin, apart with absorption band with maximum at λ_{max} = 460 nm characteristic for Z,Z-bilirubin IXα and its cis-trans photoisomer, has additional bands in UV and visible spectrum with a maximum at 337 nm and 592 nm [18, 121]. In this context, it becomes apparent that the presence in radiation source, used for phototherapy of hyperbilirubinemia, of the wavelengths of UV and visible spectrum, not corresponding with bilirubin absorption bands, may have an adverse event due to the effects of the photosensitization of lumirubin [121]. In particular, the intense mercury lines with maxima at 365, 546, 577, 585, 612 nm in the spectrum of fluorescent lamps, or lines with maxima at 334, 365, 589 nm in the spectrum of metal halide lamps, as well as a broad band emission peaking at 592 nm in the spectrum of halogen lamps (Figure 17 and Table 1) can initiate the photodynamic damage of biologically important compounds by photochemical reactions involving lumirubin.

As already noted, at a certain intensity of blue light (P~2.0 mW/cm2) the effectiveness of phototherapy is limited by the rate of excretion of bilirubin photoproducts (primarily lumirubin) of the infant. For this reason, a further increase in power density of the light flux does not lead to an acceleration of the daily reduction of bilirubin in blood [9]. For this reason, it can be expected that an excessive increase in the intensity of light used for phototherapy, can have a negative effect due to sensitizing processes involving lumirubin.

A decrease in potential side effects during phototherapy of hyperbilirubinemia, due to the sensitizing effect from lumirubin, can be promoted by alternating phototherapeutic procedures and pauses between exposures (while maintaining the overall energy load) [121]. For successful realization of this approach, the pause between phototherapy sessions must be sufficient for diffusion of lumirubin into the interior of the tissue and its subsequent excretion.

It should be noted that when exposing radiation to visible spectrum except for bilirubin and its photoproducts sensitizing (including adverse) effect may cause riboflavin (vitamin B_2), both of endogenous origin and designated as a drug for compensation due to photodecomposition during phototherapy [6].

However, there is no reason to believe that (despite a number of side effects of phototherapy for neonatal hyperbilirubinemia) a light therapy can lead to an increase in infant mortality.

7. Prospects of Using High-Brightness LEDs for Phototherapy of Neonatal Hyperbilirubinemia

There is no doubt that the most promising sources for phototherapy of neonatal hyperbilirubinemia children are high-brightness light-emitting diodes [6, 17, 20–22, 133, 156–174].

The use of LEDs is promising both in terms of improving the therapeutic efficiency and reduction (elimination) of the adverse side effects caused by the sensitizing effect of bilirubin and its photoproducts, as well as subject to the presence in the spectrum of traditional sources of intensive ultraviolet or infrared components.

Modern LEDs based on InGaN, promising for phototherapy of hyperbilirubinemia of neonatal children, are characterized by the absence of ultraviolet components and optical power in the blue or blue-green region of the spectrum up to 5 watts. Resource of such sources is about 50 000 h (for fluorescent lamps it does not exceed 2000 hours). The spectrum of light emitting diodes (unlike fluorescents) remains unchanged throughout the term of their service. The emission spectra (curves 1-4) of four types of LED light sources (based on InGaN), relevant (to some extent) to the absorption band of bilirubin (curve 5) and used in new phototherapeutic equipment for treatment of hyperbilirubinemia of neonatal children are shown in Figure 18.

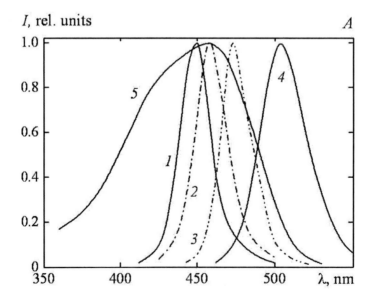

Figure 18. Normalized emission spectra of high-brightness LEDs, promising for phototherapy of neonatal hyperbilirubinemia.

The characteristic feature of the emission spectra of these LEDs is the presence of the single, relatively narrow band with maximum located at λ_{max} = 450 (curve 1), 458 (curve 2), 473 (curve 3) or 505 nm (curve 4). The half-width of the emission spectrum is $\Delta\lambda$ = 23 nm (curves 1-3) or $\Delta\lambda$ = 35 nm (curve 4). It should be noted that the devices based on the high-brightness LED for phototherapy of neonatal jaundice are recommended by the American Academy of Pediatrics, and Food and Drug Administration (FDA) [6].

It is noteworthy that if the phototherapy device Natus Neo Blue Light Phototherapy System (Natus Medical Inc., USA) is based on a 750 light-emitting diodes with a peak emission spectrum at λ_{max} = 458 nm (spot size 52x27 cm, energy irradiance E = (15±2) or (35±3.5) $\mu W/cm^2/nm$ [133,156–159], then the more recent developments are made using 18 (the unit "Malysh", Institute of Physics NASB, λ_{max} = 473 and 505 nm, P = 5.0 mW/cm^2 [165] or 5 (Bilitron 3006, FANEM, Brazil, λ_{max} = 450 nm, E =37±9 $\mu W/cm^2/nm$) [166] high-brightness LED microarrays. As follows from the given data, developers of phototherapeutic devices have no consensus on the optimal type (spectral range of radiation) LED emitter and required intensity of light output for treatment of neonatal hyperbilirubinemia.

It is clear that there is a need for further randomized supervised researches in this direction.

8. Factors Determining the Therapeutic Efficiency of Radiation Corresponding Long-Wavelength Slope of the Absorption of Bilirubin

Researches carried out using lamp (λ_{max} = 490 nm, $\Delta\lambda$ = 65 nm [53–55], λ_{max} = 504 nm, $\Delta\lambda$ = 135 nm [51, 139]), laser (argon laser, λ = 476.5; 488.0; 514.5 nm, [18, 69]) and LED (λ_{max} = 473, $\Delta\lambda$ = 23 nm and λ_{max} = 505, $\Delta\lambda$ = 35 nm [159,165]) sources, showed the presence of pronounced therapeutic effect when exposed to radiation corresponding to the long-wavelength slope of the absorption band of bilirubin. Despite the fact that for bilirubin bound to albumin, the transition from the maximum absorption band (λ_{max} =458 nm) to λ = 473, 488, 505, 515 nm the optical density of solution decreases, respectively in 1.16, 1.8, 5.1, 13.7 times, the therapeutic effect of using these sources is almost as good as the action caused by the special blue lights (λ_{max} = 452 nm) or even surpasses it. In this regard, the research [159] should be mentioned, which authors showed that with the same energy load (E = 5-8 $\mu W/cm^2/nm$) LED sources with λ_{max} = 505 nm and λ_{max} = 458 nm are comparable by therapeutic efficiency (statistically significant differences in their effects are absent). The received result is somewhat unexpected, since the long-range emission from the LED with λ_{max} = 505 nm, has almost no overlap (as indicated on Figure 18) with the absorption spectrum of bilirubin and should not cause its photoisomerization. Taking into account that moving from λ = 458 nm to λ = 505 nm quantum yield of lumirubin formation (φ_{Lr}) is increased in not more than 1.4 times, so that a slight increase of φ_{Lr} should not compensate a dramatic (in 5.1 times) reduction of bilirubin absorption rate.

In our opinion, along with the increase of φ_{Lr} there are other reasons behind high therapeutic activity of the radiation, corresponding to long-wavelength slope of the absorption band of bilirubin. One of them is that in this range the screening effect of the light from the internal filters – endogenous pigments: skin melanin (epidermis) and hemoglobin HbO_2 has low impact on bilirubin. Thus, the absorption spectrum of melanin is characterized by a non-structure band, falling with increasing radiation wavelength. That is, the screening effect of this pigment is mostly pronounced in the short wavelengths of the absorption spectrum of bilirubin. An intensive Soret band of HbO_2 (λ_{max} = 414 nm) is in the same area. Spectral dependence of the ratios of the molar extinction coefficient of bilirubin and HbO_2 is shown on Figure 19. It is illustrated that the optimal ratio between the efficiency of bilirubin absorption and hemoglobin screening effect is observed at λ = 476 nm. Let us note, that at this wavelength the absorbance of bilirubin, bound to albumin, lower than the corresponding value at the maximum absorption band in no more than 1.2 times. From this perspective LED sources with λ_{max} = 473 nm, $\Delta\lambda$ = 23 nm are exactly the most optimal for phototherapy of hyperbilirubinemia. In addition to reducing the screening effect high therapeutic efficiency of radiation of green spectral region can also be caused by a pronounced difference of spectral characteristics of bilirubin bound to albumin, in the conditions in vivo and in vitro.

These differences are due to the influence of fatty acids on the structure and, consequently, the spectral characteristics of bilirubin in the composition of its complex with HSA. According to [12] human serum albumin in the blood contains a variety of bound fatty acids that can lead to long-wavelength shift of the absorption spectrum of bilirubin in the 10–15 nm compared with commonly used *in vitro* studies of pigment complexes with high-purity preparations of the protein.

In our opinion, the above displacement is a consequence of the presence of intramolecular exciton interactions between Z,Z-bilirubin IXα chromophores [46, 67, 68]. It can be expected that the presence of fatty acids in the structure of the albumin has an impact on the conformation of tetrapyrroles, and, accordingly, on an angle between its chromophores.

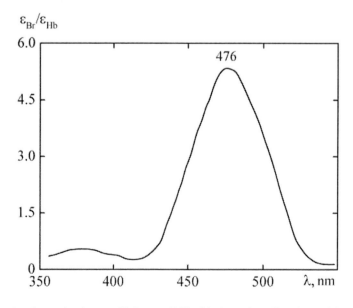

Figure 19. Ratio of molar extinction coefficients of bilirubin bound to albumin, and hemoglobin.

It leads to the change in the resulting absorption spectrum of bichromophore [46, 67, 68] in the presence of intramolecular interchromophore exciton interactions. Therefore, a more long-wave position of the absorption spectrum of bilirubin in blood compared to its spectrum in vitro conditions can be one of the reasons for the displacement range of wavelengths, optimal for phototherapy of hyperbilirubinemia from the area corresponding to the peak of the absorption spectrum of bilirubin on its long-wave slope.

It is noteworthy that the binding of fatty acids by albumin can not only affect spectral characteristics of bilirubin and its structure, but also increase the efficiency of its structural photoisomerization (lumirubin formation) in about 3 times [12].

Thus, it is expected that the in vivo maximum spectrum of light on bilirubin will comply not with the maximum absorption spectrum in vitro, but shifted to longer wavelengths. Meanwhile the wavelength, corresponding to the maximum spectrum will be determined by depth of blood vessels and subcutaneous fat that are the places of occurrence of photochemical reactions in the pigment.

Conclusion

The therapeutic effect of light during phototherapy of neonatal hyperbilirubinemia is determined by the processes of configuration (*cis-trans*) and structural photoisomerization of Z,Z-bilirubin IXα and photoproducts excretion. The main role in reducing of the level of bilirubin during phototherapy belongs to E,Z-cyclobilirubin IXα (lumirubin), characterized by a maximum speed of excretion due to the highest (among the photoproducts) hydrophilicity. Quantum yield of lumirubin formation depends on the wavelength of effecting radiation and micro-environment (such as a carrier protein) due to: (*i*) the nature bichromophore absorption (and emission) of light by Z,Z-bilirubin molecule and its photoisomers; (*ii*) spectral non-identity of the chromophores caused by the asymmetry of their structure and heterogeneity of the environment in a protein matrix; (*iii*) the presence of intramolecular interchromophore singlet-singlet energy transfer, which's probability depends on the excitation wavelength when it changes within the long-wavelength absorption band of pigment and on its conformation, determined by the type of carrier protein.Along with a pronounced therapeutic effect, long lasting influence on the infant with fluorescent, halogen and metal halide lamps light may have an adverse side effect on him. Besides the ultraviolet and infrared components, present in the emission spectrum of these lamps, negative impact on the baby can also be made by intense visible light through photosensitized processes involving endogenous pigments (including bilirubin photoproducts) and pharmacological agents.Currently there is no alternative to the use of LED sources in the devices for phototherapy of neonatal hyperbilirubinemia. Radiation sources of this type of blue-green region of the spectrum correspond to the range of absorption of bilirubin and significantly exceed the widespread lamp sources (mercury, halogen, metal halide) on the set of optical and operational characteristics. The devices for phototherapy of neonatal hyperbilirubinemia (*i*) does not contain the ultraviolet and infrared components (having side effects on the newborn); (*ii*) provides adjustment of the intensity of effecting radiation, depending on the severity of the newborn; (*iii*) allows a uniform distribution of light intensity on the surface of the child's body; (*iiii*) dozen times exceeds lamp on the work resort; (*iiiii*) resistant to mechanical damage and does not represent (in contrast to the mercury vapor lamps) environmental concerns in violation of its integrity and while disposal.

There is a need for further randomized controlled study, pointed at determining the optimal spectral range and intensity of the effecting radiation, providing an effective reduction of bilirubin in blood of newborns without adverse effects on the body of infants.

Acknowledgement

The author is grateful to Professor V. A. Mostovnikov, Professor B. M. Dzhagarov, Dr. G. R. Mostovnikova, Dr. A. L. Novakovskii, L. G. Plavskaya, A. I. Tret'yakova for fruitful discussions and critical remarks, and N. V. Bulko, I. V. Supeeva – for technical assistance in preparing this publication.

References

[1] Kivlahan, C. and James, E. J. P. (1984). The natural history of neonatal jaundice. *Pediatrics, 74(3)*, 364–370.

[2] Edwards, S. (1995). Phototherapy and neonate: providing safe and effective nursing care for jaundiced infants. *J. Neonat. Nursing, 1(5),* 9–12.

[3] Porter, M. L. and Dennis, B. L. (2002). Hyperbilirubinemia in the term newborn. *Am. Family Physician, 65(4)*, 599–606.

[4] Akobeng, A. K. (2004). Effects of treatments for unconjugated hyperbilirubinaemia in term and preterm infants. *Clin. Evid, 12(1–2),* 501–507.

[5] Bhutani, V. K., Johnson L. and Sivieri E. M. (1999). Predictive ability of a predischarge hour-specific serum bilirubin for subsequent significant hyperbilirubinemia in healthy term and near-term newborns. *Pediatrics, 103(1),* 6–14.

[6] American academy of pediatrics. Subcommittee on hyperbilirubinemia. (2004). Management of hyperbilirubinemia in the newborn infant 35 or more weeks of gestation. *Pediatrics, 114(1),* 297–316.

[7] Dennery, P. A., Seidman, D. S. and Stevenson, D. K. (2001). Drug therapy: neonatal hyperbilirubinemia. *N. Engl. J. Med., 344(8),* 581–590.

[8] Makarova, N. A. Tabolina, O. V. and Volodin, N. N. (1987). Assessment of neonatal hyperbilirubinemia and tactics therapeutic intervention. *Vopr. Okhr. Materin. Det., 6,* 59–63. [in Russian].

[9] Tan, K. L. (1982).The pattern of bilirubin response to phototherapy for neonatal hyperbilirubinaemia. *Pediatr Res., 16,* 670–674.

[10] Brown, A. K., Kim, M. H., Wu, P. Y. K. and Bryla, D. A. (1985). Efficacy of phototherapy in prevention and management of neonatal hyperbilirubinemia. *Pediatrics, 75(2),* 393–400.

[11] Ennever, J. F. (1986). Phototherapy in a new light. *Pediatr. Clin. North. Am.,. 33(3),* 603–620.

[12] Polin, R. A. (1990). Management of neonatal hyperbilirubinemia: rational use of phototherapy. *Biol. Neonate, 58(1),* 32–43.

[13] Ennever, J. F. (1990). Blue light, green light, white light, more light: treatment of neonatal jaundice. – *Clin. Perinatol., 17(2),* 467–481.

[14] Tan, K. L. (1991). Phototherapy for neonatal jaundice. *Clinics in Perinat; 18,* 423–439.

[15] Tan, K. L. (1994). Comparison of the efficacy of fiberoptic and conventional phototherapy for neonatal hyperbilirubinaemia. *J. Pediatr., 125,* 607–612.

[16] Araujo, M. C. K., Vaz, F. A. C. and Ramos, J. L. A. (1996). Progress in phototherapy. *Rev. Paul. Med., 114(2),* 1134–1140.

[17] Dicken, P., Grant, L. J. and Jones, S. (2000). An evaluation of the characteristics and performance of neonatal phototherapy equipment. *Physiol. Meas., 21(4),* 493–503.

[18] Mostovnikova, G., R., Mostovnikov, V. A., Plavskii, V. Yu., Tret'yakova, A. I., Andreev, S. P. and Ryabtsev, A. B. (2000). Phototherapeutic apparatus based on an argon laser for the treatment of hyperbilirubinemia in newborns. *J. Opt. Technol., 67(11),* 981–983.

[19] Pritchard, M. A., Beller, E. M. and Norton, B. (2004). Skin exposure during conventional phototherapy in preterm infants: a randomised controlled trial. *J. Paediatr Child Health, 40,* 270-274.

[20] Wentworth, S. D. P. (2005). Neonatal phototherapy – today's lights, lamps and devices. *Infant, 1(1),* 14–19.

[21] Stokowski, L. A. (2006). Fundamentals of phototherapy for neonatal jaundice. *Adv. Neonatal Care, 6(6),* 303–312.

[22] Maisels, M. J. and McDonagh, A. F. (2008). Phototherapy for neonatal jaundice. N. Engl. *J. Med., 358(9),* 920-928.

[23] Cremer, R. J., Perryman, P. W. and Richards, D. H. (1958). Influence of light on the hyperbilirubinaemia of infants. *Lancet, 1(24),* 1094–1097.

[24] Dobbs, R. H. and Cremer, R. J. (1985). Phototherapy (Looking back). *Arch. Dis. Child., 50(11),* 833–836.

[25] Lucey, J., Ferreiro, M. and Hewitt, J. (1968). Prevention of hyperbilirubinemia of prematurity by phototherapy. *Pediatrics, 41(6),* 1047–1054.

[26] Lucey, J. F. (1972). Neonatal jaundice and phototherapy. Pediatr. *Clin. North. Am., 19(4),* 1–7.

[27] Ennever, J. F., Knox, I., Denne, S. C. and Speck, W. T. (1985). Phototherapy for neonatal jaundice: in vivo clearance of bilirubin photoproducts. *Pediatr. Res., 19(2),* 205–208.

[28] Onishi, S., Isobe, K., Itoh, S., Manabe, M., Sasaki, K., Fukuzaki, R. and Yamakawa, T. (1986). Metabolism of bilirubin and its photoisomers in newborn infants during phototherapy. *J. Biochem., 100,* 789–795.

[29] Ennever, J. F., Costarino, A. T., Polin, R. A. and Speck, W. T. (1987). Rapid clearance of a structural isomer of bilirubin during phototherapy. *J. Clin. Invest., 79(6),* 1674–1678.

[30] Myara, A., Sender, A., Valette, V., Rostoker, C., Paumier, D., Capoulade, C., Loridon, F., Bouillie, J., Milliez, J., Brossard, Y. and Trivin, F. (1997). Early changes in cutaneous bilirubin and serum bilirubin isomers during intensive phototherapy of jaundiced neonates with blue and green light. *Biol. Neonate., 71,* 75–82.

[31] Okada, H., Masuya, K., Kurono, Y., Nagano, K., Okubo, K., Yasuda, S., Kawasaki, A., Kawada, K., Kusaka, T., Namba, M., Nishida, T., Imai, T., Isobe, K. and Itoh, S.

(2004). Change of bilirubin photoisomers in the urine and serum before and after phototherapy compared with light source. *Pediatr. Int., 46 (6),* 640-644.

[32] Okada, H., Masuya, K., Yasuda, S., Okubo, K, Kawada, K, Kusaka, T, Namba, M, Nishida,, T, Imai, T, Isobe, K and Itoh, S. (2005). Developmental changes in serum half-life of (E,Z)–cyclobilirubin. *Early Hum. Dev.*, 81(7), 619–622.

[33] Mreihil, K., McDonagh, A. F., Nakstad, B. and Hansen, T. W. (2010). Early isomerization of bilirubin in phototherapy of neonatal jaundice. *Pediatr. Res., 67(6),* 656-659.

[34] Costarino, A.T., Enneve, J.F., Baumgart, S., Speck, W.T., Paul, M. and Polin, R. A. (1985). Bilirubin photoisomerization in premature neonates under low- and high-dose phototherapy. *Pediatrics, 75(3),* 519-522.

[35] Bensasson, R. V., Land, E. J. and Truscott, T. G. (1983). *Flash Photolysis and Pulse Radiolysis: Contributions to the Chemistry of Biology and Medicine*. Oxford, New York, Toronto, Sydney, Paris, Frankfurt: Pergamon Press.

[36] Myshkin, A E and Sakharov, V N. (1982). The photochemistry of bilirubin. *Russ. Chem. Rev., 51 (1),* 40–50.

[37] Landen, G. L., Park, Y. T. and Lightner, D. A. (1983). On the role of singlet oxygen in the self–sensitized photooxygenation of bilirubin and its pyrromethenone models. *Tetrahedron, 39(11),* 1893–1907.

[38] McDonagh A F, and Assisi, F. (1972). The ready isomerization of bilirubin IXα in aqueous solution. *Biochem. J., 129(3),* 797–800.

[39] Brodersen, R. (1979). Bilirubin. Solubility and interaction with albumin and phospholipids. *J. Biol. Chem., 254(7),* 2364–2369.

[40] *Jacobsen, J. and Brodersen, R.* (1983). Albumin-bilirubin binding mechanism. Kinetic and spectroscopic studies of binding bilirubin and xanthobilirubic acid to human serum albumin. *J. Biol. Chem., 258(10),* 6319–6326.

[41] Reed, R.G. (1977). Kinetics of bilirubin binding to serum albumin and effect of palmitate. *J. Biol. Chem., 252(21),* 7483–7487.

[42] Pace, C. N., Vajdos, F., Fee, L., Grimsley, G. and Gray, T. (1995). How to measure and predict the molar absorption coefficient of a protein. *Protein. Sci., 4(11),* 2411–2423.

[43] Greenberg, J. W., Malhotra, V. and Ennever, J. F. (1987). Wavelength dependence of the quantum yield for the structural isomerization of bilirubin. *Photochem. Photobiol., 46(4),* 453–456.

[44] Kanna, Y., Arai, T. and Tokumaru, K. (1993). Effect of serum albumin from several mammals on the photoisomerization of bilirubin. *Bull. Chem. Soc. Jpn., 66(5),* 1586–1588.

[45] Chirvony, V. S., Pershukevich, P. P., Galievsky, V. A. and Shul'ga, A. M. (1997). Unusual near-IR fluorescence of ethylene-bridge-bonded porphyrin dimmers. *Opt. Spectosc., 82 (5),* 717–722.

[46] Plavskii, V. Yu., Mostovnikov, V. A., Mostovnikova G. R. and Tret'yakova A. I. (2007). Spectral fluorescence and polarization characteristics of Z,Z-bilirubin IXα. *J. Appl. Spectrosc., 74(1),* 120–132.

[47] Lamola, A. A. In: Optical properties and structure of tetrapyrroles; Blauer, G., Sund, H.; Ed.; Effect of environment on photophysical processes of bilirubin; Walter de Gruyter and Co. Inc. Berlin, New York, 1985; pp. 311–330.

[48] Greene, B. I., Lamola, A. A. and Shank, C. (1981). Picosecond primary photoprocesses of bilirubin bound to human serum albumin. *Proc. Natl. Acad. Sci. USA., 78(4),* 2008–2012.

[49] Tran, C. D. and Beddard, G. S. (1981). Excited state properties of bilirubin and its photoproduct using picosecond fluorescence and circularly polarized luminescence spectroscopy. *Biochim. Biophys. Acta, 678(3),* 497–504.

[50] Tran, C. D. and Beddard, G. S. (1982). Interaction between bilirubin and albumins using picosecond fluorescence and circularly polarized luminescence spectroscopy. *J. Am. Chem. Soc., 104(24),* 6741–6747.

[51] Vecchi, C., Donzelli, G. P., Sbrana, G. and Pratesi, R. (1986). Phototherapy for neonatal jaundice: clinical equivalence of fluorescent green and "special" blue lamps. *J. Pediatr., 108(3),* 452–426.

[52] Donzelli, G.P. (1989). Green light phototherapy : new trends. *J. Photochem. Photobiol. B.: Biol., 4(1),* 126–128.

[53] Donzelli, G.P. , Pratesi, S., Rapisardi, G., Agati, G., Fusi, F. and Pratesi, R. (1995). 1-day phototherapy of neonatal jaundice with blue-green lamp. *Lancet, 346(15),* 184–185.

[54] Ebbesen, F., Agati, G. and Pratesi, R. (2003). Phototherapy with turquoise versus blue light. *Arch. Dis. Child. Fetal Neonatal Ed., 88(5),* F.430–F.431.

[55] Ebbesen, F., Madsen, P., Støvring, S., Hundborg, H. and Agati, G. (2007). Therapeutic effect of turquoise versus blue light with equal irradiance in preterm infants with jaundice. *Acta. Paediatr., 96(6),* 837–841.

[56] Mostovnikova, G. R., Mostovnikov, V. A., Shishko, G. A., Plavskii, V. Yu., Ryabtsev, A. B., Leusenko, I. A., Ginevich, V. V., Mostovnikov, A. V., Tret'yakova, A. I. and Mikulich, A. V. Photophysical and technical basis for the efficacy of radiation technology superbright LEDs in treatment of hyperbilirubinemia. In: *Proceedings, International Conference on Laser Optics Technology in Biology and Medicine*, 14–15 October 2004, Institute of Physics NASB, Minsk (2004), Vol. 1, pp. 189–194. [in Russian].

[57] Mostovnikov, V. A., Mostovnikova, G. R., Zhdanovich, A. I., Plavskii, V. Yu., Kilina, E. B., Tret'yakov, S. A. and Samoilyukovich, V. A. (1988). Characteristic features of the effect of laser radiation in the long-wave UV region (λ = 325 nm) on enzymatic activity and spectral-luminescent characteristics of lactate dehydrogenase. *J. Appl. Spectr., 49(2),* 801 – 806.

[58] Plavskii, V. Yu., Mostovnikov, V. A., Mostovnikova, G. R., Tret'yakova, A. I. and Plavskaya, L. G. (2003). Regularities of bonding of chlorin e_6 to the oligomeric enzyme lactate dehydrogenase. *J. Appl. Spectr., 70(2),* 913–920.

[59] Chen, R.F. (1974). Fluorescence stopped-flow study of relaxation processes in the binding of bilirubin to serum albumins. *Arch. Biochem. Biophys.,160(1),* 106–112.

[60] Cu, A., Bellah, C. and Lightner, D. A. (1975). The fluorescence of bilirubin. *J. Am. Chem. Soc., 97(9),* 2579–2580.

[61] Matheson, B. C., Faini, G. J. and Lee, J. (1975). Low–temperature absorption and fluorescence spectra and quantum yields of bilirubin. *Photochem. Photobiol., 21(1),* 135–137.

[62] Bonnet, R., Dalton, J. and Hamilton, D. (1975). Luminescence of bilirubin. *J. Chem. Soc. Chem. Commun., 15,* 639–640.

[63] Barber, D. J. and Richards, J. T. (1977). Anomalous solvent effect and observation of phosphorescence in bilirubin. *Chem. Phys. Lett., 46(1),* 130–132.

[64] Dalton, J., Milgrom, L. R. and Bonnett, R. (1979). Luminescence of bilirubin. *Chem. Phys. Lett. 61(2),* 242–244.

[65] Lee, J. J. and Gillispie, G. D. (1981). The effect of pH on the fluorescence of complexes of human serum albumin and bovine serum albumin with bilirubin. *Photochem. Photobiol., 33(5),* 757–760.

[66] Lamola, A. A. and Flores, J. (1982). Effect of buffer viscosity on the fluorescence of bilirubin bound to human serum albumin. *J. Am. Chem. Soc., 104,* 2530–2534.

[67] Plavskii, V. Yu. *Photophysical aspects of laser treatment for neonatal hyperbilirubinemia.* Abstract of Ph.D. thesis. Minsk, Institute of Physics NASB, 2010. [in Russian].

[68] Plavskii, V. Yu., Mostovnikov, V. A., Tret'yakova, A. I. and Mostovnikova, G. R. (2007). Photophysical processes that determine the photoisomerization selectivity of Z,Z-bilirubin IXa in complexes with albumins. *J. Opt. Technol., 74(7),* 446–454.

[69] Mostovnikov, V. A., Mostovnikova, G. R. and Plavski, V. Y. (1994). Spectral and photochemical parameters, which define laser phototherapy hyperbilirubinemia of newborn higher efficacy. *Proc. SPIE, 2370,* 558–561.

[70] Plavskii, V. Yu., Mostovnikov, V. A., Mostovnikova, G. R. and Tret'yakova, A. I. Selective laser photoisomerization of bilirubin. In: *Materials of the International Conference on Laser Physics and the Applications of Lasers,* 14–16 May 2003, Minsk, Institute of Physics NASB, Minsk (2003), pp. 309–312. [in Russian].

[71] Zietz, B., Macpherson, A. N. and Gillbro, T. (2004). Resolution of ultrafast excited state kinetics of bilirubin in chloroform and bound to human serum albumin. *Phys. Chem. Chem. Phys., 6(19),* 4535–4537.

[72] Costarino, A. T., Ennever, J. F., Baumgart, S., Speck, W. T. and Polin, R. A. (1985). Effect of spectral distribution on isomerization of bilirubin in vivo. *J. Pediatr., 107(1),* 125–128.

[73] Onishi, S., Itoh, S. and Isobe, K. (1986). Wavelength-dependence of the relative rate constants for the main geometric and structural photoisomerization of bilirubin IX alpha bound to human serum albumin. Demonstration of green light at 510 nm as the most effective wavelength in photochemical changes from (ZZ)-bilirubin IX alpha to (EZ)-cyclobilirubin IX alpha via (EZ)-bilirubin. *Biochem. J., 236(1),* 23–29.

[74] Itoh, S., Onishi, S., Isobe, K., Manabe, M. and Yamakawa, T. (1987). Wavelength dependence of the geometric and structural photoisomerization of bilirubin bound to human serum albumin. *Biol. Neonate., 51(1),* 10–17.

[75] McDonagh, A. F., Agati, G., Fusi, F. and Pratesi, R. (1989). Quantum yields for laser photocyclization of bilirubin in the presence of human serum albumin. Dependence of quantum yield on excitation wavelength. *Photochem. Photobiol., 50(3),* 305–319.

[76] Agati, G. and Fusi, F. (1990). New trends in photobiology (invited review). Recent advances in bilirubin photophysics. *J. Photochem. Photobiol. B: Biol., 7(1),* 1–14.

[77] Ennever, J. F. and Dresing, T. J. (1991). Quantum yields for the cyclization and configurational isomerization of 4E,15Z–bilirubin. *Photochem. Photobiol., 53(1),* 25–32.

[78] Agati, G., Fusi, F., Pratesi, R. and McDonagh, A. F. (1992). Wavelength–dependent quantum yield for Z→E isomerization of bilirubin complexed with human serum albumin. *Photochem. Photobiol., 55(2),* 185–190.

[79] Agati, G., Fusi, F. and Pratesi, R. (1993). Evaluation of the quantum yield for E→Z isomerization of bilirubin bound to human serum albumin. Evidence of internal conversion processes competing with configurational photoisomerization. *J. Photochem. Photobiol. B: Biol., 17(2),* 173–180.

[80] Troup, G. J., Agati, G., Fusi, F. and Pratesi, R. (1996). Photophysics of the variable quantum yield of asymmetric bilirubin. *Aust. J. Phys., 49(3),* 673 – 681.

[81] McDonagh, A. F., Agati, G. and Lightner, D. A. (1998). Induction of wavelength-dependent photochemistry in bilirubins by serum albumin. *Monatsh. Chem., 129(6-7),* 649 – 660.

[82] Mazzoni, M., Agati, G., Troup, G. J. and Pratesi, R. (2003). Analysis of wavelength-dependent photoisomerization quantum yields in bilirubins by fitting two exciton absorption bands. *J. Opt. A: Pure Appl. Opt., 5(5),* S374-S380.

[83] Mazzoni, M., Agati, G., Pratesi, R. and Persico, M. A (2005). spectroscopic study of the wavelength-dependent photoisomerizations of bilirubins bound to human serum albumin. *J. Opt. A: Pure Appl. Opt., 7(12),* 742- 747.

[84] McDonagh, A. F. and Lightner, D. A. In Optical properties and structure of tetrapyrroles; Blauer, G., Sund, H.; Ed.; Mechanism of phototherapy of neonatal jaundice. Regiospecific photoisomerization of bilirubins; Walter de Gruyter and Co. Inc. Berlin, New York, 1985; pp. 297–307.

[85] Onishi, S., Itoh, S., Yamakawa, T., Isobe, K., Manabe, M., Toyota, S. and Imai, T. (1985). Comparison of kinetic study of the photochemical changes of (ZZ)-bilirubin IX α bound to human serum albumin with that bound to rat serum albumin. *Biochem. J., 230(3),* 561–567.

[86] Ennever, J. F. (1988). Phototherapy for neonatal jaundice. *Photochem. Photobiol., 47(6),* 871–876.

[87] Kanna, Y., Arai, T. and Tokumaru, K. (1993). Photoisomerization of bilirubins and the role of intramolecular hydrogen bonds. *Bull. Chem. Soc. Jpn., 66(5),* 1482–1489.

[88] Kanna, Y., Arai, T. and Tokumaru, K. (1994). Effect of binding sites of human serum albumin on the efficiency and photostationary state isomer rations of the photoisomerization of bilirubin. *Bull. Chem. Soc. Jpn., 67,* 2758–2762.

[89] Sloper, R. W. and Truscott, T. G. (1980). Excited states of bilirubin. *Photochem. Photobiol., 31(5),* 445–450.

[90] Plavski, V. Y., Mostovnikov, V. A. and Mostovnikova, G. R. In Spectroscopy of Biological Molecules; Merlin, J. C., Turrell, S., Huvenne, J. P.; Ed.; Bichromophoric character of the absorption and emission of light by molecules of bilirubin; Kluwer Acad. Publ.: Dordrecht, Boston, London, 1995, pp. 271–272.

[91] Mugnoli, A., Manito, P., and Monti, D. (1983). Structure of bilirubin IX (isopropylammonium salt) chloroform solvate, $C_{33}H_{34}N_4O_6^{2-}.2C_3H_{10}N^+.2CHCl_3$. *Acta. Cryst., C39(9),* P. 1287–1291.

[92] Lightner, D. A., Gawronski, J. K. and Wijekoon, W. M. D. (1987). Complementarity and chiral recognition: enantioselective complexation of bilirubin. *J. Am. Chem. Soc., 109(21),* 6354–6362.

[93] Kuz'mitskii, V. A. (2001). Quantum–chemical calculations of the spatial structure of bilirubin molecule fragments. *J. Appl. Spectrosc., 68(1),* 45–54.

[94] Heirwegh, K. P., Fevery, J. and Blanckaert, N. (1989). Chromatographic analysis and structure determination of biliverdins and bilirubins. *J. Chromatogr., 496(1),* 1–26.

[95] Zunszain, P. A., Ghuman, J., McDonagh, A. F. and Curry, S. (2008). Crystallographic analysis of human serum albumin complexed with 4Z,15E-bilirubin-IXα. *J. Mol. Biol., 381(2),* 394–406.

[96] Lightner, D. A., Wijekoon, W. M. D. and Zhang, M. H. (1988). Understanding bilirubin conformation and binding. Circular dichroism of human serum albumin complexes with bilirubin and its esters. *J. Biol. Chem., 263(32),* 16669–16676.

[97] Rubinov, A. N. and Tomin, V. I. (1970). Bathochromic luminescence in solutions of organic dyes at low temperatures. *Opt. Spectrosc., 29(6),* 578–580.

[98] Tomin, V. I. and Rubinov, A. N. (1980). Spectroscopy of inhomogeneous configurational broadening in dye solutions. *J. Appl. Spectrosc., 35(2),* 855 – 865.

[99] Rubinov, A. N., Tomin, V. I. and Bushuk, B. A. (1982). Kinetic spectroscopy of orientational states of solvated dye molecules in polar solutions. *J. Luminesc., 26(4),* 377–391.

[100] Weber, G. and, Shinitzky, M. (1970). Failure of energy transfer between identical aromatic molecules on excitation at the long wave edge of the absorption spectrum. *Proc. Nat. Acad. Sci. USA, 65,* 823–830.

[101] Lakowicz, J. R. (1999). *Principles of Fluorescence Spectroscopy.* New York: Kluwer Academic/Plenum Publishers.

[102] Iwase, T., and Kusaka, T. and Itoh, S. (2010). (*EZ*)-Cyclobilirubin formation from bilirubin in complex with serum albumin derived from various species. *J. Photochem. Photobiol. B: Biol., 98(2),* 138–143.

[103] McDonagh, A. F., Lightner, D. A., Resinger, M. and, Palma, L. A. (1986). Human serum albumin as a chiral template. Stereoselective photocyclization of bilirubin. *J. Chem. Soc. Chem. Commun., 2,* 249–250.

[104] Blauer, G., Harmatz, D. and Snir, J. (1972). Optical properties of bilirubin-serum albumin complexes in and aqueous solution. I. Dependence on pH. *Biochim. Biophys. Acta, 278(1),* 68–88.

[105] Blauer, G. and King, T. E. (1970). Interactions of bilirubin with bovine serum albumin in aqueous solution. *J. Biol. Chem., 245(2),* 372–381.

[106] Davydov, A. S. (1971). *Theory of Molecular Excitons.* New York: Plenum Press.

[107] Kasha, M., Rawls, H., and Ashraf El-Bayoumi, M. (1965). The exciton model in molecular spectroscopy. *Pure. Appl. Chem., 11(3-4),* 371-392.

[108] Broude, V. L., Rashba, E. I. and Sheka, E. F. (1985). *Spectroscopy of molecular excitons.* Berlin; New York: Springer-Verlag.

[109] Plavskii V. Yu., Mostovnikov V. A., Tret'yakova A. I. and Mostovnikova G. R. Exciton interchromophore interactions in the Z,Z-bilirubin IXα molecule, in *Proc. Intern. Conf on Laser Physics and Optical Technology,* 25–29 September 2006, Minsk, 2006, Grodno, 2006, vol. 2, pp. 336–339. [in Russian].

[110] Lamola, A. A., Braslavsky, S. E., Schaffner, K. and Lightner, D. A. (1983). Spectral study of the photochemistry of dipyrrole models for bilirubin bound to human serum albumin. *Photochem. Photobiol., 37(3),* 263–270.

[111] Rubaltelli, F. F. and Jori, G. (1976). Visible light irradiation of human and bovine serum albumin-bilirubin complex. *Photochem. Photobiol., 29(5)*, 991–1000.

[112] Hulea, S. A., Smith, T. L., Wasowicz, E. and Kummerow, F. A. (1996). Bilirubin sensitized photooxidation of human plasma low density lipoprotein. *Biochim. Biophys. Acta., 1304(3)*, 197–209.

[113] Rosenstein, B. S., Ducore, J. M. and Cummings, S. W. (1983). The mechanism of bilirubin-photosensitized DNA strand breakage in human cells exposed to phototherapy light. *Mutation. Res., 112(6)*, 397–406.

[114] Christensen, T., Reitan, J. B. and Kinn, G. (1990). Single-strand breaks in the DNA of human cells exposed to visible light from phototherapy lamps in the presence and absence of bilirubin. *J. Photochem. Photobiol. B.: Biol., 7(2–4)*, 337–346.

[115] Girotti, A.W. (1976). Bilirubin-sensitized photoinactivation of enzymes in the isolated membrane of the human erythrocyte. *Photochem. Photobiol., 24(6)*, 525–532.

[116] Deziel, M.R. and Girotti, A W. (1980). Photodynamic action of bilirubin on liposomes and erythrocyte membranes. *J. Biol. Chem., 255(17)*, 8192–8198.

[117] Bohm, F., Drygalla, F. and Charlesworth, P. (1995). Bilirubin phototoxicity to human cells by green light phototherapy in vitro. *Photochem. Photobiol., 62(6)*, 980–983.

[118] Roll, E. (2005). Bilirubin-induced cell death during continuous and intermittent phototherapy and in the dark. *Acta. Paediatrica., 94(10)*, 1437–1442.

[119] Roll, E. B. and Christensen, T. (2005). Formation of photoproducts and cytotoxicity of bilirubin irradiated with turquoise and blue phototherapy light. *Acta. Paediatr., 94(10)*, 1448–1454.

[120] Plavskii, V. Yu., Mostovnikov, V. A., Tret'yakova, A. I. and Mostovnikova, G. R. (2008). Sensitizing effect of Z,Z-bilirubin IXα and its photoproducts on enzymes in model solutions. *J. Appl. Spectrosc., 75(3)*, 407–419.

[121] Vorobey, A., Plavsky, V., Vorobey, P., Steindal, A.H. and Moan, J. Photochemical reactions of bilirubin and folic acid in different solutions. 12th congress European society for photobiology. ESP 2007, Bath, England, 1st – 6th September 2007, *Programme and Book of Abstracts*. P. 161–162.

[122] Shapovalenko, E. P. and Kolosov, I. V. (1978) A study of thermodynamics and mechanism of bilirubin and albumin interaction in neutral and weakly alkaline media. *Rus. J. Bioorg. Chem.*, 4, 514–522.

[123] Malhotra, V. and Ennever, J. F. (1986). Determination of the relative detector response for unstable bilirubin photoproducts without isolation. *J. Chromatogr., 383(1)*, 153–157.

[124] Migliorini, M. G., Galvan, P., Sbrana, G., Donzelli, G. P. and Vecchi, C. (1988). Bilirubin photoconversion induced by monochromatic laser radiation. Comparison between aerobic and anaerobic experiments in vitro. *Biochem. J., 256(3)*, 841–846.

[125] Bacci, M., Linari, R., Agati, G. and Fusi, F. (1989). UV excitable fluorescence of lumirubin. *J. Photochem. Photobiol. B.: Biol., 3(3)*, 419–427.

[126] Lightner, D. A., Wooldridge, T. A. and McDonagh, A. F. (1979). Configurational isomerization of bilirubin and the mechanism of jaundice phototherapy. *Biochem. Biophys. Res. Commun., 86(2)*, 235–243.

[127] Yasuda, S., Itoh, S., Imai, T., Isobe, K. and Onishi, S. (2001). Cyclobilirubin formation by in vitro photoirradiation with neonatal phototherapy light. *Pediatr. Int., 43(3)*, 270–275.

[128] Onishi, S., Itoh, S., Isobe, K., Ochi, M., Kunikata, T. and Imai, T. (1989). Effect of the binding of bilirubin to either the first class or the second class of binding sites of the human serum albumin molecule on its photochemical reaction. *Biochem. J., 257(3),* 711–714.

[129] Moroi, Y., Matuura, R. and Hisadome, T. (1985). Bilirubin in aqueous solution. Absorption spectrum, aqueous solubility, and dissociation constants. *Bull. Chem. Soc. Jpn., 58(5),* 1426–1431.

[130] Agati, G., Fusi, F., Pratesi, R., Galvan P. and Donzelli, G. P. (1998). Bilirubin photoisomerization products in serum and urine from a Crigler–Najjar type I patient treated by phototherapy. *J. Photochem. Photobiol. B., 47(2–3),* 181–189.

[131] Dobretsov, G. E., Miller, Yu. I. and Kolchin, Yu. A. (1990). Interaction of N-phenyl-1-amino-8-sulfonaphthalene (ANS) with albumin in blood in hyperbilirubinemia. *Ukr Biokhim Zh., 62(5),* 29–33 [in Russian].

[132] Volodin, N. N., Antonov, A. G., Aronskind, E. V., Baybarina, E. N., Degtyarev, D. N., Degtyareva, A. V., Kovtun, O. P., Mukhametshin, F. G. and Parshikova, O. V. (2006). Protocol of diagnosing and treatment of hyperbilirubinemia in neonates. *Woprosy Pract. Pediatr., 1(6),* 9–18 [in Russian].

[133] Lightner, D. A., Linnane, W. P. and Ahlfors, C. E. (1984). Bilirubin photooxidation products in the urine of jaundiced neonates receiving phototherapy. *Pediatr. Res., 18(8),* 696–700.

[134] Vreman, H. J., Wong, R. J. and Stevenson, D. K. (2004). Phototherapy: current methods and future direction. *Semin. Perinatol., 28(5),* 326–333.

[135] Use of phototherapy for neonatal hyperbilirubinemia. Fetus and Newborn Committee, Canadian Paediatric Society (1986) *Can. Med. Assoc. J., 134(11),* 1237–1245.

[136] Vogl, T. P., Hegyi, T., Hiat, I. M., Polin, R. A. and Indyk, L. (1978). Intermittent phototherapy in the treatment of jaundice in the premature infant. *J. Pediatr., 92(4),* 627–630.

[137] Rubaltelli, F. F. Zanardo, V. and Granati, B. (1978). Effect of various phototherapy regimens on bilirubin decrement. *Pediatrics., 61(6),* 838–841.

[138] Rudenko, E. B. and Kalinicheva, V. I. (1990). Comparative evaluation of the effectiveness of various schedules of phototherapy in premature newborn infants with hyperbilirubinemia. *Pediatriia, 4,* 58–61 [in Russian].

[139] Peirce, S. C., Hedges, A. J. and Crawford, D. C. (2006). *Market survey: infant warming and phototherapy.* Report 06046. London: Purchasing and Supply Agency. Centre for evidence-based purchasing.

[140] Tan, K. L. (1989). Efficacy of fluorescent daylight, blue and green lamps in the management of non-hemolytic hyperbilirubinemia. *J. Pediatrics,* 114(1), 132–137.

[141] Gies, H. P. and Roy, C. R. (1990). Bilirubin phototherapy and potential UVR hazards. *Health Physics, 58(3),* 313–320.

[142] Jacobson, E. D. and Krell, K. (1982). Genetic effects of fluorescent lamp radiation on eukaryotic cells in culture. *Photochem. Photobiol., 35(6),* 875–879.

[143] Speck, W. T., (1981). Rosenkranz P.G., Behrman M. and Rosenkranz H. S. Embryotoxic effect of phototherapy: Separation of therapeutic and gametotoxic activities. *Photochem. Photobiol., 33(1),* 121–122.

[144] Rubaltelli, F. F., Da Riol, R., D'Amore, E. S. and Jori, G. (1996). The bronze baby syndrome: evidence of increased tissue concentration of copper porphyrins. *Acta. Paediatr., 85(3),* 381–384.

[145] Zammarchi, E., La Rosa, S., Pierro, U., Lenzi, G., Bartolini, P. and Falorni, S. (1989). Free tryptophan decrease in jaundiced newborn infants during phototherapy. *Biol. Neonate., 55(4/5),* 224–227.

[146] Hovi, L., Hekali, R. and Siimes, M. A. (1979). Evidence of riboflavin depletion in breast-fed newborns and its further acceleration during treatment of hyperbilirubinemia by phototherapy. *Acta. Paediatr. Scand., 68(4),* 567–570.

[147] Kehyayan, E., Galdi, I., Pellicciotta, G., Girardi, A.M. and Caviezel, F. (1985). Effect of phototherapy on plasma levels of GH, LH and FSH in the newborn. *J. Endocrinol. Invest., 8(6),* 561–565.

[148] Vanhaesebrouck, P., De Bock, F., Zecic, A., De Praeter, C., Smets, K, De Coen, K. and Goossens, L. (2005). Phototherapy-mediated syndrome of inappropriate secretion of antidiuretic hormone in an in utero selective serotonin reuptake inhibitor–exposed newborn infant. *Pediatrics, 115(5),* 508–511.

[149] Gounaris, A., Alexiou, N., Costalos, C., Daniilidou, M., Bakoleas, V. and Constantellou, E. (1998). Gut hormone levels in neonates under phototherapy. *Early. Hum. Dev., 51(1),* 57–60.

[150] Kurt, A., Aygun, A. D., Kurt, A. N. C., Godekmerdan, A., Akarsu, S. and Yilmaz, E. (2009). Use of phototherapy for neonatal hyperbilirubinemia affects cytokine production and lymphocyte subsets. *Neonatology, 95(3),* 262–266.

[151] Aouthmany, M. M. (1999). Phototherapy increases hemoglobin degradation and bilirubin production in preterm infants. *J. Perinatol., 19(4),* 271–274.

[152] Grunhagen, D. J., De Boer, M. G., De Beaufort, A. J. and Walther, F. J. (2002). Transepidermal water loss during halogen spotlight phototherapy in preterm infants. *Pediatr. Res., 51(3),* 402–405.

[153] Matichard, E., Hénanff, A. L., Sanders, A., Leguyadec, J., Crickx, B. and Descamps, V. (2006). Effect of neonatal phototherapy on melanocytic nevus count in children. *Arch. Dermatol.,* 142(12), 1599–1604.

[154] Csoma, Z., Kemeny, L. and Olah, J. (2008). Phototherapy for neonatal jaundice. *N. Engl. J. Med., 358(23),* 2523–2524.

[155] Csoma, Z., Hencz, P., Orvos, H., Kemeny, L., Dobozy, A., Dosa-Racz, E., Erdei, Z., Bartusek, D. and Olah, J. (2007). Neonatal blue-light phototherapy could increase the risk of dysplastic nevus development. *Pediatrics, 119(6),* 1269.

[156] Vreman, H. J., Wong, R. J., Stevenson, D. K., Route R. K., Reader S. D., Fejer M. M., Gale R. and Seidman D. S. (1998). Light-emitting diodes: a novel light source for phototherapy. *Pediatr. Res., 44(5),* 804–809.

[157] Seidman, D.S., Moise, J., Ergaz, Z., Laor A., Vreman H. J., Stevenson D. K. and Gale R. (2000). A new blue light-emitting phototherapy device: a prospective randomized controlled study. *J. Pediatr., 136(6),* 771–774.

[158] Maisels, M. J. (2001). Phototherapy Traditional and Nontraditional. *J. Perninatol., 21(1),* S93–S97.

[159] Seidman, D.S., Moise, J., Ergaz, Z., Laor A., Vreman H. J., Stevenson D. K. and Gale R. (2003). A prospective randomized controlled study of phototherapy using blue and

blue-green light-emitting devices, and conventional halogen-quartz phototherapy. *J. Perinatol., 23(2),* 123–127.

[160] Rosen, H., Rosen, A., Rosen, D., Onaral B. and Hiatt M. (2005). Use of a light emitting diode (LED) array for bilirubin phototransformation. *Conf Proc IEEE Eng. Med. Biol. Soc. (7):* 7266–7268.

[161] Chang, Y. S., Hwang, J. H., Kwon, H. N., Choi, C. W., Ko, S. Y., Park, W. S., Shin, S.M. and Lee, M. (2005). In vitro and in vivo efficacy of new blue light emitting diode phototherapy compared to conventional halogen quartz phototherapy for neonatal jaundice. *J. Korean Med. Sci., 20(1),* 61–64.

[162] Sarin, M., Dutta, S. and Narang A. (2006). Randomized controlled trial of compact fluorescent lamp versus standard phototherapy for the treatment of neonatal hyperbilirubinemia. *Indian Pediatr., 43(7),* 583–590.

[163] Maisels, M. J., Kring, E. A. and DeRidder, J. (2007). Randomized controlled trial of light-emitting diode phototherapy. *J. Perinatol., 27(9),* 565–567.

[164] Okada, H., Abe, T., Etoh, Y., Yoshino, S., Kato, I., Iwaki, T., Okubo, K., Yasuda, S., Kawada, K., Kusaka, T., Namba, M., Nishida, T., Imai, T., Isobe, K. and Itoh S. (2007). In vitro production of bilirubin photoisomers by light irradiation using neoBLUE. *Pediatr. Int. 49(3),* 318–321.

[165] Mostovnikova, G. R., Vil'chuk, K.U., Ryabtsev, A. B., Mostovnikov, A. V., Gned'ko, T. V., Leusenko, I. A., Mostovnikov, V. A. and Plavskii V. Yu. (2008). Use of radiation superbright LEDs to increase efficacy of phototherapy for hyperbilirubinemia (jaundice) in newborn. *Proc. VII Int. Conf. Laser Physics and Optics Technology. June 17–19, 2008, Minsk,* Institute of Physics NASB, Minsk (2008), Vol. 2, p. 443 – 447. [in Russian].

[166] Martins, B. M., De Carvalho, M., Moreira, M. E. and Lopes, J. M. (2007). Efficacy of new micro processed phototherapy system with five high intensity light emitting diodes (Super LED). *J. Pediatr. (Rio J.), 83(3),* 253–258.

[167] Bertini, G., Perugi, S., Elia, S., Pratesi, S., Dani, C. and Rubaltelli, F. F. (2008). Transepidermal water loss and cerebral hemodynamics in preterm infants: conventional versus LED phototherapy. *Eur. J. Pediatr., 167(1),* 37–42.

[168] Tanaka, K., Hashimoto, H., Tachibana, T., Ishikawa, H. and Ohki, T. (2008). Apoptosis in the small intestine of neonatal rat using blue light-emitting diode devices and conventional halogen-quartz devices in phototherapy. *Pediatr. Surg. Int., 24(7),* 837–842.

[169] Uras N., Karadag A., Tonbul A., Karabel M., Dogan G. and Tatli M. M. (2009). Comparison of light emitting diode phototherapy and double standard conventional phototherapy for nonhemolytic neonatal hyperbilirubinemia. *Turk. J. Med. Sci, 39(3),* 337–341.

[170] Tzeng, C. B., Wey, T. S. and Young M. S. (2009). A new phototherapy apparatus designed for the curing of neonatal jaundice. *IFMBE Proc, 23(3),* 1132–1135.

[171] Sebbe, P. F., Villaverde, A. B., Moreira, L. M., Barbosa, A. M. and Veissid, N. (2009). Characterization of a novel LEDs device prototype for neonatal jaundice and its comparison with fluorescent lamps sources: Phototherapy treatment of hyperbilirubinemia in Wistar rats. *Spectroscopy, 23(5–6),* 243–255.

[172] Bhutani. V. K. (2009). Performance evaluation for neonatal phototherapy. *Indian Pediatr., 46(1),* 19–21.

[173] Vreman, H. J. (2010). Phototherapy: the challenge to accurately measure irradiance. *Indian Pediatr., 47,* 127–128.

[174] Kumar, P., Murki, S., Malik, G. K., Chawla, D., Deorari, A. K., Karthi, N., Subramanian, S., Sravanthi, J., Gaddam, P. and Singh S. N. (2010). Light emitting diodes versus compact fluorescent tubes for phototherapy in neonatal jaundice: a multi center randomized controlled trial. *Indian Pediatr., 47(2),* 131–137.

In: Bilirubin: Chemistry, Regulation and Disorder
Editors: J. F. Novotny and F. Sedlacek
ISBN: 978-1-62100-911-5
© 2012 Nova Science Publishers, Inc.

Chapter II

Genetic Variants in Bilirubin Metabolism Pathway, Serum Bilirubin and Cardiovascular Disease

Rong Lin[1,2,*] *and Li Jin*[2]

[1]Department of Biology, Hainan Medical College, Haikou, Hainan, China
[2]State Key Laboratory of Genetic Engineering and Ministry of Education Key Laboratory of Contemporary Anthropology, School of Life Sciences and Institutes of Biomedical Sciences, Fudan University, Shanghai, China

Abstract

Due to the potent antioxidant property of bilirubin, increasing attention has been drawn to the potential protective effects of bilirubin against cardiovascular disease (CVD), and the polymorphisms that may play roles in influencing serum total bilirubin (TBIL) levels and CVD. This article evaluates associations between serum bilirubin, bilirubin metabolism gene polymorphisms and CVD. A series of variants in the bilirubin metabolism genes, including polymorphisms in heme oxygenase-1(*HMOX1*), uridine diphosphate glycosyltransferase 1 (*UGT1A1*) and solute carrier organic anion transporter family member 1B3 (*SLCO1B3*), were found to be associated with TBIL levels. Serum bilirubin has consistently been shown to be inversely associated with various CVD, mostly among men in different populations. However, the findings on bilirubin-related genetic polymorphisms are controversial in association with CVD. These previous association studies of bilirubin metabolism gene polymorphisms with CVD typically focus on single-locus analysis, especially on the *HMOX1* (GT)n repeat or *UGT1A1* (TA)n repeat polymorphism. In addition to the *HMOX1* (GT)n repeat and *UGT1A1* (TA)n repeat polymorphisms, other variants, which have been identified associated with TBIL levels by genome-wide association studies (GWAS) and non-GWAS studies, should be

---

[*] Corresponding author: Dr Rong Lin, Hainan Medical College, 3 Xueyuan Road, Haikou, Hainan Province 571101, China. Tel: 86-898-66893739; Fax: 86-898-31350713; E-mail: xianronglin@ gmail.com.

examined in the genetic association studies of CVD. And examination beyond individual variant hits, by focusing on bilirubin metabolism pathway, is also important to unleashing the true power of association studies.

Bilirubin metabolism is a critical pathway which is required for the degradation of ageing erythrocytes. For many years, bilirubin was considered to be a toxic waste product of bilirubin metabolism. Today, bilirubin is regarded as an effective physiological antioxidant and, ultimately, may have relevant clinical implications in the prevention of cardiovascular disease (CVD). Recent evidence also indicates that some enzymes in bilirubin metabolism may play multifunctional roles in the cardiovascular system. Here, we firstly present a review of the vasoprotective effects of bilirubin metabolism. Later, we provide detailed information on the association between serum bilirubin, bilirubin metabolism gene polymorphisms, and CVD.

1. Bilirubin Metabolism Pathway in Human

Serum bilirubin is derived primarily from hemoglobin in aging red blood cells, which are removed by macrophages. Within macrophages, hemoglobin is broken down to heme and globin. Heme oxygenases (HMOXs) convert heme to biliverdin, free ferrous iron, and carbon monoxide (CO) (Figure 1). Biliverdin is subsequently reduced by biliverdin reductases (BLVRs) to bilirubin. Bilirubin is largely carried by albumin in the blood and transported into the liver by solute carrier organic anion transporter (SLCO) family members, such as SLCO1B1 and SLCO1B3, which are located in the hepatocyte sinusoidal membrane. [1] Within hepatocytes, the solubility of bilirubin is increased by the addition of one or two molecules of glucuronic acid, a process that is catalyzed by uridine diphosphate glycosyltransferase 1 family, polypeptide A1 (UGT1A1). Bilirubin monoglucuronide and diglucuronide metabolites, collectively referred to as conjugated bilirubin, are then actively transported into the bile by ATP-binding cassette, sub-family C (CFTR/MRP), member 2 (ABCC2) at the hepatocyte canalicular membrane. [2]

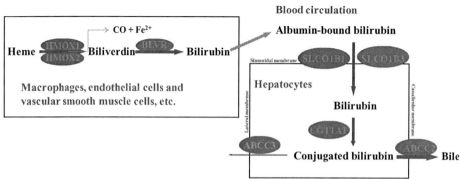

HMOX, heme oxygenase; BLVR, biliverdin reductase; SLCO, solute carrier organic anion transporter; UGT1A1, uridine-diphosphate glucuronosyltransferase 1A1; ABCC2, ATP-binding cassette, sub-family C (CFTR/MRP), member 2; CO, carbon monoxide.

Figure 1. Bilirubin metabolism pathway.

When biliary bilirubin excretion is impaired, bilirubin glucuronides are transported into blood by ABCC3 which is located in the hepatocyte lateral membrane. [3, 4]

In healthy subjects, unconjugated bilirubin is a predominant form of serum bilirubin. Serum conjugated bilirubin and total bilirubin levels are measured in the routine laboratory assay. Conjugated bilirubin reacts directly with the diazo reagent, while the reaction of unconjugated bilirubin with the diazo reagent requires an accelerator detergent. Thus, unconjugated bilirubin is also called 'indirect bilirubin' and conjugated bilirubin is called 'direct bilirubin'. Serum total bilirubin (TBIL) equals direct bilirubin plus indirect bilirubin. In this article, serum bilirubin represents TBIL sometimes for convenience, unless otherwise stated.

2. Biological Properties of Bilirubin

Bilirubin, which was considered to be a toxic waste product for many years, causes hyperbilirubinemia when in excess. Hyperbilirubinemia subsequently increases the risk of schizophrenia, [5] and may lead to acute unconjugated bilirubin encephalopathy, even kernicterus. [6] However, increasing experimental evidence indicated that bilirubin, including free bilirubin, albumin-bound bilirubin and conjugated bilirubin, is an effective antioxidant within normal/mild-increased range. Bilirubin scavenges peroxyl radicals efficiently, and suppresses the oxidation of lipids and lipoproteins, especially low density lipoprotein (LDL). [7] LDL, which is known as "bad" cholesterol, becomes more dangerous when it is oxidized. Oxidation of LDL renders it "sticky" and facilitates its deposition on the internal lining of blood vessels. Oxidized LDL initiates and contributes to the development of atherosclerosis by inducing inflammation in the blood vessel wall, atherosclerotic plaque development and profound immune response. Serum bilirubin shows a direct, highly significant association with the total serum antioxidant capacity in humans. [8-10] In addition to its property as a radical scavenger, bilirubin has an inhibitory effect on the activity of nicotinamide adenine dinucleotide phosphate (NADPH) oxidase, which is an important source of reactive oxygen species (ROS) production. [11, 12]

Moreover, despite of the low concentration of intracellular bilirubin (≈20-50 nmol/L [<0.3% normal serum TBIL levels]), the antioxidant ability of intracellular bilirubin is also powerful. Intracellular bilirubin protects against almost 10,000-fold higher concentrations of hydrogen peroxide (H_2O_2) by the bilirubin redox cycle in which bilirubin oxidized to biliverdin by ROS is rapidly reduced back to bilirubin by biliverdin reductase A (BLVRA). [13] It is well known that glutathione (GSH) is a principal endogenous intracellular small molecule antioxidant cytoprotectant. Although GSH is present in millimolar concentrations whereas bilirubin levels are well below 100 nM in most cells, bilirubin redox cycle is of comparable, or greater, importance to the cycling of GSH in physiologic cytoprotection. [13] Bilirubin and GSH provide physiologic antioxidant activity for distinct types of intracellular molecules. [14] Lipophilic bilirubin is associated intimately with cell membranes, where it might prevent lipid peroxidation and protect membrane proteins. On the other hand, water-soluble GSH primarily protect cytoplasmic constituents. However, recent studies have raised suspicion regarding the role of the bilirubin redox cycle for a sustained antioxidant effect of bilirubin. [15, 16]

Some evidence, both in vitro and vivo, implies that bilirubin has anti-inflammatory properties. [17] Bilirubin inhibits tumor necrosis factor-α (TNF-α)-induced upregulation of E-selectin, vascular cell adhesion molecule 1 (VCAM-1), and intercellular adhesion molecule 1 (ICAM-1) in vitro. [18] Subjects with Gilbert syndrome (GS), a very mild form of hyperbilirubinemia, had significantly lower concentrations of soluble forms of CD40 ligand and P-selectin. [19]

Bilirubin and biliverdin have also been shown to modulate immune effector functions. Biliverdin and bilirubin inhibit complement-dependent reactions in vitro and biliverdin administration inhibits Forssman anaphylaxis in guinea pigs, indicating that bile pigments may be endogenous tissue protectors partially by virtue of their anti-complement activity. [20] Because of its antioxidant, anti-inflammatory, and other biological properties, higher bilirubin levels could possibly prevent plaque formation and subsequent atherosclerosis.

3. Vasoprotective Effects of HMOX1

Two isoforms of HMOX exists in human: HMOX1 and HMOX2. [21] *HMOX2* is constitutively expressed and abundant in testes, brain, liver and vasculature. Compared to *HMOX2*, *HMOX1* is highly induced by many factors, such as heavy metals, cytokines, endotoxin, heme, nitric oxide (NO), hypoxia, and ultraviolet (UV) irradiation. *HMOX1* expression is highest in the reticuloendothelial system (spleen, liver and bone marrow), which are the responsible organs for the removal of aging red blood cells.

In tissues that are not directly responsible for hemoglobin degradation, the basal expression of *HMOX1* is very low but can be rapidly induced by stimulation. *HMOX1* expression is subjected to regulation by many cellular signaling pathways through the multiple response elements present in *HMOX1* gene promoter. [22] All findings suggest that HMOXs, especially HMOX1, may have broader roles besides hemoglobin degradation.

HMOX metabolism was viewed as a potentially toxicological pathway for many years because of its destruction of hemoproteins [23, 24] and various "waste" products. [25, 26] Interestingly, numerous studies have supported the multifunctional roles of HMOX1 in the cardiovascular system during the past decade. The vasoprotection afforded by HMOX1 is largely attributable to its end products: biliverdin, bilirubin (its effect has been described above), Fe^{2+} and CO (Reviewed in Reference [27]). CO could affect cardiovascular function through activation of soluble guanylate cyclase and the consequent increase in intracellular cGMP concentrations. [28]

CO is known to have anti-apoptotic, anti-inflammatory and anti-proliferative properties. [11, 12, 29] CO is also an active vasodilator involved in the regulation of vasomotor tone, platelet aggregation, and vascular smooth muscle cell proliferation. [28, 30, 31] The free iron released through HMOX activity drives the synthesis of ferritin. Ferritin can provide protection to endothelial cells against oxidative damages and prevention of oxidative modifications of LDL by virtue of its iron-binding capacity. [27, 32, 33]

Furthermore, HMOX-mediated consumption of heme may reduce heme-induced toxic cell injury, and decreased hemoglobin concentrations may enhance vasodilatation. Hemoglobin is a scavenger of NO that blunts NO-dependent vasodilatation.

4. Protective Effects of BLVRA

BLVRA was known for a long time only as an enzyme converting biliverdin to bilirubin. However, recent data disclosed new important features of this protein which are not limited to its reductase activity. BLVRA is a powerful intracellular antioxidant not only for it is a regulator for the induction of *HMOX1* and activating transcription factor-2 (*ATF-2*) expression by oxidative stress,[34, 35] but also for its roles in the bilirubin redox cycle. [13] In addition, BLVRA is an antagonist to insulin-mediated glucose uptake by cells,[36] and insulin resistance (IR) is an important risk factor for coronary artery disease (CAD) [37] and other CVD. [38, 39]

5. Genetic Polymorphisms of Bilirubin Metabolism Genes and Serum Bilirubin

In humans, serum bilirubin is highly heritable. [40-42] The study of two clinic visits in 84 Utah pedigrees indicated that a major gene explains 27% and 28% of the variance in bilirubin levels at visit 1 and visit 2, respectively, and 22% of the variance in bilirubin levels could be explained by other genes. [40] A Framingham Heart study, another pedigree-based study, also showed that the heritability of TBIL levels is estimated to be 49%±6% and *UGT1A1* might be a major gene controlling TBIL levels. [42]

(1) HMOX

Heme oxygenases are the first and rate-limiting enzymes in heme degradation. A guanine-thymine (GT)n repeat polymorphism and single nucleotide polymorphism (SNP) rs2071746 (T-413A) in the promoter region of the *HMOX1* gene are putatively functional. The number of (GT)n repeats shows a trimodal distribution with 23, 30 and 34 repeats at peaks in many populations such as the Japanese, Han Chinese, Uyghur, Kazak populations. [43, 44] The peak at 34 repeats is much lower and thus does not occur in some populations such as German. [45] Though the exact cut-off is still unknown, the allele type is usually classified into three classes, namely, short (S: <27 GT), middle (M: 27-32 GT), and long alleles (L: ≥33 GT). *HMOX1* expressions and HMOX activities induced by H_2O_2 stimulation were significantly higher in lymphoblastoid cell lines from Japanese subjects with the S/S genotype than those with the L/L genotype. [46] In addition, a cell with the S/M genotype had higher *HMOX1* expression than that with the M/M genotype under ultraviolet A irradiation. [47] The basal transcriptional activity of the *HMOX1* promoter/luciferase fusion genes with the S allele is higher than those with the M or L alleles. [43, 48] H_2O_2 exposure up-regulated the transcriptional activity of the *HMOX1* promoter/luciferase fusion genes with S allele but did not do so with M or L alleles. [43] The difference of transcriptional activity between M and L alleles remains uncertain. Therefore, the allele type is also classified into two classes, namely, short (S: <27 GT) and long alleles (L: ≥27 GT), in some studies.

The A-413T polymorphism is 154 nucleotides apart from the (GT)n repeat polymorphism. The A-30 and T-23 haplotypes are the two major haplotypes of A-413T-(GT)n repeat. [44, 49] Luciferase reporter assays conducted by Ono et al. [49, 50] showed that the promoter activities of the A-30 and A-23 haplotypes were significantly higher than those of the T-23 and T-30 haplotypes, which led the conclusion that the T-413A polymorphism might be responsible for the promoter activity. Notably, unlike the functional experiments of the (GT)n repeat polymorphism, Ono et al. only measured the basal promoter activities without any stimulation and only performed luciferase reporter assays which are less perfect than the assays using human cells carrying different (GT)n repeats. Further experiments are needed to measure the promoter activities of the A-30 and T-23 haplotypes under H_2O_2 or other stimulation.

Seven non-genome-wide association studies (GWAS) population-based studies have examined variants of *HMOX* related to TBIL levels. [44, 51-56] The association between the (GT)n repeat polymorphism and serum bilirubin levels was inconsistent. [44, 51-53, 55, 56] In Caucasians, carriers of short alleles (<25 GT) of the (GT)n repeat polymorphism had higher bilirubin levels compared with non-carriers. [51] However, this association was not observed in another two Caucasian populations. [52, 53] In healthy Indian adults, carriers of short alleles (<20 GT) had significantly higher bilirubin levels compared with non-carriers. [55] The (GT)n repeat polymorphism showed significant association with TBIL levels in the Uyghur population, but not in the Han and Kazak populations. [44] It can explained 1.1% of the variation of bilirubin in the Uyghur population (P=0.006). Chinese subjects with the L/L(≥27 GT) genotype had significantly lower bilirubin levels than those with the S/S (<27 GT) and S/L genotypes. [56] The (GT)n repeat polymorphism also showed no relation to neonatal hyperbilirubinemia in Japanese [57] and Turks,[58] but there was a relation to prolonged jaundice associated with breast milk in Turks. [58] The A-413T polymorphism has consistently been found not to be associated with TBIL levels. [44, 52, 54]

Until now, four GWASs have focused on serum bilirubin levels, three of which were conducted in Caucasians [59-61] and one in Koreans. [62] However, all of them failed to identify any SNPs in *HMOX* associated with serum bilirubin. Lin et al. [54] investigated four tagging SNPs in *HMOX1* and also failed to detect any SNPs in *HMOX1* associated with TBIL levels. Lin et al. [63] estimated gene variants influencing serum bilirubin levels in the Framingham Heart Study, using a low genome coverage GeneChip—Affymetrix 100K SNP GeneChip. Later, individuals from the Framingham Heart Study were included the GWAS by Johnson et al.,[60] which was performed with Affymetrix 500K SNP GeneChip and an additional genefocused 50K SNP GeneChip. Therefore, in this article, we did not group the study of Lin et al. [63] into the GWASs on serum bilirubin levels.

(2) BLVRA

Two isoforms of BLVR exists in human: BLVRA and BLVRB. In early fetuses, BLVRB is expressed and thus the predominant bilirubin isomer is bilirubin IXβ. Adults express BLVRA which produces bilirubin IXα exclusively. [64, 65] BLVRA functions in bilirubin metabolism pathway through three distinct tracks: (1) a reductase that catalyzes the conversion of biliverdin to bilirubin, (2) a regulator of HMOX1 enzyme activities[66] by binding to HMOX1 protein,[67] and (3) a bzip-type transcription factor for *HMOX1* regulation. [34] Three non-GWAS population-based studies have pursued the role of genetic variants of *BLVRA* underlying serum bilirubin levels and not found any variants of *BLVRA* significantly associated with serum bilirubin. [44, 54, 68] The four GWASs on serum bilirubin levels also failed to identify any SNPs in *BLVRA* associated with serum bilirubin levels.

(3) SLCO1B1 and SLCO1B3

SLCO1B1 is exclusively expressed at the hepatocyte basolateral membrane and transports unconjugated bilirubin from the blood circulation into the liver. [69] The drugs, such as indinavir, saquinavir, cyclosporin A, and rifamycin SV, can inhibit SLCO1B1 transport activity and induce hyperbilirubinemia in humans. [70] Recent data suggested that SNP rs2306283 (Asp130Asn) and rs4149056 (Val174Ala), two common non-synonymous SNPs within *SLCO1B1*, are associated with altered pharmacokinetics of pravastatin. [71-73] Japanese subjects with the *SLCO1B1*15* (Asp130Ala174) allele had higher serum unconjugated bilirubin levels and lower pravastin clearance. [74] In 42 healthy Chinese volunteers, serum bilirubin in both the *SLCO1B1*1b/*15* (Asp130Val174/Asp130Ala174) and *15/*15* (Asp130Ala174/Asp130Ala174) groups was significantly higher than that in the *SLCO1B1*1b/*1b* group. [75] Those two studies suggested that SNP rs4149056 is associated with serum bilirubin but both were carried out in very small Asian samples. In an association study with serum TBIL levels in 752 healthy Japanese, three SNPs within the *SLCO1B1* gene (rs4149025, rs4149014 and rs4149018) showed P-values between 0.01 and 0.001. [76] However, two non-GWAS population-based studies in large Asian populations did not observe any variants of *SLCO1B1*, including rs4149014, rs4149018 and rs4149056, associated with serum bilirubin levels. [44, 54] A small study (69 Caucasians and 38 Blacks) did not replicate the finding that subjects with the *SLCO1B1*15* allele had higher serum unconjugated bilirubin levels. [77]

The GWAS in 8,841 Koreans also showed that variants in *SLCO1B1* did not reach genome-wide significance [62]. SNP rs4149056 was filtered out by low SNP call rate in this GWAS. The three GWASs in Caucasians indicated that SNP rs4149056 is associated with serum bilirubin levels (Table 1). [59-61] SNP rs4149056 has the most significant P-value in the *SLCO1B1* locus ($P = 6.7 \times 10^{-13}$) and accounts for approximately 0.6% of the variation in TBIL levels. [60] The different contribution of *SLCO1B1* may be owing to population-specific genetic features, and sample size might also contribute to controversial results.

The effects of SNP rs2306283 and rs4149056 on hyperbilirubinemia are also controversial. The minor allele of SNP rs2306283 and rs4149056 in the *SLCO1B1* gene increases the risk of unconjugated hyperbilirubinemia amongst Taiwanese adults. [78] Taiwanese neonates with the minor allele of rs2306283 were at high risk to develop severe hyperbilirubinemia whereas those with the minor allele of rs4149056 were not. [79] Neither SNP rs2306283 nor rs4149056 was associated with neonatal hyperbilirubinemia in a multi-ethnic population,[80] a Thai population,[81] and in two Chinese populations. [82, 83]

SLCO1B3 shares greater than 80% amino acid identity with SLCO1B1 [84] and also takes up bilirubin from the blood into the hepatocytes. Significant association with TBIL levels were observed at *SLCO1B3* in all three GWASs using Affymetrix arrays [59, 60] [62] but not in the GWAS using Illumina arrays. [61] Those GWASs showed significant differences in the associated variations of *SLCO1B3* between Koreans and Caucasians (Table 1): (1) SNP rs2417940 at Intron 7 of *SLCO1B3* was significantly associated in Koreans (P=1.03×10^{-17}) [62] but not in Caucasians; (2)SNP rs17680137 (P=0.584) and rs2117032 (P=2.76×10^{-5}), which were identified as the top-ranked SNPs in Caucasians, did not reach the genome-wide significance level in Koreans (Table 1). Of note, the minor allele frequency (MAF) of SNP rs17680137 is 0.00373 and 0.293 in Koreans and Caucasians (SardiNIA stage 1), respectively, and it is associated with bilirubin levels only in Caucasians. The GWAS by Bielinski et al. [61] used Illumina arrays, which did not include the probes for SNP rs2117032 and rs17680137. But the Illumina arrays included the probes for SNP rs10841712 (r^2=0.965) and rs7306033 (r^2=1), which are in tight linkage with rs2117032 and rs17680137, respectively. However, this GWAS showed that no associations of SNPs in *SLCO1B3* with bilirubin levels reached a significant level of P=1.0×10^{-6}. [61] More data are warranted to support the association between the polymorphisms in *SLCO1B1* and *SLCO1B3* and bilirubin levels.

(4) UGT1A1

UGT1A1 is another rate-limiting enzyme in bilirubin metabolism. Of the genes involved in bilirubin metabolism, *UGT1A1* has been the most widely studied. A TA insertion in the TATAA box in the promoter of *UGT1A1* [normal (TA)$_6$TAA], results in the sequence (TA)$_7$TAA. *UGT1A1*28* is characterized by seven TA repeats, whereas six TA repeats represent the common wild allele *UGT1A1*1*. The (TA)$_5$ and (TA)$_8$ alleles have rarely been found among Asians, African-Americans and Caucasians. [44, 85, 86] The number of the (TA)n repeat is inversely associated with the *UGT1A1* promoter activity. [85] The (TA)$_7$ allele (*UGT1A1*28*) reduces *UGT1A1* expression by 70 to 80%. [87] Although a few studies showed that the (TA)n repeat polymorphism of *UGT1A1* was not associated with hyperbilirubinemia,[88, 89] the (TA)n repeat polymorphism in the *UGT1A1* gene promoter has been studied extensively and *UGT1A1*28* has been proved to be associated with higher risk of hyperbilirubinemia. Some non-GWAS population-based studies have examined the (TA)n repeat polymorphism associated with serum bilirubin levels. [44, 86, 90-100] All of them consistently showed that (TA)$_7$/(TA)$_7$ carriers had highest TBIL levels, followed by(TA)$_6$/(TA)$_7$ carriers, and then (TA)$_6$/(TA)$_6$ carriers. SNP rs887829, which has been shown to be tightly associated with the (TA)n repeat polymorphism (r^2≈1) in Japanese, Europeans and Africans,[90, 101] is strongly associated with TBIL levels in all four GWASs (Table 1) [59, 60, 62] and non-GWAS. [54] The r^2 value is either 1 or very close to 1, showing that the variants are almost in absolute (or perfect) linkage disequilibrium (LD). These results further confirmed the role of the (TA)n repeat polymorphism on serum bilirubin levels.

The frequency of (TA)$_7$ allele is high in Caucasians (0.357-0.415) [85, 102] and Indians (0.351). [103] The (TA)$_7$ allele with the frequency 0.323 is associated with high bilirubin levels, which can explain 18.6% of the total variation of bilirubin in Caucasians. [86] In contrast, the frequency of (TA)$_7$ allele is lower in Asian populations (0.143 in Taiwanese, 0.188 in Malays, and 0.100-0.168 in Japanese). [92, 103-106]

Table 1. Ethnic differences in *SLCO1B1*, *SLCO1B3* and *UGT1A1* variants associated with serum total bilirubin levels in the four genome-wide association studies

Gene	SNP	Location	Korean MAF	LD (r^2)	P (Stage I)	Caucasian MAF	LD (r^2)	Sanna et al. [59] P	Johnson et al. [60] P	Bielinski et al. [61] P
SLCO1B1	rs4149056	Exon 5 (Val174Ala)	0.13	0.137[a]	NA	0.16	0.08[a]	0.0015	6.7×10⁻¹³	1.26×10⁻⁷[b]
SLCO1B3	rs2417940	Intron 7	0.21	reference	1.03×10⁻¹⁷	0.13	reference	NS	NA	NA[c]
	rs2117032	Near-3'	0.44	0.28	2.76×10⁻⁵	0.37	0.073	2.91×10⁻¹⁴	1.30×10⁻⁶	N[d]
	rs17680137	Intron 7	0.0037	0	0.584	0.19	0.031	4.41×10⁻¹⁴	1.20×10⁻⁶	N[d]
UGT1A1	rs11891311	-30 kb	0.12	reference	4.78×10⁻¹⁴⁸	0.31	reference	NA	1.4×10⁻²³⁵	N[d] *
	rs887829	-311 bp	0.12	0.96	5.37×10⁻¹⁴⁸	0.28	0.67	1.44×10⁻⁶⁹	5.0×10⁻³²⁴	1.73×10⁻⁵⁵
	rs6742078	Intron 1	0.12	0.96	7.19×10⁻¹⁴⁸	0.28	0.67	NA	5.0×10⁻³²⁴	5.81×10⁻⁶²
	rs4148323	Exon 1 (Gly71Arg)	0.19	0.032	2.56×10⁻⁷⁰	0	-	-	-	-

SNP, single nucleotide polymorphism; MAF, minor allele frequency; LD, linkage disequilibrium; NA, not available; NS, not significant.
[a] r^2 between rs4149056 and rs2417940.
[b] P value available from the data of SNP rs1871395, which is in perfect LD with rs4149056 (r^2=1) in Caucasians.
[c] Original data was not available from the literature.
[d] Not included in the GeneChips.
* P= 1.84×10⁻²² for SNP rs6725478, which is in tight linkage with rs11891311 (r^2= 0.926) in Caucasians.

The (TA)$_7$ allele with low (TA)$_7$ allele frequencies (0.134 in Han Chinese, 0.256 in Uyghur, and 0.277 in Kazak) is associated with high bilirubin levels, which can explain 7.1% (P=1.47×10^{-12}), 7.0% (P=1.90×10^{-11})and 5.9% (P=4.93×10^{-7}) of the total variation of bilirubin, respectively in the Han Chinese, Uyghur and Kazak populations. [44]

In addition to the (TA)n repeat polymorphism, SNP rs4148323 (Gly74Arg) at Exon 1 of *UGT1A1* is also important as a genetic determinant of serum bilirubin. Several studies have indicated that SNP rs4148323 is associated with hyperbilirubinemia [79, 81, 107-112] and serum bilirubin levels [44, 54, 76] in East Asians. SNP rs4148323 is virtually monomorphic in Caucasians and Africans while it is common in East Asians (Japanese, MAF=0.130; Taiwanese, MAF=0.109). [92, 104] The MAFs of rs4148323 for the Han, Uyghur and Kazak populations were 0.211, 0.168 and 0.211, respectively, which explained 9.8% (P=4.15×10^{-18}), 4.4% (P=4.12×10^{-8}) and 3.9% (P=6.95×10^{-5}) of the total variation of TBIL levels, respectively. [44] This strong association between SNP rs4148323 and TBIL levels was also observed in the GWAS in Koreans (P=2.56×10^{-70}) (Table 1). However, a few studies indicated that rs4148323 was not responsible for the variation of TBIL levels. [90, 94]

In addition to SNP rs887829 and rs4148323, rs11891311 and rs6742078 have been reported to be strongly associated with TBIL levels in the GWASs (Table 1). SNP rs4148324, in perfect LD with rs6742078 in the HapMap Han Chinese, European and Japanese populations, is associated with TBIL levels (P=1.99×10^{-10}). [76] Lin et al. [54] did not estimate rs11891311, rs6742078 and any SNPs in strong LD with them (r^2≥0.8). However, they did observe that SNP rs4399719 in the *UGT1A1* promoter, other than SNP rs887829 and rs4148323, is also associated with TBIL levels (P=5.28×10^{-5}) in Chinese Han population. [54] SNP rs4124874 (*UGT1A1* T-3279G), which is perfectly linked with rs4399719 in the HapMap Han Chinese, European, Japanese and Yoruba populations, alone has a significant effect on TBIL levels in Japanese [113], Chinese [94] and Europeans. [114] SNP rs3755319, also in perfect LD with rs4399719 in the HapMap Han Chinese, European and Japanese populations, is associated with TBIL levels (P=4.47×10^{-9}). [76]

(5) ABCC2 and ABCC3

Conjugated bilirubin-glucuronides are excreted into bile by ABCC2. As a compensatory pathway, the glucuronidated bilirubin may also be secreted back to blood by ABCC3 under conditions of impaired biliary excretion. Although *ABCC2* is well known as a gene responsible for Dubin-Johnson syndrome, characterized by conjugated hyperbilirubinemia,[115] a study on serum bilirubin levels revealed no association between *ABCC2* variants and serum bilirubin levels. [74] The four GWASs on serum bilirubin levels also failed to identify any SNPs in *ABCC2* and *ABCC3* associated with serum bilirubin levels. Three SNPs in *ABCC2* (rs4148386, rs4148396 and rs2145852) showed P-values between 0.01 and 0.001 in Japanese, however, far above the significance level after Bonferroni correction. [76]

6. Ethnic Differences in Serum Bilirubin and Genetic Polymorphisms of Bilirubin Metabolism Genes

There are striking ethnic differences in TBIL levels. Mean TBIL levels of the Han participants (12.6 μmol/L, 95% confidence interval [CI] 12.2-12.9 μmol/L) were lower than that of the Kazak participants (13.4 μmol/L, 95% CI 12.9-14.0 μmol/L) (P=0.004) whereas higher than that of the Uyghur participants (11.8 μmol/L, 95% CI 11.4-12.1 μmol/L) (P=0.002). [44] The frequency of (TA)$_7$ allele (0.134) in the Han population was lower than that in the Kazak (0.277) (P=4.38×10^{-19}) and Uyghur (0.256) (P=1.59×10^{-17}) populations. The distribution of the (TA)n repeat polymorphism of *UGT1A1* can explain the difference between the Han and Kazak populations but not explain the difference between the Han and Uyghur populations.

African Americans had lower bilirubin levels but had higher frequencies of (TA)$_7$ and (TA)$_8$ than Caucasians. [85] One of the reasons for this may be that African Americans have lower hemoglobin levels than Caucasians. [116, 117] Asians had higher bilirubin levels but had lower frequencies of (TA)$_7$ than Caucasians. [85] These studies explained ethnic differences in TBIL levels mostly by the distribution of the (TA)n repeat polymorphism and SNP rs4148323 of *UGT1A1* among ethnic groups.

Non-GWAS population-based studies and GWASs on serum bilirubin have identified more polymorphisms involving the bilirubin metabolism genes, in addition to the (TA)n repeat polymorphism and SNP rs4148323 of *UGT1A1*, associated with TBIL levels. These polymorphisms may also account for the differences among populations and need to be further explored to explain the differences.

7. Serum Bilirubin and Coronary Artery Disease

The first indication that serum bilirubin levels might be related to CAD was reported in 1994. [118] In that study, low serum bilirubin levels were found to be associated with an increased risk of CAD, whereas high-normal levels were found to be associated with a decreased risk of CAD. Of note, the study was conducted only in men. In 1996, Hopkins et al. [119] first confirmed that serum bilirubin levels were inversely correlated with the severity of CAD in both men and women.

Later, a fairly large number of descriptive (cross-sectional and case-control studies) and prospective epidemiological studies have examined the association between serum bilirubin and CAD in different populations. Serum bilirubin has consistently been shown to be inversely associated with CAD, mostly among men. Early in 2003, a meta-analysis involving eleven studies has been performed and the results showed that serum bilirubin levels are inversely related to the severity of atherosclerosis in men (r =-0.31, P<0.0001). [120] Women were not included in the meta-analysis because of the limited number of women involving the published studies. Non-parametric, regression, and stratified analyses all showed an inverse and dose-dependent association between serum bilirubin levels and atherosclerotic process from subclinical to clinical outcomes. [120] The results of regression analysis indicated that serum bilirubin of 10 μmol/L was the cut point for discrimination of cardiovascular risk. [120]

However, the association of serum bilirubin with CVD was not clear in women so far. Prospective association studies of serum bilirubin with CVD have displayed inconsistent results in women. A reverse relationship between serum bilirubin and myocardial infarction (MI), CAD, and any CVD event was observed in men, but not in women, in the Framingham Study. [121] The limited number of MI (n=61) and CVD (n=103) female cases and the relatively young age of the women at baseline (39.5 years) in the Framingham Study might account for the differences in men and women. Another prospective case-control study by Ekblom et al. [98] investigated the association of serum bilirubin with MI and showed the reverse relationship in both men and women but the relationship did not remain statistically significant after correction for systolic blood pressure, smoking, Apolipoprotein B/Apolipoprotein A1, high-sensitivity C-reactive protein (hs-CRP), albumin and *UGT1A1* (TA)n genotypes.

Results from the retrospective association studies of serum bilirubin with CVD have also displayed inconsistent in women. Hopkins et al. [119] and Lingenhel et al. [114] observed that low bilirubin levels were significantly associated with CAD both in men and women whereas Lin et al. [54] and Endler et al. [122] observed inverse relationship only in men but not in women.

In a cross-sectional study involving 398 men and 239 women, serum bilirubin levels were found to be strongly associated with coronary artery calcification (CAC) scores, a marker of the atherosclerotic process, and independent determinants of CAC in both men and women (P<0.0001). [123] An increment in serum bilirubin of 1 μmol/L led to a 14% decrease in odds for having a CAC score ≥400 after adjustment for several other major risk factors. [123]

Women have lower serum bilirubin than men in adolescent and adult life. Lower bilirubin levels in women may be due to lower hemoglobin levels [124] and hormonal effects before menopause on UGT activity. Evidence from in vitro, animal and clinical studies showed that (1) UGT1A1 glucuronidates estriol, 17b-estradiol, ethinylestradiol and catechol estrogens, and (2) estrogen and progesterone significantly enhanced UGT1A1 activities. [125-127] Lower bilirubin levels may be a possible reason for a weak or null relationship between serum bilirubin and CVD in women. [128] Another possible reason is that estrogen may offset an increased risk of CVD associated with low serum bilirubin. [51, 121] Estrogen can decrease LDL and increase high-density lipoprotein, reduce the oxidation of LDL, increase the local production of nitric oxide in the vascular wall, and subsequently reduce the risk of CVD. [129]

8. Serum Bilirubin, Stroke and Other Vascular Diseases

(1) Stroke

In 2009, a large prospective study on 1,137 male and 827 female stroke cases found an association between high bilirubin levels and reduced risk for stroke in men but in women. [130] However, in 2010, another prospective study on 127 male and 104 female stroke cases showed risk reduction for higher bilirubin levels only in women. [97] There are only a few previous cross-sectional studies on the issue of serum bilirubin and stroke risk. Bilirubin was included in the National Health and Nutrition Examination Survey (NHANES) 1999-2004 study (n=13,214), and a history of stroke was recorded in 453 subjects. [131] In this cross-sectional study, higher bilirubin levels showed association with reduced stroke prevalence, and gender-stratified analysis was not performed.

(2) Carotid Artery Atherosclerosis

Several studies have investigated the association between bilirubin and carotid artery atherosclerosis (CAA). In these studies, the degree of CAA was measured by the presence of plaque or intima-media thickness (IMT) of the carotid artery. A prospective case-control study, which nested in the Atherosclerosis Risk in Community (ARIC) cohort study in Caucasians, reported no statistically significant difference in serum bilirubin between cases and controls of CAA. [132] In this study, cases of CAA were 150 individuals with elevated mean IMT (mean: 1.04 mm; range: 0.88-1.59 mm), and controls were 150 age-gender-matched individuals with low mean IMT (mean: 0.62 mm; range: 0.43-0.74 mm). Contrary to this observation, a study involving 1,741 Japanese men and women combined, those with carotid plaques (n=330), which was defined as focal thickening of the intimal-medial layer with an IMT of 1.3 mm or greater, were found to have lower levels of bilirubin. [133] In that study, each 0.1 mg/dL (1.7μmol/L) increase in serum bilirubin was found to decrease the risk of carotid artery plaque by 3.7%. The study subjects were expanded in the second Japanese study including 5,473 men and 2,671 women,[134] in which an inverse association between serum bilirubin and the prevalence of carotid plaque were found to be stronger in women than in men. Subsequent cross-sectional studies consistently showed an inverse relationship between serum bilirubin and CAA. [135-137]

(3) Peripheral Artery Disease

Peripheral artery disease (PAD) is another atherosclerosis-related morbidity. Ankle brachial index (ABI), the ratio of ankle and brachial systolic blood pressure, is the common tests currently used in screening and population studies. A small study has been conducted on 31 PAD patients (19 men and 12 women) without elevated liver enzymes. In this study, male and female PAD patients both showed lower serum bilirubin. [138] The NHANES,[139] a cross-sectional cohort study that consisted of 7,075 adults, showed the inverse association between serum bilirubin and PAD in men and smokers but not in women and nonsmokers. In this study, a 1.7 μmol/L increase in serum bilirubin was associated with a multivariate Odds Ratio (OR) of 0.94 (95% CI 0.90-0.98) after adjustment for age, gender, smoking, diabetes and other cardiovascular risk factors. A similar decrease in the multivariate-adjusted prevalence odds associated with serum bilirubin (OR 0.91, 95% CI 0.86-0.97 per 1.7 μmol/L increment) was observed in the Cardiovascular Disease in Intermittent Claudication (CAVASIC) Study of 255 male patients with intermittent claudication and 255 community controls matched for age and presence of Type 2 diabetes. [100] The Bogalusa Heart Study showed that serum bilirubin was beneficially associated with arterial compliance of small arteries (but not of large arteries), as assessed by radial artery pressure pulse contour analysis, independent of cardiovascular risk factors in asymptomatic young adults. [140]

(4) Essential Hypertension

Few studies have examined the association between serum bilirubin levels and hypertension. [68, 122, 124, 141, 142] Madhavan et al. [124] reported that the young offspring with parental history of hypertension (n=1,178) had lower bilirubin levels than those without such parental history (n=1,367). Papadakis et al. [141] have shown that serum bilirubin levels are significantly lower in the untreated hypertensives (n=34) when compared with normotensives (n=272) or the treated hypertensives (n=89). Chin et al. [142] followed 1,208 Koreans for 10 years and found the effect of serum bilirubin on the development of hypertension, which was more evident in females and in non-smokers. But, in contrast, Endler et al. [122] and Lin et al. [68]showed no association between serum bilirubin and hypertension.

9. Genetic Polymorphisms of Bilirubin Metabolism Genes and CVD Risk

(1) HMOX1

Because numerous studies have supported the multifunctional roles of HMOX1 in the cardiovascular system, the association between *HMOX1* polymorphisms, especially the (GT)n repeat polymorphism, and CVD has been of a recent interest. The studies conducted to date are outlined in Table 2. The studies investigating the inflammation response after coronary surgery are also presented in Table 2, for the effects of *HMOX1* polymorphisms on restenosis might be related to their effects on the inflammation response. In Table 2, studies are sorted by CVD end point, ethnicity and publication time.

I. HMOX1 and CAD

As shown in Table 2, of all 28 studies involving *HMOX1* polymorphisms and CVD, 8 studies have investigated the association between *HMOX1* polymorphisms and CAD. Of these 8 studies, five studies (Kaneda et al.,[143] Ono et al.,[49] Chen et al.,[48] Chen et al. [56] and Lin et al. [54]) were conducted in Orientals and three (Schillinger et al.,[144] Endler et al. [51] and Lublinghoff et al. [52]) were conducted in Caucasians. Three of five in Orientals (Kaneda et al.,[143] Chen et al. [48] and Chen et al. [56]) only investigated the *HMOX1* (GT)n repeat polymorphism and reported that the *HMOX1* (GT)n repeat polymorphism might protect from CAD. Of note, in all those three studies in Orientals, the effect of the *HMOX1* (GT)n repeat polymorphism was restricted to selected subgroups of patients. While there was no overall association between the (GT)n repeat polymorphism and CAD (≥75% stenosis) among 796 Chinese and 577 Japanese patients undergoing coronary angiography,[48, 143] homozygosity of the long allele (≥32 GT [48] or ≥27 GT [143]) was associated with an increased prevalence of CAD in Chinese individuals with diabetes mellitus (n=214) [48] and in Japanese subjects with diabetes (n=205), hypercholesterolemia (n=189) or a smoking habit (n=305) [143]. Chen et al. [56] expanded the number of patients (n=986) and confirmed an increased prevalence odds of CAD associated with homozygosity of the long allele (≥27 GT) among diabetic patients, but did not replicate the Japanese observation among those with Hypercholesterolemia or smokers.

Another two studies in Orientals (Ono et al. [49] and Lin et al. [54]) did not include the *HMOX1* (GT)n repeat polymorphism. Ono et al. [49] investigated two *HMOX1* polymorphisms, T-413A and G-1135A, and indicated that the A/A genotype of the T-413A polymorphism might protect from CAD in Japanese, but the finding was not reproduced in Caucasians [52] and Chinese. [54] Lin et al. [54] investigated four SNPs in *HMOX1*, including T-413A, and did not find any SNPs associated with CAD.

In contrast to Orientals, three studies in Caucasians (Schillinger et al.,[144] Endler et al. [51] and Lublinghoff et al. [52]) all indicated that the *HMOX1* (GT)n repeat polymorphism was unrelated to CAD. The observation of Schillinger et al. [144] made in a relatively small patient cohort (70 CAD patients and 61 healthy controls) may be a chance finding and give rise to numerous possibilities for error. Endler et al. [51] did not observe an association between CAD or MI and the *HMOX1* (GT)n repeat polymorphism neither in the total population nor in patients' subgroups with diabetes or hyperlipidemia, although they observed potentially beneficial effects of the short allele (<25 GT) on lipid profile and serum bilirubin. Likewise, in the study of Lublinghoff et al.,[52] no association of the *HMOX1* (GT)n repeat and T-413A polymorphisms with CAD or MI was found, even in a subgroup analysis of type 2 diabetic patients.

It is preferable to test whether the associations may be conveyed through their effects on TBIL levels by association analysis between *HMOX1* polymorphisms and bilirubin levels. However, of the 8 studies, only 4 (Lublinghoff et al.,[52] Endler et al.,[51] Lin et al. [54] and Chen et al. [56]) have performed this association analysis. Lublinghoff et al. [52] investigated the A-413T and (GT)n repeat polymorphisms and did not observe any significant association between *HMOX1* genotypes and serum bilirubin levels as well as CAD in Caucasians. Endler et al. [51] observed that carriers of the short allele (<25 GT) had significantly higher bilirubin levels than non-carriers but did not have lower risk of CAD or MI. Lin et al. [54] investigated four SNPs in *HMOX1*, including A-413T, and did not find any significant association between *HMOX1* polymorphisms and serum bilirubin levels as well as CAD in Chinese. Chen et al. [56] reported that subjects with the L/L (≥27 GT) genotype had significantly lower bilirubin levels than those with S/S and S/L genotypes, and had an increased CAD risk among diabetic patients, first suggesting that the association of the (GT)n repeat polymorphism and CAD may be conveyed through its influence on serum bilirubin.

II. HMOX1 and Restenosis

Further, in a small population, a lower number of *HMOX1* (GT)n repeats was associated with a lower risk of restenosis after balloon dilatation of the femoral artery [145] suggesting a protective effect of HMOX1 on restenosis. The longer (GT)n repeats have been related to higher risk of restenosis after coronary stenting. [146, 147] However, Schillinger et al. [148] observed that short (<25) (GT)n repeats carriers had lower inflammatory response, which was indicated by the change of C-reactive protein (CRP) and serum amyloid A (SAA) after 24 and 48 hours, after balloon angioplasty but not after coronary stenting. Later, Schillinger et al. [149] renewed their study and observed that the lower number (<25) of (GT)n repeats was associated with a lower risk of restenosis and a lower postintervention CRP at 24 h and 48 h after balloon dilatation but not after coronary stenting. Moreover, Li et al. [150] showed that the *HMOX1* (GT)n repeat polymorphism was not associated with the risk of restenosis and the change of CRP and inerleukin-6 (IL-6) after 24 and 48 hours after coronary stenting. Li et al. [151] also found no significant relationship between the *HMOX1* (GT)n repeat polymorphism and peak IL-6, CRP and fibrinogen levels after coronary artery bypass surgery. Of note, all these studies comprised less than 500 patients.

Recently, Wijpkema et al. [152] followed a group of 3,104 patients for 10 months after successful percutanous coronary intervention (PCI), and did not find significant association between the (GT)n repeat polymorphism and restenosis after PCI. Tiroch et al. [153] studied 1,357 patients 6 months after coronary stenting and, again, did not find significant association between the (GT)n repeat polymorphism and restenosis after coronary stenting. In addition, two small studies both showed no association of the (GT)n repeat polymorphism with the development of acute rejection, transplant coronary artery disease and cardiac allograft vasculopathy after heart transplantation in Caucasians. [154, 155] In order to assess the impact of the (GT)n repeat polymorphism on restenosis, more studies based on larger populations are needed, and studies with limited number of patients should be included together with studies based on larger populations in a meta-analysis.

III. HMOX1 and Hypertension

HMOX1 induction provides potent protection against oxidative stress-induced damage. HMOX1-derived CO plays a role in the pathology of hypertension. Therefore, the effect of *HMOX1* polymorphisms on hypertension and blood pressure has become one of the research topics. Lin et al. [156] reported that the longer (GT)n repeat group had a lower risk of hypertension, lower systolic and diastolic blood pressures in a Chinese Han population. SNP rs9607267 in Intron 2 of *HMOX1* might also affect the susceptibility to hypertension. [157] Ono et al. [50] observed that the A/A genotype of T-413A was associated with an increased incidence of hypertension in Japanese women, but not in men. However, this association could not be validated in Han Chinese. [156, 157] As mentioned above, the same group--Ono et al. [49] also observed that the A/A genotype of the T-413A polymorphism may reduce the incidence of CAD, was not repeated in Caucasians [52] and Chinese. [54] Until now, no published GWASs of CVD showed that the SNPs in *HMOX1* reached genome-wide significance.

Ono et al. [49, 50] showed that the basal promoter activities of the A-30 and A-23 haplotypes of A-413T-(GT)n repeat were significantly higher than those of the T-23 and T-30 haplotypes by luciferase reporter assays, and concluded that the T-413A polymorphism might be responsible for the *HMOX1* promoter activity. The (GT)n repeat polymorphism have been demonstrated to be responsible for the *HMOX1* promoter activity under stimulations. [43, 46, 47] But no functional experiments were conducted to test the role of the T-413A polymorphism under stimulations. Further experiments are required to measure the promoter activities of the A-30 and T-23 haplotypes in human cell lines with or without stimulations. We conjecture that the T-413A polymorphism might be irresponsible for the *HMOX1* promoter activity under stimulations and thus does not play a role in the development of CVD.

IV. HMOX1 and Other Vascular Diseases

Dick et al. [158] followed 472 Caucasian PAD patients for median 21 months for the occurrence of coronary events (MI, percutaneous coronary interventions and coronary artery bypass graft), cerebrovascular events (stroke or carotid revascularization) and all-cause mortality, and observed a significant association of the (GT)n repeat polymorphism with coronary events, but not with cerebrovascular events, mortality and overall major adverse cardiovascular events.

As stated above, in Caucasians, two studies (Endler et al. [51] and Lublinghoff et al. [52]) showed no association between the (GT)n repeat polymorphism and CAD. But here Dick et al. [158] showed significant association between the (GT)n repeat polymorphism and coronary events in PAD patients.

As mentioned above, the effect of the *HMOX1* (GT)n repeat polymorphism on CAD was restricted to selected subgroups of patients in Orients. [48, 56, 143] Similarly, Funk et al. [159] showed that the effect of the *HMOX1* (GT)n repeat polymorphism on cerebrovascular events (ischaemic stroke events or transient ischaemic attack) was restricted to the subgroup of patients without hyperlipidemia, although Dick et al. [158] found no association between the (GT)n repeat polymorphism and cerebrovascular events (stroke or carotid revascularization) in Caucasian PAD patients.

The S/S (<25 GT) genotype exerted a protective effect on the development of cerebrovascular events in patients without hyperlipidemia, while this effect was no longer detectable in hyperlipidemic patients. [159] Bai et al. [160] also showed no overall association between the (GT)n repeat polymorphism and ischemic stroke. But they indicated that the long genotype (average (GT)n repeats of two alleles >26) was associated with an increased prevalence of ischemic stroke in Chinese individuals with poor high-density lipoprotein cholesterol (HDL-C) status (<40 mg/dL for men/50 mg/dL for women), but not among subgroups stratified by hypercholesterolemia, hypertriglyceridemia, or high LDL-C level.

Of note, Funk et al. [159] did not include HDL-C measure, and hyperlipidemia was defined as hypercholesterolemia or a history of hyperlipidemia, or lipid-lowering medication. That is to say, Funk et al. [159] could not estimate the abnormal HDL-C levels, especially in patients with normal total cholesterol levels. Therefore, the protective effect of the *HMOX1* (GT)n repeat polymorphism on cerebrovascular events dependent on the presence of lipid conditions should be confirmed further in population-based studies. The results imply that the longer (GT)n repeats alone may not be sufficient to cause coronary or cerebrovascular diseases. [48, 56, 143, 158-160] But it may contribute to the onset or progression of coronary or cerebrovascular diseases when certain conditions of enhanced oxidative stress coexist.

Recently, the relationship of *HMOX1* polymorphisms with arsenic-associated cardiovascular disease was first studied by Wu et al. [161, 162]. Wu et al. [161] studied 367 participants with an indication of carotid atherosclerosis and an additional 420 controls without the indication from two arsenic exposure areas in Taiwan, a low arsenic-exposed Lanyang cohort and a high arsenic-exposed LMN cohort.

In this study, the arsenic's effect on carotid atherosclerosis differed between carriers of the short (GT)n allele (<27 GT) (OR 1.39, 95% CI 0.86-2.25, P=0.181) and non-carriers (OR 2.65, 95% CI 1.03-6.82, P= 0.044) in the high-exposure LMN cohort. In contrast, no such results were found in the low-exposure Lanyang cohort.

Table 2. Published studies evaluating the association of *HMOX1* polymorphisms with cardiovascular end points

Study	Year	Ethnicity	Study design	CVD end point	Number of cases/controls	Polymorphism	OR*[a] (95% CI), P value, other findings
Kaneda et al.[143]	2002	Japanese	Retrospective	CAD	287/290 155/50 with diabetes	(GT)n S <27, L ≥27	S/S vs L/L Total 0.65(0.36-1.2), NS Diabetes 0.23(0.076-0.71), P<0.01 Hypertension OR not shown, NS HC 0.23(0.07-0.72), P<0.05 Smokers 0.40 (0.17-0.95), P<0.05
Ono et al.[49]	2004	Japanese	Retrospective	CAD	MI/AP/controls 393/204/ 1972	T-413A G-1135A	MI OR[b] not provided, P=0.0468 AP OR[b] not provided, P= 0.0096 OR not shown, NS
Chen et al.[48]	2002	Chinese	Retrospective	CAD	474/322 151/63 with diabetes	(GT)n S <23, M 23-31, L ≥32	Total no OR shown, NS Diabetes L/L+M/L+S/L vs M/M+S/M+S/S 4.7 (1.9-12.0), P=0.001
Chen et al.[56]	2008	Chinese	Retrospective	CAD	664/322 200/63 with diabetes	(GT)n S <27, L ≥27	L/L vs L/S+S/S Total OR not shown, NS Diabetes 2.81(1.22-6.47), P= 0.015 Hypertension OR not shown, NS HC OR not shown, NS Smoking OR not shown, NS
Lin et al.[54]	2009	Chinese	Retrospective	CAD	1311/1000	4 polymorphisms[c]	OR not shown, NS
Schillinger et al.[144]	2002	Caucasian	Prospective	AAA/CAD/ PAD	AAA/CAD/ PAD/healthy controls 70/70/70/61	(GT)n S <25, L ≥25	S/S+S/L vs L/L AAA/Controls OR not shown, P= 0.04 CAD/Controls OR not shown, NS PAD/Controls OR not shown, NS AAA/CAD 0.38(0.17-0.78), P=0.006 AAA/PAD 0.35(0.14-0.81), P= 0.01
Endler et al.[51]	2004	Caucasian	Retrospective	MI and stable CAD	MI/CAD/controls 258/180/ 211	(GT)n S <25, L ≥25	OR not shown, NS
LURIC [Lublinghoff et al.[52]]	2009	Caucasian	Prospective	CAD CAD plus MI	2526/693 1339/693	A-413T (GT)n S <25 or 27, L ≥25 or 27	OR not shown, NS OR not shown, NS

Study	Year	Ethnicity	Study design	CVD end point	Number of cases/controls	Polymorphism	OR *[a] (95% CI), P value, other findings
Exner et al.[145]	2001	Caucasian	Retrospective (6 months follow-up)	Restenosis after balloon angioplasty	Men 23/73	(GT)n S <25, M 25-28, L ≥29	S/S+S/M+S/L vs M/M+M/L+L/L 0.2 (0.06-0.70), P=0.007
Schillinger et al.[148]	2002	Caucasian	Prospective	Inflammation response after balloon angioplasty or stenting	Balloon angioplasty:150 Stenting: 61	(GT)n S <25, M 25-28, L ≥29	Balloon angioplasty: Delta CRP(24h) (P<0.0001) Delta CRP(48h) (P<0.0001) Delta SAA(24h) (P=0.02) Delta SAA(48h) (P=0.006) Delta fibrinogen (NS) Stenting: NS for all inflammatory markers
Chen et al.[146]	2004	Chinese	Prospective (6 months follow-up)	Restenosis and adverse cardiac events after coronary stenting	Restenosis 111/212 Adverse cardiac events 59/264	(GT)n S <26, L ≥26	Restenosis L/L+S/L vs S/S 3.74 (1.61-8.70), P=0.002 Adverse cardiac events L/L vs S/L+S/S 3.26(1.58-6.72), P=0.001
Schillinger et al.[149]	2004	Caucasian	Prospective (6 months follow-up)	Restenosis and inflammation response after balloon angioplasty or stenting	Balloon angioplasty: 74/136 Stenting: 21/47	(GT)n S <25, L ≥25	Balloon angioplasty: Restenosis S/S+S/L vs L/L 0.43(0.24-0.71), P<0.001 CRP at 24 h (P =0.009) and 48 h (P <0.001) Stenting: Restenosis OR not shown, NS CRP at 24 h and 48 h: NS
Gulesserian et al.[147]	2005	Caucasian	Prospective (6-9 months follow-up)	Restenosis after coronary stenting	102/97	(GT)n S <29, L ≥29 G+99C (rs2071747)	L/L+L/S vs S/S 1.9(1.0-3.4), P=0.04 OR not shown, NS
Li et al.[150]	2005	88.8% Caucasian	Prospective (6 months follow-up)	Restenosis after coronary stenting Inflammation response after coronary stenting	52/135	(GT)n S <30, M 30-37, L ≥38	OR not shown, NS The change of CRP and IL-6 after 24 and 48 hours: NS
Li et al[151]	2005	Caucasian	Prospective	Inflammatory response after CABG	275 patients undergoing CABG	(GT)n S <30, M 30-37, L ≥38	Peak IL-6, CRP and fibrinogen levels after CABG: NS

Table 2. Published studies evaluating the association of *HMOX1* polymorphisms with cardiovascular end points (Continued)

Study	Year	Ethnicity	Study design	CVD end point	Number of cases/controls	Polymorphism	OR* a (95% CI), P value, other findings
Wijpkema et al.[152]	2006	97% Caucasian	Prospective (10 months follow-up)	Restenosis after percutaneous coronary intervention	324/2601	(GT)n S <25, L ≥25	L/L vs S/S 0.83 (0.53-1.29), NS
Tiroch et al.[153]	2007	Caucasian	Prospective (6 months follow-up)	Restenosis after coronary stenting	401/956	(GT)n S <25, L ≥25	OR not shown, NS
Holweg et al.[154]	2005	Caucasian	Prospective	AR or TCAD after cardiac transplantation	AR 241/63 TCAD 85/207	(GT)n S <27, L ≥27	OR not shown, NS
Ullrich et al.[155]	2005	Caucasian	Retrospective (9±4 years follow-up)	CAV after cardiac transplantation	95/57	(GT)n S <25, L ≥25	OR not shown, NS
Ono et al.[50]	2003	Japanese	Retrospective	Essential hypertension	Women 369/572 Men 390/572	T-413A G-1135A	A/A vs T/T+A/T Women 1.59(1.14-2.20), P=0.0058 Men OR not shown, NS OR not shown, NS
Yun et al.[157]	2009	Chinese	Retrospective	Essential hypertension	503/490	rs9607267 rs2071749	T/T+T/C vs C/C 1.41(1.02-1.95), P=0.040 G/G+G/A vs A/A 1.18(0.69-2.02), NS
Lin et al.[156]	2011	Chinese	Retrospective	Essential hypertension	312/369	T-413A (GT)n S <27, M 27-32, L ≥33 rs2071749	A/A vs T/T 0.98(0.62-1.55), NS M/M+S/L+M/L+L/L vs S/S+S/M 0.58(0.41-0.82), P=0.002 A/A vs G/G 1.02(0.57-1.82), NS
Dick et al.[158]	2005	Caucasian	Prospective (median 21 months follow-up)	MACE[d] in PAD patients	173/299	(GT)n S <25, L ≥25	Coronary events S/S+S/L vs L/L 0.46(0.24-0.87), P=0.016
Funk et al.[159]	2004	Caucasian	Retrospective	Ischaemic stroke events and transient ischaemic attack	399/398	(GT)n S <25, L ≥25	Without hyperlipidemia S/S vs L/L 0.2(0.1-0.6), P<0.05 S/L vs L/L 0.8(0.4-1.4), NS With hyperlipidemia OR not shown, NS
Bai et al.[160]	2010	Chinese	Retrospective	Ischemic stroke	183/164 Low HDL-C[e] 131/68	(GT)n Genotype S ≤26, Genotype L >26[f]	Total OR not shown, NS Low HDL-C Genotype L vs S 2.07(1.07-4.01), P=0.0303

Table 2. (Continued)

Study	Year	Ethnicity	Study design	CVD end point	Number of cases/controls	Polymorphism	OR*[a] (95% CI), P value, other findings
Lanyang and LMN cohorts [Wu et al.[161]]	2010	Chinese	Prospective	CAA	367/420	(GT)n S <27, L ≥27	Arsenic's effect on CAA S/S+S/L : L/L Low arsenic-exposure (Lanyang): 1.12(0.91-1.38), NS : 1.13(0.81-1.57), NS High arsenic-exposure(LMN): 1.39(0.86-2.25), NS : 2.65(1.03-6.82), P=0.044
Lanyang cohort [Wu et al.[162]]	2010	Chinese	Prospective(median of 10.7 years follow-up)	Cardiovascular deaths (CAD, CBVD and PAD)	22/482	(GT)n S <27, L ≥27	L/S+S/S vs L/L 0.38(0.16-0.90), P=0.028
Kanai et al.[163]	2003	Japanese	Retrospective	Kawasaki disease	61/122	(GT)n S <27, M 27-32, L ≥33	OR not shown, NS

CVD, cardiovascular disease; CAD, coronary artery disease; HC, hypercholesterolemia; MI, myocardial infarction; AP, angina pectoris; CBVD, cerebrovascular disease; PAD, peripheral arterial disease; CAA, carotid artery atherosclerosis; CRP, C-reactive protein; SAA, Serum Amyloid A; IL-6, inerleukin-6; CABG, coronary artery bypass surgery; AAA, abdominal aortic aneurysms; TCAD, transplant coronary artery disease; AR, acute rejection; CAV, cardiac allograft vasculopathy; MACE, major adverse cardiovascular event; HDL-C, high density lipoprotein cholesterin; OR, odds ratio; CI, confidence interval; NS: not significant.

* A measure for the effects of polymorphisms on CVD, except for the study of Wu et al. [161], which is a measure for the Arsenic's effect on CAA.
[a] OR except for the study of Wijpkema et al. [152] which is relative risk (RR), and the study of Dick et al. [158] which is hazard ratio (HR).
[b] Adjusted ORs are wrongly provided in this study and then we do not provide in this table.
[c] Four polymorphisms, which are tag SNPs from the phase II HapMap Han Chinese, including T-413A.
[d] MACEs consist of coronary events (fatal and nonfatal myocardial infarction, percutaneous coronary interventions and coronary artery bypass grafting), cerebrovascular events (fatal and nonfatal stroke, carotid revascularization (carotid stenting or carotid endarterectomy)), all-cause mortality and a combination of all these MACEs.
[e] Low HDL-C level is defined by HDL-C level <40 mg/dL (for men)/<50 mg/dL (for women).
[f] Averaged length of (GT)n repeats of the HMOX1 gene promoter was calculated for each participant and the averaged repeat number >26 and ≤26 are defined as the long (L) and short (S) genotype, respectively.

The results suggested that at a relatively high arsenic exposure level, carriers of the short (GT)n allele may have a smaller carotid atherosclerosis risk than non-carriers. Wu et al. [162] also evaluated the association between the *HMOX1* (GT)n repeat polymorphism and cardiovascular mortality in the Lanyang cohort. They observed that carriers of the short (GT)n allele (<27 GT) may have lower rates of cardiovascular mortality than non-carriers in this arsenic-exposed population.

In addition to inflammation [148] or restenosis [145, 149] after balloon angioplasty or stenting, CAD,[51] cerebrovascular events,[159] cardiac allograft vasculopathy after heart transplantation,[155] cardiovascular adverse events in PAD patients,[158] Schillinger et al. at the Medical University of Vienna-- addressed abdominal aortic aneurysm (AAA) and found a significant association between the *HMOX1* (GT)n repeat polymorphism and the presence of AAA in a matched case-control study including a total of 271 subjects (70 AAA patients, 70 CAD patients, 70 PAD patients and 61 healthy controls). [144] AAA patients were less frequently carriers of the short (<25 GT) repeats, suggesting the short repeats may have a protective anti-inflammatory effect against the development of AAA. Kanai et al. [163] reported a lack of association between the *HMOX1* (GT)n repeat polymorphism and Kawasaki disease, a systemic vasculitis, in Japanese children.

In summary, in cardiovascular disease, the role of the (GT)n repeat polymorphism remains less clear and needs further investigation.

(2) BLVRA

Lin et al. [54] investigated the association of *BLVRA* polymorphisms with CAD and found additive effects of rs2877262 (*BLVRA* G+1238/in6C) (age-adjusted OR 0.73, 95% CI 0.59-0.89, P=0.0021) and rs2690381 (*BLVRA* G+2613/in6A) (age-adjusted OR 0.70, 95% CI 0.56-0.86, P=0.0008) on female CAD but not on male CAD. [54] Neither rs2877262 nor rs2690381 was associated with TBIL levels. That is to say, these two effects were not conveyed through the effect on TBIL levels. Lin et al. [68] also found that the minor allele of SNP rs699512 (Thr3Ala), the only common non-synonymous SNP within *BLVRA*, was associated with lower risk of essential hypertension, lower systolic and diastolic blood pressures. Most Affymetrix arrays used in the published GWASs of CVD were not included the three *BLVRA* SNPs and no published GWASs of CVD indicated that the *BLVRA* polymorphisms reached genome-wide significance until now.

(3) SLCO1B1

SLCO1B1 functions as not only the transporter of bilirubin and its glucuronide conjugates but also the transporter of other CAD-related endogenous compounds such as thyroid hormones, leukotriene C4, prostaglandin E2, and bile acids. [164] Among them, thyroid hormones and prostaglandin E2 are also related to blood pressure regulation. SNP rs4149013 (*SLCO1B1* A-12099G) has been shown to be associated with male CAD (age-adjusted OR 0.70, 95% CI 0.55-0.91, P=0.0069) but not with female CAD under a dominant mode of inheritance. [54] This association was not conveyed through the effect on TBIL levels. Another SNP (rs4149014, *SLCO1B1* T-11556G) was found to be correlated with essential hypertension in Uyghurs. [165] All GWASs of CVD failed to detect SNP rs4149013 and rs4149014 as associated with CVD. We consider the two following factors might contribute to this failure. Firstly, of the 11 HapMap populations, the minor allele frequency of SNP rs4149013 and rs4149014 is higher than 0.100 only in the Chinese and Japanese populations and very low in other populations, while most GWASs of CVD were conducted in Caucasians. Secondly, some GeneChips used in the GWASs of CVD did not include SNP rs4149013 and rs4149014, especially rs4149013. For example, neither Affymetrix Human SNP array 500K nor 5.0 included SNP rs4149013.

In addition, Lin et al. [165] and Kivisto et al. [166] both showed that SNP rs4149056 (Val174Ala) was not associated with hypertension. SNP rs4149056, which is common in Caucasians and Asians but rare in Africans, were also undetected in the GWASs of hypertension.

SLCO1B1 has also been shown to regulate the hepatic uptake of statins, which is widely used to lower LDL cholesterol and then substantially reduce many cardiovascular events. But in rare cases, myopathy occurs in association with statin therapy, especially when the statins are administered at higher doses or with certain other medications. A genomewide scan yielded a single strong association of statin-induced myopathy with SNP rs4363657 in *SLCO1B1* ($P=4\times10^{-9}$). [167] The noncoding SNP rs4363657 is in nearly complete LD with nonsynonymous SNP rs4149056 ($r^2=0.97$), which has been linked to statin metabolism. [168]

(4) SLCO1B3

No non-GWAS studies were performed to estimate the association of the polymorphisms in *SLCO1B3* with CVD, although several polymorphisms in *SLCO1B3* have been found to be associated with TBIL levels and TBIL levels are inversely associated with the risk of CVD. And no GWASs of CVD indicated that the polymorphisms in *SLCO1B3* achieved genome-wide significance.

(5) UGT1A1

Because individuals with higher bilirubin levels have been found to have lower risk of CVD, and *UGT1A1* has a key effect on serum bilirubin levels, the relationship between the *UGT1A1* polymorphisms and CVD became a subject of interest. The studies conducted to date are summarized in Table 3. In Table 3, studies are sorted by ethnic populations and publication time. Seven were performed in Caucasians. [86, 97-100, 114, 169] Three were performed in Chinese. [54, 91, 170]

The Rotterdam Study examined the risk of MI in individuals with the *UGT1A1*28* allele, however, did not found a protective effect the *UGT1A1*28* allele on MI, even stratified by gender. [99] Moreover, this study also showed that serum bilirubin is not associated with MI. The association might have been missed because of the relatively advanced population age (almost 70 years at baseline) at entry into the study, the lack of adjustment for liver disease, the use of non-standardized blood collection procedures, lack of sub-stratification of individual genotypes according to serum bilirubin levels, low number of cases and controls, or population substructure influences. [86, 99] Genetic factors for CVD may be more difficult to detect in an elderly group than in a young group because an elderly group was affected by many more other factors, including environmental factors, which might have contributed to a high baseline incidence of atherosclerosis.

The Etude Cas-te'moins de L'Infarctus du Myocarde (ECTIM) case-control study involving 2 Caucasian male populations showed an opposite result in that the *UGT1A1*28* allele carriers appear to have an increased risk of MI. [169] Serum bilirubin levels were not available in the ECTIM study.

An association of serum bilirubin, *UGT1A1* genotypes, and CVD was observed in the Framingham Offspring Study involving 1780 individuals. [86] The mean age of the participants was 36 years at entry, and CVD and CAD events were followed for 24 years. This was the first time that a significantly decreased risk of CVD and CAD was found for the *UGT1A1*28* homozygosity $(TA)_7/(TA)_7$ genotype carriers. The $(TA)_7/(TA)_7$ genotype carriers had significantly higher serum bilirubin levels and one-third risk of CVD and CAD than the $(TA)_6/(TA)_6$ and $(TA)_6/(TA)_7$ genotypes carriers. However, in the 59 cases with MI, no effect on risk was seen for *UGT1A1*28*, possibly because of the inadequate power associated with the low number of incident MI. Bilirubin levels were also not significantly lower in subjects with MI than in individuals free of those complications, but significantly lower in subjects with a CVD or CAD event. This association was of borderline significance for the MI event group. CVD, CAD, and MI risks decreased by 10%, 13%, and 13%, respectively, for each 0.1 mg/dL (1.7 µmol/L) increase in serum bilirubin when the genotype was not included in the Cox regression model. The $(TA)_7$ allele was significantly associated with higher serum bilirubin levels, and could explain 18.6% of the variation in serum bilirubin. When both serum bilirubin and genotype were included, only the bilirubin effect remained in the model and the *UGT1A1* (TA)n repeat polymorphism was no longer significant. This result suggests that serum bilirubin levels are probably more closely associated with CVD and CAD than the genotype, and may be an intermediate phenotype for CVD and CAD events.

The Cardiovascular Disease in Patients with Intermittent Claudication (CAVASIC) study was a case-control study of PAD. [100] There was a clear association between low bilirubin levels and PAD, and the *UGT1A1*28* carriers had significantly higher bilirubin levels in both the case and control groups. However, the association between *UGT1A1*28* and PAD was not observed. Another case-control study was performed on bilirubin levels, *UGT1A1*28* as well as SNP rs4124874 (*UGT1A1* T-3279G), and CAD. [114] Both variants were found to be associated with bilirubin levels but not with CAD although low bilirubin levels were found to be significantly associated with CAD both in men and women. [114]

A prospective case-control study, including 571 cases with first-time MI (3.5-year median lag time until event) and 1,074 matched control subjects was performed by Ekblom et al. to evaluate a possible protective effect of serum bilirubin and *UGT1A1*28* against MI. [98] Serum bilirubin was lower in cases than in controls. The $(TA)_7$ allele showed a strong gene-dosage effect on bilirubin levels both in controls and cases. However, the *UGT1A1*28* allele did not influence the risk of MI.

Ekblom et al. [97] also investigated if bilirubin and *UGT1A1*28* are protective against ischemic stroke in a prospective case-control setting. Cases with first-ever ischemic stroke (n=231; median lag time 4.9 years) and 462 matched referents were enrolled. Serum bilirubin was lower in cases than in controls, but the difference reached significance only in women.

Table 3. Published studies evaluating the association of *UGT1A1* polymorphisms with bilirubin levels and cardiovascular end points[a]

Study	Year	Ethnicity	Study design	CVD end point	Number of cases/controls	Polymorphism	OR[b] (95% CI), P value, other findings
Rotterdam [Bosma et al. [99]]	2003	Caucasian	Prospective	MI	185/255	UGT1A1*28	$(TA)_7/(TA)_7$ vs $(TA)_6/(TA)_6$ Ajusted 1.3(0.8-2.2), $(TA)_6/(TA)_7$ vs $(TA)_6/(TA)_6$ Adjusted 0.9(0.7-1.3), NS $(TA)_7/(TA)_7$ vs $[(TA)_6/(TA)_7+(TA)_6/(TA)_6]$ Unadjusted 1.7(0.9-3.1), NS Significant association was also not found when stratified by gender (data not shown).
ECTIM [Gajdos et al. [169]]	2006	Caucasian	Retrospective	MI	Men 366/314	UGT1A1*28	$(TA)_7/(TA)_7$ vs $(TA)_6/(TA)_6$ Adjusted 1.8(0.9-3.5), $(TA)_6/(TA)_7$ vs $(TA)_6/(TA)_6$ Adjusted 1.5 (1.0-2.1), P=0.017
Framingham [Lin et al. [86]]	2006	Caucasian	Prospective (24 years follow-up)	CVD[c]	156/1583	UGT1A1*28	0.36(0.18-0.74), P=0.005
				CAD[d]	117/1583		0.30 (0.12-0.74), P=0.009
				MI	59/1583		0.52 (0.19-1.43), P=0.202 Gender-stratified analysis was not performed.
CAVASIC [Rantner et al. [100]]	2008	Caucasian	Retrospective	PAD	Men 255/255	UGT1A1*28	OR not shown, NS
Lingenhel et al. [114]	2008	Caucasian	Retrospective	CAD			The results in full study group were not shown.
					Men 365/397	UGT1A1*28	$(TA)_7/(TA)_7$ vs $[(TA)_6/(TA)_7+(TA)_6/(TA)_6]$ 0.97(0.61-1.54), NS
						T-3279G	G/G vs T/T+T/G 0.88 (0.61-1.28), NS
					Women 112/222	UGT1A1*28	$(TA)_7/(TA)_7$ vs $[(TA)_6/(TA)_7+(TA)_6/(TA)_6]$ 0.93(0.45-1.93), NS
						T-3279G	G/G vs T/T+T/G 0.92 (0.52-1.62), NS
MONICA, NSHDSC and the local Mammography Screening Project [Ekblom et al. [98]]	2010	Caucasian	Prospective	First-ever MI	618/1184	UGT1A1*28	$(TA)_7/(TA)_7$ vs $[(TA)_6/(TA)_7+(TA)_6/(TA)_6]$ 0.95 (0.65-1.38), NS
					Men 447/862	UGT1A1*28	$(TA)_7/(TA)_7$ vs $[(TA)_6/(TA)_7+(TA)_6/(TA)_6]$ 0.83 (0.52-1.30), NS
					Women 171/322	UGT1A1*28	$(TA)_7/(TA)_7$ vs $[(TA)_6/(TA)_7+(TA)_6/(TA)_6]$ 1.33 (0.66-2.68), NS

Study	Year	Ethnicity	Study design	CVD end point	Number of cases/controls	Polymorphism	OR[b] (95% CI), P value, other findings
MONICA and NSHDSC [Ekblom et al. [97]]	2010	Caucasian	Prospective	First-ever ischemic stroke	231/462	UGT1A1*28	$(TA)_7/(TA)_7$ vs $[(TA)_6/(TA)_7+(TA)_6/(TA)_6]$ 1.12 (0.64-1.97), NS
					Men 127/254	UGT1A1*28	$(TA)_7/(TA)_7$ vs $[(TA)_6/(TA)_7+(TA)_6/(TA)_6]$ 1.77 (0.88-3.56), NS

Table 3. (Continued)

Study	Year	Ethnicity	Study design	CVD end point	Number of cases/controls	Polymorphism	OR[b] (95% CI), P value, other findings
					Women 104/208	UGT1A1*28	$(TA)_7/(TA)_7$ vs $[(TA)_6/(TA)_7+(TA)_6/(TA)_6]$ 0.48 (0.17-1.36), NS
Hsieh et al. [170]	2008	Chinese	Retrospective	CAD	Men 61/74	5 polymorphisms[e]	OR not shown, NS
Lin et al. [54]	2009	Chinese	Retrospective	CAD	1311/1000	8 polymorphisms[f] rs887829(UGT1A1 G-364A)	AA vs (GG+GA) 0.35(0.16-0.78), P= 0.0069
					Men 926/501	rs887829(UGT1A1 G-364A)	AA vs (GG+GA) 0.24(0.10-0.60), P=0.0014
					Women 385/499		(GA+AA) vs GG 1.29(0.94-1.76), NS
Chen et al. [91]	2011	Chinese	Prospective (12 years follow-up)	CVE[g]	186/475	UGT1A1*28	$(TA)_7/(TA)_7$ vs $(TA)_6/(TA)_6$ 0.10(0.03-0.36), P<0.001 $(TA)_6/(TA)_7$ vs $(TA)_6/(TA)_6$ 0.64(0.45-0.89), P=0.008 $(TA)_7/(TA)_7$ vs $[(TA)_6/(TA)_7+(TA)_6/(TA)_6]$ 0.12(0.04-0.43), P=0.001 Gender-stratified analysis was not performed.

CVD, cardiovascular disease; MI, myocardial infarction; PAD, peripheral arterial disease; CAD, coronary artery disease; CVE, cardiovascular event; OR, odds ratio; CI, confidence interval; NS, not significant.

[a] There are significant associations between UGT1A1 polymorphisms and bilirubin levels as well as between bilirubin levels and CVD end points in all studies except for (1) Rotterdam, in which bilirubin levels were not measured at a standardized time and showed no association with MI, (2) ECTIM, in which bilirubin levels were not measured, (3) the study of Ekblom et al. [97], in which association between bilirubin levels and CVD end points was observed only in women, and (4) the study of Lin et al. [54], in which association between bilirubin levels and CVD end points was observed only in men. Because different studies used different units to measure bilirubin levels and compared bilirubin levels in different genotypes of the different polymorphisms, the values of bilirubin levels are not listed in the table. [b] OR except for the studies of Lin et al. [86] and Chen et al. [91], which are hazard ratios. [c] CVD comprises fatal and nonfatal myocardial infarction, angina pectoris, coronary insufficiency, stroke, transient ischemic attack, intermittent claudication, congestive heart failure, and death from coronary artery disease. [d] CAD comprises fatal and nonfatal myocardial infarction, angina pectoris, and coronary insufficiency. [e] Five polymorphisms, including UGT1A1*28 and 4 SNPs in exons. [f] Eight polymorphisms, which are tag SNPs from the phase II HapMap Han Chinese. Of them, only one SNP (i.e. rs887829) showed association with CAD and thus the results are listed just for SNP rs887829. [g] CVE comprises fatal and nonfatal myocardial infarction, stroke, congestive heart failure and arrhythmia, coronary artery disease, transient ischemic attack, peripheral arterial disease, and sudden death.

The *UGT1A1*28* allele showed a strong gene-dose association with serum bilirubin levels both in controls and cases, but no association with risk for stroke. There are some GWASs on stroke,[171-175] which have recently been reviewed. [176, 177] One study has reported associations on chromosome 2, but these SNPs were not close to *UGT1A1* at 2q37. [174]

It should be noted that the power and genome coverage of several of these GWASs might have been insufficient.

A recent small Chinese case-control study on CAD demonstrated similar results in that CAD was associated with bilirubin levels but not with the *UGT1A1* variants. [170] It is notable that the sample size was so small (n=135 men) that homozygous carriers of the *UGT1A1*28* allele could not be found in this study.

In another study, *UGT1A1* genotypes and CAD risk was investigated in 1,311 CAD patients and 1,000 controls in a Chinese Han population who underwent coronary angiography. [54] In this study, eight common tag SNPs in *UGT1A1* were selected from the phase II HapMap Han Chinese population, which captured 30 of 30 (100%) common SNPs of *UGT1A1* at $r^2 \geq 0.8$, while most of other 9 studies just estimated the (TA)n repeat polymorphism. In this study, the A/A genotype of SNP rs887829 (*UGT1A1* G-364A), which is located upstream of the (TA)n repeat (-53bp) and in almost perfect LD with the (TA)n repeat ($r^2 \approx 1$) in Japanese, Europeans and Africans,[90, 101] showed association with increased TBIL levels in both men and women and decreased risk only for male CAD (OR 0.24, 95% CI 0.10-0.60, P=0.0014).

An inverse relationship between serum bilirubin levels and CAD was observed in men but not in women, which is consistent with other studies. [51, 122] The difference of bilirubin levels between male CAD cases and controls remained significant after adjustment for age and all TBIL-related SNPs [rs4399719 (*UGT1A1* T-2473G), rs887829 (*UGT1A1* G-364A) and rs4148323 (*UGT1A1* G211A)] in this study. These results suggested that other factors might be contributive to the difference between male CAD cases and controls. The authors noted that the difference might be partly because of increased oxidative activity in CAD-prone individuals, causing increased consumption of bilirubin, however, cannot exclude that other polymorphisms not in this study may account for this difference. These results also point out that serum bilirubin levels are probably more closely associated with CAD than the genotype, which have been indicated in the Framingham Offspring Study. This study had adequate statistical power because it had 2-20 times the number of CAD patients as the other 9 studies. It also had adequate statistical power for the gender-stratified analysis. This study showed that the A/A genotype of SNP rs887829 (*UGT1A1* G-364A) showed association with decreased risk in full study group (OR 0.35, 95% CI 0.16-0.78, P= 0.0069). However, in the gender-stratified analysis, this association remained only in men. It is required to confirm this association in more studies with large sample size, especially with large case sample size.

A prospective cohort study in Chinese was carried out to explore the association between serum bilirubin and *UGT1A1*28* polymorphism and their effect on cardiovascular events (CVEs) in chronic hemodialysis patients. [91] This study comprised 661 chronic hemodialysis patients who were prospectively followed for 12 years. A reverse association was noted between serum bilirubin and CVEs among chronic hemodialysis patients. For each 0.1-mg/dl decrease in bilirubin levels, risks for CVEs increased 9%. Moreover, the *UGT1A1*28* polymorphism had strong effects on bilirubin levels and the $(TA)_7/(TA)_7$ genotype might have an important effect on reducing CVEs in chronic hemodialysis patients.

Compared with subjects with the $(TA)_6/(TA)_6$ genotype, subjects with the $(TA)_7/(TA)_7$ genotype had a significantly lower risk for CVEs (hazard ratio 0.10, 95% CI 0.03-0.36, P<0.001).

Of note, of the three studies in Chinese,[54, 91, 170] only one study [170] in small sample size did not observe significant association between *UGT1A1*28* polymorphism and CVD. The results of the 7 studies in Caucasians are controversial. [86, 97-100, 114, 169] Only the long-term prospective Framingham Heart Study found a strong association of *UGT1A1*28* with CVD outcomes and, because of the lower power, a nonsignificant trend with MI. [86] Of the 7 studies in Caucasians, 4 have estimated the association between *UGT1A1*28* and MI and none of them detect this association. [86, 98, 99, 169]

In all studies involving bilirubin, *UGT1A1*, and CVD, clear associations between UGT1A1*28 and bilirubin levels were observed. Serum bilirubin levels also showed association with CVD outcomes in most of the studies; however, the UGT1A1*28 genotype was found to be associated with CVD only in the prospective Framingham Heart Study, the large Han Chinese case-control study and the Han Chinese prospective cohort study. In addition, multiple GWASs on CAD showed no significant disease association with SNPs in the proximity of the *UGT1A1* locus at chromosome 2q37. [178-185]

Most published non-GWAS studies have estimated the association between *HMOX1* (GT)n repeat or *UGT1A1* (TA)n repeat polymorphism, serum bilirubin and CVD. No non-GWAS studies have synchronously estimated both *HMOX1* (GT)n repeat and *UGT1A1* (TA)n repeat polymorphisms except for the large Han Chinese case-control study. [54] Recent GWASs and non-GWAS studies have revealed some other polymorphisms associated with serum bilirubin, especially in other genes such *SLCO1B3*. It is necessary to investigate the association between those polymorphisms, in addition to *HMOX1* (GT)n repeat and *UGT1A1* (TA)n repeat polymorphisms, serum bilirubin and CVD. That is to say, comprehensive pathway-based association study of bilirubin metabolism gene variants is required to detect novel signatures for association with TBIL levels and CVD. Maybe we can find that other polymorphisms and their interactions can explain the difference of bilirubin levels between the CVD cases and controls, and show association with CVD. To our knowledge, the large Han Chinese case-control study [54] is the first and systematic association study to detect SNPs across four main genes involved in bilirubin metabolism - namely *HMOX1*, *BLVRA*, *UGT1A1*, and *SLCO1B1* associated with TBIL levels and CAD.

(6) ABCC2 and ABCC3

No studies have estimated the association of *ABCC2* and *ABCC3* polymorphisms with CVD.

Summary

Abundant evidence indicates multifunctional roles of bilirubin metabolism in the cardiovascular system. The vasoprotection afforded by bilirubin metabolism is largely attributable to its products (biliverdin, bilirubin, Fe^{2+} and CO) and enzymes such as HMOX1 and BLVRA. The retrospective and prospective epidemiological studies also indicate that high normal and moderately elevated serum bilirubin levels protect against CVD over the past 16 years. Although GWASs and non-GWAS studies have identify a series of variants in bilirubin metabolism gene that are associated with TBIL levels such as SNP rs4149056 in *SLCO1B1* and rs2117032 in *SLCO1B3*, published association studies of bilirubin metabolism gene polymorphisms with CVD mostly focus on individual single locus, mainly on the *HMOX1* (GT)n repeat or *UGT1A1* (TA)n repeat polymorphism. The results of these genetic studies for CVD are inconsistent. Further studies are warranted to investigate the associations between those polymorphisms, in addition to *HMOX1* (GT)n repeat and *UGT1A1* (TA)n repeat polymorphisms, serum bilirubin and CVD. Comprehensive pathway-based association study of bilirubin metabolism gene variants is also required to detect novel signatures for association with CVD.

Acknowledgments

This work was supported by grants from the Research Start-Up Fund in Hainan Medical College, and the Young Scientists Fund of the National Natural Science Foundation of China (Grant No. 31100904).

References

[1] Trauner M, Boyer JL. Bile salt transporters: molecular characterization, function, and regulation. *Physiol. Rev.* Apr 2003;83:633-671.

[2] Akita H, Suzuki H, Ito K, Kinoshita S, Sato N, Takikawa H, Sugiyama Y. Characterization of bile acid transport mediated by multidrug resistance associated protein 2 and bile salt export pump. *Biochimica et biophysica acta*. Mar 9 2001;1511:7-16.

[3] Konig J, Rost D, Cui Y, Keppler D. Characterization of the human multidrug resistance protein isoform MRP3 localized to the basolateral hepatocyte membrane. *Hepatology (Baltimore, Md*. Apr 1999;29:1156-1163.

[4] Lee YM, Cui Y, Konig J, Risch A, Jager B, Drings P, Bartsch H, Keppler D, Nies AT. Identification and functional characterization of the natural variant MRP3-Arg1297His of human multidrug resistance protein 3 (MRP3/ABCC3). *Pharmacogenetics*. Apr 2004;14:213-223.

[5] Miyaoka T, Seno H, Itoga M, Iijima M, Inagaki T, Horiguchi J. Schizophrenia-associated idiopathic unconjugated hyperbilirubinemia (Gilbert's syndrome). *J. Clin. Psychiat*. Nov 2000;61:868-871.

[6] Alexandra Brito M, Silva RF, Brites D. Bilirubin toxicity to human erythrocytes: a review. *Clin. Chim. Acta*. Dec 2006;374:46-56.

[7] Mayer M. Association of serum bilirubin concentration with risk of coronary artery disease. *Clin. Chem*. Nov 2000;46:1723-1727.

[8] Gopinathan V, Miller NJ, Milner AD, Rice-Evans CA. Bilirubin and ascorbate antioxidant activity in neonatal plasma. *FEBS letters*. Aug 1 1994;349:197-200.

[9] Belanger S, Lavoie JC, Chessex P. Influence of bilirubin on the antioxidant capacity of plasma in newborn infants. *Biology of the neonate*. 1997;71:233-238.

[10] Vitek L, Jirsa M, Brodanova M, Kalab M, Marecek Z, Danzig V, Novotny L, Kotal P. Gilbert syndrome and ischemic heart disease: a protective effect of elevated bilirubin levels. *Atherosclerosis*. Feb 2002;160:449-456.

[11] McCarty MF. "Iatrogenic Gilbert syndrome"--a strategy for reducing vascular and cancer risk by increasing plasma unconjugated bilirubin. *Medical hypotheses*. 2007;69:974-994.

[12] Abraham NG, Kappas A. Pharmacological and clinical aspects of heme oxygenase. *Pharmacological reviews*. Mar 2008;60:79-127.

[13] Baranano DE, Rao M, Ferris CD, Snyder SH. Biliverdin reductase: a major physiologic cytoprotectant. *Proc. Natl. Acad. Sci. USA*. Dec 10 2002;99:16093-16098.

[14] Sedlak TW, Saleh M, Higginson DS, Paul BD, Juluri KR, Snyder SH. Bilirubin and glutathione have complementary antioxidant and cytoprotective roles. *Proc. Natl. Acad. Sci.USA*. Mar 31 2009;106:5171-5176.

[15] Maghzal GJ, Leck MC, Collinson E, Li C, Stocker R. Limited role for the bilirubin-biliverdin redox amplification cycle in the cellular antioxidant protection by biliverdin reductase. *J. Biol. Chem.* Oct 23 2009;284:29251-29259.

[16] McDonagh AF. The biliverdin-bilirubin antioxidant cycle of cellular protection: Missing a wheel? *Free Radic. Biol. Med*. Sep 1 2010;49:814-820.

[17] Vitek L, Schwertner HA. The heme catabolic pathway and its protective effects on oxidative stress-mediated diseases. *Advances in clinical chemistry*. 2007;43:1-57.

[18] Mazzone GL, Rigato I, Ostrow JD, Bossi F, Bortoluzzi A, Sukowati CH, Tedesco F, Tiribelli C. Bilirubin inhibits the TNFalpha-related induction of three endothelial adhesion molecules. *Biochem. Biophys. Res. Commun.* Aug 21 2009;386:338-344.

[19] Tapan S, Dogru T, Tasci I, Ercin CN, Ozgurtas T, Erbil MK. Soluble CD40 ligand and soluble P-selectin levels in Gilbert's syndrome: a link to protection against atherosclerosis? *Clinical biochemistry*. Jun 2009;42:791-795.

[20] Nakagami T, Toyomura K, Kinoshita T, Morisawa S. A beneficial role of bile pigments as an endogenous tissue protector: anti-complement effects of biliverdin and conjugated bilirubin. *Biochimica et biophysica acta*. Oct 3 1993;1158:189-193.

[21] Abraham NG, Kappas A. Heme oxygenase and the cardiovascular-renal system. *Free Radic. Biol. Med*. Jul 1 2005;39:1-25.

[22] Wang CY, Chau LY. Heme oxygenase-1 in cardiovascular diseases: molecular mechanisms and clinical perspectives. *Chang Gung medical journal*. Jan-Feb 2010;33: 13-24.

[23] Kumar S, Bandyopadhyay U. Free heme toxicity and its detoxification systems in human. *Toxicology letters*. Jul 4 2005;157:175-188.

[24] Sassa S. Why heme needs to be degraded to iron, biliverdin IXalpha, and carbon monoxide? *Antioxid. Redox Signal*. Oct 2004;6:819-824.

[25] Mense SM, Zhang L. Heme: a versatile signaling molecule controlling the activities of diverse regulators ranging from transcription factors to MAP kinases. *Cell research*. Aug 2006; 16:681-692.

[26] Ryter SW, Alam J, Choi AM. Heme oxygenase-1/carbon monoxide: from basic science to therapeutic applications. *Physiol. Rev*. Apr 2006;86:583-650.

[27] Dulak J, Loboda A, Jozkowicz A. Effect of heme oxygenase-1 on vascular function and disease. *Current opinion in lipidology.* Oct 2008;19:505-512.
[28] Durante W, Schafer AI. Carbon monoxide and vascular cell function (review). *International journal of molecular medicine.* Sep 1998;2:255-262.
[29] Breimer LH, Mikhailidis DP. Could carbon monoxide and bilirubin be friends as well as foes of the body? *Scandinavian journal of clinical and laboratory investigation.* Feb 2010;70:1-5.
[30] Siow RC, Sato H, Mann GE. Heme oxygenase-carbon monoxide signalling pathway in atherosclerosis: anti-atherogenic actions of bilirubin and carbon monoxide? *Cardiovascular research.* Feb 1999;41:385-394.
[31] Johnson RA, Kozma F, Colombari E. Carbon monoxide: from toxin to endogenous modulator of cardiovascular functions. *Brazilian journal of medical and biological research = Revista brasileira de pesquisas medicas e biologicas / Sociedade Brasileira de Biofisica ... [et al.* Jan 1999;32:1-14.
[32] Balla G, Jacob HS, Balla J, Rosenberg M, Nath K, Apple F, Eaton JW, Vercellotti GM. Ferritin: a cytoprotective antioxidant strategem of endothelium. *J. Biol. Chem.* Sep 5 1992; 267:18148-18153.
[33] Ujhelyi L, Balla G, Jeney V, Varga Z, Nagy E, Vercellotti GM, Agarwal A, Eaton JW, Balla J. Hemodialysis reduces inhibitory effect of plasma ultrafiltrate on LDL oxidation and subsequent endothelial reactions. *Kidney international.* Jan 2006;69:144-151.
[34] Ahmad Z, Salim M, Maines MD. Human biliverdin reductase is a leucine zipper-like DNA-binding protein and functions in transcriptional activation of heme oxygenase-1 by oxidative stress. *J. Biol. Chem.* Mar 15 2002;277:9226-9232.
[35] Kravets A, Hu Z, Miralem T, Torno MD, Maines MD. Biliverdin reductase, a novel regulator for induction of activating transcription factor-2 and heme oxygenase-1. *J. Biol. Chem.* May 7 2004;279:19916-19923.
[36] Lerner-Marmarosh N, Shen J, Torno MD, Kravets A, Hu Z, Maines MD. Human biliverdin reductase: a member of the insulin receptor substrate family with serine/threonine/tyrosine kinase activity. *Proc. Natl. Acad. Sci. USA* May 17 2005;102: 7109-7114.
[37] LeWinter MM. Association of syndromes of insulin resistance with coronary artery disease. *Coron. Artery Dis.* Dec 2005;16:477-480.
[38] Schulman IH, Zhou MS. Vascular insulin resistance: a potential link between cardiovascular and metabolic diseases. *Current hypertension reports.* Feb 2009;11:48-55.
[39] Chahwala V, Arora R. Cardiovascular manifestations of insulin resistance. *American journal of therapeutics.* Sep-Oct 2009;16:e14-28.
[40] Hunt SC, Wu LL, Hopkins PN, Williams RR. Evidence for a major gene elevating serum bilirubin concentration in Utah pedigrees. *Arterioscler. Thromb. Vasc. Biol.* Aug 1996;16:912-917.
[41] Kronenberg F, Coon H, Gutin A, Abkevich V, Samuels ME, Ballinger DG, Hopkins PN, Hunt SC. A genome scan for loci influencing anti-atherogenic serum bilirubin levels. *Eur. J. Hum. Genet.* Sep 2002;10:539-546.
[42] Lin JP, Cupples LA, Wilson PW, Heard-Costa N, O'Donnell CJ. Evidence for a gene influencing serum bilirubin on chromosome 2q telomere: a genomewide scan in the Framingham study. *American journal of human genetics.* Apr 2003;72:1029-1034.

[43] Yamada N, Yamaya M, Okinaga S, Nakayama K, Sekizawa K, Shibahara S, Sasaki H. Microsatellite polymorphism in the heme oxygenase-1 gene promoter is associated with susceptibility to emphysema. *American journal of human genetics*. Jan 2000;66:187-195.

[44] Lin R, Wang X, Wang Y, Zhang F, Wang Y, Fu W, Yu T, Li S, Xiong M, Huang W, Jin L. Association of polymorphisms in four bilirubin metabolism genes with serum bilirubin in three Asian populations. *Human mutation*. Apr 2009;30:609-615.

[45] Funke C, Tomiuk J, Riess O, Berg D, Soehn AS. Genetic analysis of heme oxygenase-1 (HO-1) in German Parkinson's disease patients. *J. Neural Transm*. Jul 2009;116:853-859.

[46] Hirai H, Kubo H, Yamaya M, Nakayama K, Numasaki M, Kobayashi S, Suzuki S, Shibahara S, Sasaki H. Microsatellite polymorphism in heme oxygenase-1 gene promoter is associated with susceptibility to oxidant-induced apoptosis in lymphoblastoid cell lines. *Blood*. Sep 1 2003;102:1619-1621.

[47] Okamoto I, Krogler J, Endler G, Kaufmann S, Mustafa S, Exner M, Mannhalter C, Wagner O, Pehamberger H. A microsatellite polymorphism in the heme oxygenase-1 gene promoter is associated with risk for melanoma. *Int. J. Cancer*. Sep 15 2006;119:1312-1315.

[48] Chen YH, Lin SJ, Lin MW, Tsai HL, Kuo SS, Chen JW, Charng MJ, Wu TC, Chen LC, Ding YA, Pan WH, Jou YS, Chau LY. Microsatellite polymorphism in promoter of heme oxygenase-1 gene is associated with susceptibility to coronary artery disease in type 2 diabetic patients. *Human genetics*. Jul 2002;111:1-8.

[49] Ono K, Goto Y, Takagi S, Baba S, Tago N, Nonogi H, Iwai N. A promoter variant of the heme oxygenase-1 gene may reduce the incidence of ischemic heart disease in Japanese. *Atherosclerosis*. Apr 2004;173:315-319.

[50] Ono K, Mannami T, Iwai N. Association of a promoter variant of the haeme oxygenase-1 gene with hypertension in women. *J. Hypertens*. Aug 2003;21:1497-1503.

[51] Endler G, Exner M, Schillinger M, Marculescu R, Sunder-Plassmann R, Raith M, Jordanova N, Wojta J, Mannhalter C, Wagner OF, Huber K. A microsatellite polymorphism in the heme oxygenase-1 gene promoter is associated with increased bilirubin and HDL levels but not with coronary artery disease. *Thromb Haemost*. Jan 2004;91:155-161.

[52] Lublinghoff N, Winkler K, Winkelmann BR, Seelhorst U, Wellnitz B, Boehm BO, Marz W, Hoffmann MM. Genetic variants of the promoter of the heme oxygenase-1 gene and their influence on cardiovascular disease (the Ludwigshafen Risk and Cardiovascular Health study). *BMC medical genetics*. 2009;10:36.

[53] Vasavda N, Menzel S, Kondaveeti S, Maytham E, Awogbade M, Bannister S, Cunningham J, Eichholz A, Daniel Y, Okpala I, Fulford T, Thein SL. The linear effects of alpha-thalassaemia, the UGT1A1 and HMOX1 polymorphisms on cholelithiasis in sickle cell disease. *British journal of haematology*. Jul 2007;138:263-270.

[54] Lin R, Wang Y, Wang Y, Fu W, Zhang D, Zheng H, Yu T, Wang Y, Shen M, Lei R, Wu H, Sun A, Zhang R, Wang X, Xiong M, Huang W, Jin L. Common variants of four bilirubin metabolism genes and their association with serum bilirubin and coronary artery disease in Chinese Han population. *Pharmacogenet Genomics*. Apr 2009;19:310-318.

[55] D'Silva S, Borse V, Colah RB, Ghosh K, Mukherjee MB. Association of (GT)n Repeats Promoter Polymorphism of Heme Oxygenase-1 Gene with Serum Bilirubin Levels in Healthy Indian Adults. *Genetic testing and molecular biomarkers.* Apr 2011;15:215-218.

[56] Chen YH, Chau LY, Chen JW, Lin SJ. Serum bilirubin and ferritin levels link heme oxygenase-1 gene promoter polymorphism and susceptibility to coronary artery disease in diabetic patients. *Diabetes Care.* Aug 2008;31:1615-1620.

[57] Kanai M, Akaba K, Sasaki A, Sato M, Harano T, Shibahara S, Kurachi H, Yoshida T, Hayasaka K. Neonatal hyperbilirubinemia in Japanese neonates: analysis of the heme oxygenase-1 gene and fetal hemoglobin composition in cord blood. *Pediatr. Res.* Aug 2003;54:165-171.

[58] Bozkaya OG, Kumral A, Yesilirmak DC, Ulgenalp A, Duman N, Ercal D, Ozkan H. Prolonged unconjugated hyperbilirubinaemia associated with the haem oxygenase-1 gene promoter polymorphism. *Acta Paediatr.* May 2010;99:679-683.

[59] Sanna S, Busonero F, Maschio A, McArdle PF, Usala G, Dei M, Lai S, Mulas A, Piras MG, Perseu L, Masala M, Marongiu M, Crisponi L, Naitza S, Galanello R, Abecasis GR, Shuldiner AR, Schlessinger D, Cao A, Uda M. Common variants in the SLCO1B3 locus are associated with bilirubin levels and unconjugated hyperbilirubinemia. *Human molecular genetics.* Jul 15 2009;18:2711-2718.

[60] Johnson AD, Kavousi M, Smith AV, Chen MH, Dehghan A, Aspelund T, Lin JP, van Duijn CM, Harris TB, Cupples LA, Uitterlinden AG, Launer L, Hofman A, Rivadeneira F, Stricker B, Yang Q, O'Donnell CJ, Gudnason V, Witteman JC. Genome-wide association meta-analysis for total serum bilirubin levels. *Human molecular genetics.* Jul 15 2009;18:2700-2710.

[61] Bielinski SJ, Chai HS, Pathak J, Talwalkar JA, Limburg PJ, Gullerud RE, Sicotte H, Klee EW, Ross JL, Kocher JP, Kullo IJ, Heit JA, Petersen GM, de Andrade M, Chute CG. Mayo genome consortia: a genotype-phenotype resource for genome-wide association studies with an application to the analysis of circulating bilirubin levels. *Mayo Clinic proceedings.* Jul 2011;86:606-614.

[62] Kang TW, Kim HJ, Ju H, Kim JH, Jeon YJ, Lee HC, Kim KK, Kim JW, Lee S, Kim JY, Kim SY, Kim YS. Genome-wide association of serum bilirubin levels in Korean population. *Human molecular genetics.* Sep 15 2010;19:3672-3678.

[63] Lin JP, Schwaiger JP, Cupples LA, O'Donnell CJ, Zheng G, Schoenborn V, Hunt SC, Joo J, Kronenberg F. Conditional linkage and genome-wide association studies identify UGT1A1 as a major gene for anti-atherogenic serum bilirubin levels--the Framingham Heart Study. *Atherosclerosis.* Sep 2009;206:228-233.

[64] Yamaguchi T, Komoda Y, Nakajima H. Biliverdin-IX alpha reductase and biliverdin-IX beta reductase from human liver. Purification and characterization. *J. Biol. Chem.* Sep 30 1994;269:24343-24348.

[65] Yamaguchi T, Nakajima H. Changes in the composition of bilirubin-IX isomers during human prenatal development. *Eur. J. Biochem.* Oct 15 1995;233:467-472.

[66] Liu Y, Ortiz de Montellano PR. Reaction intermediates and single turnover rate constants for the oxidation of heme by human heme oxygenase-1. *J. Biol. Chem.* Feb 25 2000;275:5297-5307.

[67] Wang J, de Montellano PR. The binding sites on human heme oxygenase-1 for cytochrome p450 reductase and biliverdin reductase. *J. Biol. Chem.* May 30 2003;278:20069-20076.

[68] Lin R, Wang X, Zhou W, Fu W, Wang Y, Huang W, Jin L. Association of a BLVRA Common Polymorphism with Essential Hypertension and Blood Pressure in Kazaks. *Clin. Exp. Hypertens.* Jul 1 2011.

[69] Cui Y, Konig J, Leier I, Buchholz U, Keppler D. Hepatic uptake of bilirubin and its conjugates by the human organic anion transporter SLC21A6. *J. Biol. Chem.* Mar 30 2001;276:9626-9630.

[70] Campbell SD, de Morais SM, Xu JJ. Inhibition of human organic anion transporting polypeptide OATP 1B1 as a mechanism of drug-induced hyperbilirubinemia. *Chemico-biological interactions*. Nov 20 2004;150:179-187.

[71] Nishizato Y, Ieiri I, Suzuki H, Kimura M, Kawabata K, Hirota T, Takane H, Irie S, Kusuhara H, Urasaki Y, Urae A, Higuchi S, Otsubo K, Sugiyama Y. Polymorphisms of OATP-C (SLC21A6) and OAT3 (SLC22A8) genes: consequences for pravastatin pharmacokinetics. *Clin. Pharmacol. Ther.* Jun 2003;73:554-565.

[72] Kivisto KT, Niemi M. Influence of drug transporter polymorphisms on pravastatin pharmacokinetics in humans. *Pharm. Res.* Feb 2007;24:239-247.

[73] Mwinyi J, Johne A, Bauer S, Roots I, Gerloff T. Evidence for inverse effects of OATP-C (SLC21A6) 5 and 1b haplotypes on pravastatin kinetics. *Clin. Pharmacol. Ther.* May 2004;75:415-421.

[74] Ieiri I, Suzuki H, Kimura M, Takane H, Nishizato Y, Irie S, Urae A, Kawabata K, Higuchi S, Otsubo K, Sugiyama Y. Influence of common variants in the pharmacokinetic genes (OATP-C, UGT1A1, and MRP2) on serum bilirubin levels in healthy subjects. *Hepatol. Res.* Oct 2004;30:91-95.

[75] Zhang W, He YJ, Gan Z, Fan L, Li Q, Wang A, Liu ZQ, Deng S, Huang YF, Xu LY, Zhou HH. OATP1B1 polymorphism is a major determinant of serum bilirubin level but not associated with rifampicin-mediated bilirubin elevation. *Clinical and experimental pharmacology and physiology*. Dec 2007;34:1240-1244.

[76] Saito A, Kawamoto M, Kamatani N. Association study between single-nucleotide polymorphisms in 199 drug-related genes and commonly measured quantitative traits of 752 healthy Japanese subjects. *J. Hum. Genet.* Jun 2009;54:317-323.

[77] Ho RH, Choi L, Lee W, Mayo G, Schwarz UI, Tirona RG, Bailey DG, Michael Stein C, Kim RB. Effect of drug transporter genotypes on pravastatin disposition in European- and African-American participants. *Pharmacogenet Genomics*. Aug 2007;17:647-656.

[78] Huang CS, Huang MJ, Lin MS, Yang SS, Teng HC, Tang KS. Genetic factors related to unconjugated hyperbilirubinemia amongst adults. *Pharmacogenet Genomics*. Jan 2005;15:43-50.

[79] Huang MJ, Kua KE, Teng HC, Tang KS, Weng HW, Huang CS. Risk factors for severe hyperbilirubinemia in neonates. *Pediatr. Res.* Nov 2004;56:682-689.

[80] Watchko JF, Lin Z, Clark RH, Kelleher AS, Walker MW, Spitzer AR. Complex multifactorial nature of significant hyperbilirubinemia in neonates. *Pediatrics*. Nov 2009;124:e868-877.

[81] Prachukthum S, Nunnarumit P, Pienvichit P, Chuansumrit A, Songdej D, Kajanachumpol S, Pakakasama S, Hongeng S. Genetic polymorphisms in Thai neonates with hyperbilirubinemia. *Acta Paediatr.* Jul 2009;98:1106-1110.

[82] Wong FL, Boo NY, Ainoon O, Wang MK. Variants of organic anion transporter polypeptide 2 gene are not risk factors associated with severe neonatal hyperbilirubinemia. *The Malaysian journal of pathology*. Dec 2009;31:99-104.

[83] Chang PF, Lin YC, Liu K, Yeh SJ, Ni YH. Risk of Hyperbilirubinemia in Breast-Fed Infants. *J. Pediatr*. May 16 2011.

[84] Smith NF, Figg WD, Sparreboom A. Role of the liver-specific transporters OATP1B1 and OATP1B3 in governing drug elimination. *Expert opinion on drug metabolism and toxicology*. Oct 2005;1:429-445.

[85] Beutler E, Gelbart T, Demina A. Racial variability in the UDP-glucuronosyltransferase 1 (UGT1A1) promoter: a balanced polymorphism for regulation of bilirubin metabolism? *Proc. Natl. Acad. Sci. USA*. Jul 7 1998;95:8170-8174.

[86] Lin JP, O'Donnell CJ, Schwaiger JP, Cupples LA, Lingenhel A, Hunt SC, Yang S, Kronenberg F. Association between the UGT1A1*28 allele, bilirubin levels, and coronary heart disease in the Framingham Heart Study. *Circulation*. Oct 3 2006;114:1476-1481.

[87] Bosma PJ, Chowdhury JR, Bakker C, Gantla S, de Boer A, Oostra BA, Lindhout D, Tytgat GN, Jansen PL, Oude Elferink RP, et al. The genetic basis of the reduced expression of bilirubin UDP-glucuronosyltransferase 1 in Gilbert's syndrome. *N. Engl. J. Med*. Nov 2 1995;333:1171-1175.

[88] Muslu N, Turhan AB, Eskandari G, Atici A, Ozturk OG, Kul S, Atik U. The frequency of UDP-glucuronosyltransferase 1A1 promoter region (TA)7 polymorphism in newborns and it's relation with jaundice. *J. Trop. Pediatr*. Feb 2007;53:64-68.

[89] Babaoglu MO, Yigit S, Aynacioglu AS, Kerb R, Yurdakok M, Bozkurt A. Neonatal jaundice and bilirubin UDP-glucuronosyl transferase 1A1 gene polymorphism in Turkish patients. *Basic Clin. Pharmacol. Toxicol*. Apr 2006;98:377-380.

[90] Sai K, Saeki M, Saito Y, Ozawa S, Katori N, Jinno H, Hasegawa R, Kaniwa N, Sawada J, Komamura K, Ueno K, Kamakura S, Kitakaze M, Kitamura Y, Kamatani N, Minami H, Ohtsu A, Shirao K, Yoshida T, Saijo N. UGT1A1 haplotypes associated with reduced glucuronidation and increased serum bilirubin in irinotecan-administered Japanese patients with cancer. *Clin. Pharmacol. Ther*. Jun 2004;75:501-515.

[91] Chen YH, Hung SC, Tarng DC. Serum Bilirubin Links UGT1A1*28 Polymorphism and Predicts Long-Term Cardiovascular Events and Mortality in Chronic Hemodialysis Patients. *Clin. J. Am. Soc. Nephrol*. Mar 2011;6:567-574.

[92] Huang CS, Luo GA, Huang ML, Yu SC, Yang SS. Variations of the bilirubin uridine-diphosphoglucuronosyl transferase 1A1 gene in healthy Taiwanese. *Pharmacogenetics*. Aug 2000;10:539-544.

[93] Ki CS, Lee KA, Lee SY, Kim HJ, Cho SS, Park JH, Cho S, Sohn KM, Kim JW. Haplotype structure of the UDP-glucuronosyltransferase 1A1 (UGT1A1) gene and its relationship to serum total bilirubin concentration in a male Korean population. *Clin. Chem*. Dec 2003;49:2078-2081.

[94] Zhang A, Xing Q, Qin S, Du J, Wang L, Yu L, Li X, Xu L, Xu M, Feng G, He L. Intra-ethnic differences in genetic variants of the UGT-glucuronosyltransferase 1A1 gene in Chinese populations. *Pharmacogenomics J*. Oct 2007;7:333-338.

[95] Italia KY, Jijina FF, Jain D, Merchant R, Nadkarni AH, Mukherjee M, Ghosh K, Colah RB. The effect of UGT1A1 promoter polymorphism on bilirubin response to

hydroxyurea therapy in hemoglobinopathies. *Clinical biochemistry*. Nov 2010;43:1329-1332.

[96] Borucki K, Weikert C, Fisher E, Jakubiczka S, Luley C, Westphal S, Dierkes J. Haplotypes in the UGT1A1 gene and their role as genetic determinants of bilirubin concentration in healthy German volunteers. *Clinical biochemistry*. Nov 2009;42:1635-1641.

[97] Ekblom K, Marklund SL, Johansson L, Osterman P, Hallmans G, Weinehall L, Wiklund PG, Hultdin J. Bilirubin and UGT1A1*28 are not associated with lower risk for ischemic stroke in a prospective nested case-referent setting. *Cerebrovascular diseases (Basel, Switzerland)*. 2010;30:590-596.

[98] Ekblom K, Marklund SL, Jansson JH, Osterman P, Hallmans G, Weinehall L, Hultdin J. Plasma bilirubin and UGT1A1*28 are not protective factors against first-time myocardial infarction in a prospective, nested case-referent setting. *Circ. Cardiovasc. Genet.* Aug 2010;3:340-347.

[99] Bosma PJ, van der Meer IM, Bakker CT, Hofman A, Paul-Abrahamse M, Witteman JC. UGT1A1*28 allele and coronary heart disease: the Rotterdam Study. *Clin. Chem.* Jul 2003;49:1180-1181.

[100] Rantner B, Kollerits B, Anderwald-Stadler M, Klein-Weigel P, Gruber I, Gehringer A, Haak M, Schnapka-Kopf M, Fraedrich G, Kronenberg F. Association between the UGT1A1 TA-repeat polymorphism and bilirubin concentration in patients with intermittent claudication: results from the CAVASIC study. *Clin. Chem.* May 2008;54:851-857.

[101] Horsfall LJ, Zeitlyn D, Tarekegn A, Bekele E, Thomas MG, Bradman N, Swallow DM. Prevalence of clinically relevant UGT1A alleles and haplotypes in African populations. *Annals of human genetics*. Mar 2011;75:236-246.

[102] Monaghan G, Foster B, Jurima-Romet M, Hume R, Burchell B. UGT1*1 genotyping in a Canadian Inuit population. *Pharmacogenetics*. Apr 1997;7:153-156.

[103] Balram C, Sabapathy K, Fei G, Khoo KS, Lee EJ. Genetic polymorphisms of UDP-glucuronosyltransferase in Asians: UGT1A1*28 is a common allele in Indians. *Pharmacogenetics*. Jan 2002;12:81-83.

[104] Akaba K, Kimura T, Sasaki A, Tanabe S, Ikegami T, Hashimoto M, Umeda H, Yoshida H, Umetsu K, Chiba H, Yuasa I, Hayasaka K. Neonatal hyperbilirubinemia and mutation of the bilirubin uridine diphosphate-glucuronosyltransferase gene: a common missense mutation among Japanese, Koreans and Chinese. *Biochem. Mol. Biol. Int.* Sep 1998;46:21-26.

[105] Ando Y, Chida M, Nakayama K, Saka H, Kamataki T. The UGT1A1*28 allele is relatively rare in a Japanese population. *Pharmacogenetics*. Aug 1998;8:357-360.

[106] Saeki M, Saito Y, Jinno H, Tohkin M, Kurose K, Kaniwa N, Komamura K, Ueno K, Kamakura S, Kitakaze M, Ozawa S, Sawada J. Comprehensive UGT1A1 genotyping in a Japanese population by pyrosequencing. *Clin. Chem.* Jul 2003;49:1182-1185.

[107] Chou HC, Chen MH, Yang HI, Su YN, Hsieh WS, Chen CY, Chen HL, Chang MH, Tsao PN. 211 G to a variation of UDP-glucuronosyl transferase 1A1 gene and neonatal breastfeeding jaundice. *Pediatr. Res.* Feb 2011;69:170-174.

[108] Chang PF, Lin YC, Liu K, Yeh SJ, Ni YH. Prolonged unconjugated hyperbiliriubinemia in breast-fed male infants with a mutation of uridine diphosphate-glucuronosyl transferase. *J. Pediatr.* Dec 2009;155:860-863.

[109] Boo NY, Wong FL, Wang MK, Othman A. Homozygous variant of UGT1A1 gene mutation and severe neonatal hyperbilirubinemia. *Pediatr. Int.* Aug 2009;51:488-493.

[110] Huang CS, Chang PF, Huang MJ, Chen ES, Hung KL, Tsou KI. Relationship between bilirubin UDP-glucuronosyl transferase 1A1 gene and neonatal hyperbilirubinemia. *Pediatr. Res.* Oct 2002;52:601-605.

[111] Huang CS, Chang PF, Huang MJ, Chen ES, Chen WC. Glucose-6-phosphate dehydrogenase deficiency, the UDP-glucuronosyl transferase 1A1 gene, and neonatal hyperbilirubinemia. *Gastroenterology.* Jul 2002;123:127-133.

[112] Sugatani J, Yamakawa K, Yoshinari K, Machida T, Takagi H, Mori M, Kakizaki S, Sueyoshi T, Negishi M, Miwa M. Identification of a defect in the UGT1A1 gene promoter and its association with hyperbilirubinemia. *Biochem. Biophys. Res. Commun.* Mar 29 2002;292:492-497.

[113] Saeki M, Saito Y, Sai K, Maekawa K, Kaniwa N, Sawada J, Kawamoto M, Saito A, Kamatani N. A combinatorial haplotype of the UDP-glucuronosyltransferase 1A1 gene (#60-#IB) increases total bilirubin concentrations in Japanese volunteers. *Clin. Chem.* Feb 2007;53:356-358.

[114] Lingenhel A, Kollerits B, Schwaiger JP, Hunt SC, Gress R, Hopkins PN, Schoenborn V, Heid IM, Kronenberg F. Serum bilirubin levels, UGT1A1 polymorphisms and risk for coronary artery disease. *Exp. Gerontol.* Dec 2008;43:1102-1107.

[115] Wada M, Toh S, Taniguchi K, Nakamura T, Uchiumi T, Kohno K, Yoshida I, Kimura A, Sakisaka S, Adachi Y, Kuwano M. Mutations in the canilicular multispecific organic anion transporter (cMOAT) gene, a novel ABC transporter, in patients with hyperbilirubinemia II/Dubin-Johnson syndrome. *Human molecular genetics.* Feb 1998;7:203-207.

[116] Perry GS, Byers T, Yip R, Margen S. Iron nutrition does not account for the hemoglobin differences between blacks and whites. *J. Nutr.* Jul 1992;122:1417-1424.

[117] Beutler E, West C. Hematologic differences between African-Americans and whites: the roles of iron deficiency and alpha-thalassemia on hemoglobin levels and mean corpuscular volume. *Blood.* Jul 15 2005;106:740-745.

[118] Schwertner HA, Jackson WG, Tolan G. Association of low serum concentration of bilirubin with increased risk of coronary artery disease. *Clin. Chem.* Jan 1994;40:18-23.

[119] Hopkins PN, Wu LL, Hunt SC, James BC, Vincent GM, Williams RR. Higher serum bilirubin is associated with decreased risk for early familial coronary artery disease. *Arterioscler. Thromb. Vasc. Biol.* Feb 1996;16:250-255.

[120] Novotny L, Vitek L. Inverse relationship between serum bilirubin and atherosclerosis in men: a meta-analysis of published studies. *Exp. Biol. Med. (Maywood).* May 2003;228: 568-571.

[121] Djousse L, Levy D, Cupples LA, Evans JC, D'Agostino RB, Ellison RC. Total serum bilirubin and risk of cardiovascular disease in the Framingham offspring study. *Am. J.Cardiol.* May 15 2001;87:1196-1200; A1194, 1197.

[122] Endler G, Hamwi A, Sunder-Plassmann R, Exner M, Vukovich T, Mannhalter C, Wojta J, Huber K, Wagner O. Is low serum bilirubin an independent risk factor for coronary artery disease in men but not in women? *Clin. Chem.* Jul 2003;49:1201-1204.

[123] Tanaka M, Fukui M, Tomiyasu K, Akabame S, Nakano K, Hasegawa G, Oda Y, Nakamura N. Low serum bilirubin concentration is associated with coronary artery calcification (CAC). *Atherosclerosis.* Sep 2009;206:287-291.

[124] Madhavan M, Wattigney WA, Srinivasan SR, Berenson GS. Serum bilirubin distribution and its relation to cardiovascular risk in children and young adults. *Atherosclerosis*. May 1997;131:107-113.

[125] Walden CE, Knopp RH, Johnson JL, Heiss G, Wahl PW, Hoover JJ. Effect of estrogen/progestin potency on clinical chemistry measures. The Lipid Research Clinics Program Prevalence Study. *American journal of epidemiology*. Mar 1986;123:517-531.

[126] Buckley DB, Klaassen CD. Tissue- and gender-specific mRNA expression of UDP-glucuronosyltransferases (UGTs) in mice. *Drug Metab Dispos*. Jan 2007;35:121-127.

[127] Muraca M, Fevery J. Influence of sex and sex steroids on bilirubin uridine diphosphate-glucuronosyltransferase activity of rat liver. *Gastroenterology*. Aug 1984;87:308-313.

[128] Breimer LH. Mortality associated with bilirubin levels. *Journal of insurance medicine (New York, N.Y.* 2009;41:230-231; author reply 232.

[129] Freeman R. Hormone replacement therapy (estrogen and progesterone): is it necessary for heart disease prevention? *Prev. Cardiol*. Winter 2000;3:21-23.

[130] Kimm H, Yun JE, Jo J, Jee SH. Low serum bilirubin level as an independent predictor of stroke incidence: a prospective study in Korean men and women. *Stroke*. Nov 2009; 40:3422-3427.

[131] Perlstein TS, Pande RL, Creager MA, Weuve J, Beckman JA. Serum total bilirubin level, prevalent stroke, and stroke outcomes: NHANES 1999-2004. *The American journal of medicine*. Sep 2008;121:781-788 e781.

[132] Nieto FJ, Iribarren C, Gross MD, Comstock GW, Cutler RG. Uric acid and serum antioxidant capacity: a reaction to atherosclerosis? *Atherosclerosis*. Jan 2000;148:131-139.

[133] Ishizaka N, Ishizaka Y, Takahashi E, Yamakado M, Hashimoto H. High serum bilirubin level is inversely associated with the presence of carotid plaque. *Stroke*. Feb 2001;32: 580-583.

[134] Ishizaka N, Ishizaka Y, Toda E, Nagai R, Yamakado M. Association between serum uric acid, metabolic syndrome, and carotid atherosclerosis in Japanese individuals. *Arterioscler. Thromb. Vasc. Biol*. May 2005;25:1038-1044.

[135] Erdogan D, Gullu H, Yildirim E, Tok D, Kirbas I, Ciftci O, Baycan ST, Muderrisoglu H. Low serum bilirubin levels are independently and inversely related to impaired flow-mediated vasodilation and increased carotid intima-media thickness in both men and women. *Atherosclerosis*. Feb 2006;184:431-437.

[136] Vitek L, Novotny L, Sperl M, Holaj R, Spacil J. The inverse association of elevated serum bilirubin levels with subclinical carotid atherosclerosis. *Cerebrovascular diseases (Basel, Switzerland)*. 2006;21:408-414.

[137] Yang XF, Chen YZ, Su JL, Wang FY, Wang LX. Relationship between serum bilirubin and carotid atherosclerosis in hypertensive patients. *Internal medicine (Tokyo, Japan)*. 2009;48:1595-1599.

[138] Breimer LH, Spyropolous KA, Winder AF, Mikhailidis DP, Hamilton G. Is bilirubin protective against coronary artery disease? *Clin. Chem*. Oct 1994;40:1987-1988.

[139] Perlstein TS, Pande RL, Beckman JA, Creager MA. Serum total bilirubin level and prevalent lower-extremity peripheral arterial disease: National Health and Nutrition Examination Survey (NHANES) 1999 to 2004. *Arterioscler. Thromb. Vasc. Biol*. Jan 2008;28:166-172.

[140] Bhuiyan AR, Srinivasan SR, Chen W, Sultana A, Berenson GS. Association of serum bilirubin with pulsatile arterial function in asymptomatic young adults: the Bogalusa Heart Study. *Metabolism: clinical and experimental*. May 2008;57:612-616.

[141] Papadakis JA, Ganotakis ES, Jagroop IA, Mikhailidis DP, Winder AF. Effect of hypertension and its treatment on lipid, lipoprotein(a), fibrinogen, and bilirubin levels in patients referred for dyslipidemia. *American journal of hypertension*. Jul 1999;12:673-681.

[142] Chin HJ, Song YR, Kim HS, Park M, Yoon HJ, Na KY, Kim Y, Chae DW, Kim S. The bilirubin level is negatively correlated with the incidence of hypertension in normotensive Korean population. *Journal of Korean medical science*. Jan 2009;24 Suppl:S50-56.

[143] Kaneda H, Ohno M, Taguchi J, Togo M, Hashimoto H, Ogasawara K, Aizawa T, Ishizaka N, Nagai R. Heme oxygenase-1 gene promoter polymorphism is associated with coronary artery disease in Japanese patients with coronary risk factors. *Arterioscler. Thromb. Vasc. Biol.* Oct 1 2002;22:1680-1685.

[144] Schillinger M, Exner M, Mlekusch W, Domanovits H, Huber K, Mannhalter C, Wagner O, Minar E. Heme oxygenase-1 gene promoter polymorphism is associated with abdominal aortic aneurysm. *Thrombosis research*. Apr 15 2002;106:131-136.

[145] Exner M, Schillinger M, Minar E, Mlekusch W, Schlerka G, Haumer M, Mannhalter C, Wagner O. Heme oxygenase-1 gene promoter microsatellite polymorphism is associated with restenosis after percutaneous transluminal angioplasty. *J. Endovasc. Ther.* Oct 2001;8:433-440.

[146] Chen YH, Chau LY, Lin MW, Chen LC, Yo MH, Chen JW, Lin SJ. Heme oxygenase-1 gene promotor microsatellite polymorphism is associated with angiographic restenosis after coronary stenting. *European heart journal*. Jan 2004;25:39-47.

[147] Gulesserian T, Wenzel C, Endler G, Sunder-Plassmann R, Marsik C, Mannhalter C, Iordanova N, Gyongyosi M, Wojta J, Mustafa S, Wagner O, Huber K. Clinical restenosis after coronary stent implantation is associated with the heme oxygenase-1 gene promoter polymorphism and the heme oxygenase-1 +99G/C variant. *Clin. Chem.* Sep 2005;51:1661-1665.

[148] Schillinger M, Exner M, Mlekusch W, Ahmadi R, Rumpold H, Mannhalter C, Wagner O, Minar E. Heme oxygenase-1 genotype is a vascular anti-inflammatory factor following balloon angioplasty. *J. Endovasc. Ther.* Aug 2002;9:385-394.

[149] Schillinger M, Exner M, Minar E, Mlekusch W, Mullner M, Mannhalter C, Bach FH, Wagner O. Heme oxygenase-1 genotype and restenosis after balloon angioplasty: a novel vascular protective factor. *J. Am. Coll. Cardiol.* Mar 17 2004;43:950-957.

[150] Li P, Elrayess MA, Gomma AH, Palmen J, Hawe E, Fox KM, Humphries SE. The microsatellite polymorphism of heme oxygenase-1 is associated with baseline plasma IL-6 level but not with restenosis after coronary in-stenting. *Chinese medical journal*. Sep 20 2005;118:1525-1532.

[151] Li P, Sanders J, Hawe E, Brull D, Montgomery H, Humphries S. Inflammatory response to coronary artery bypass surgery: does the heme-oxygenase-1 gene microsatellite polymorphism play a role? *Chinese medical journal*. Aug 5 2005;118: 1285-1290.

[152] Wijpkema JS, van Haelst PL, Monraats PS, Bruinenberg M, Zwinderman AH, Zijlstra F, van der Steege G, de Winter RJ, Doevendans PA, Waltenberger J, Jukema JW, Tio

RA. Restenosis after percutaneous coronary intervention is associated with the angiotensin-II type-1 receptor 1166A/C polymorphism but not with polymorphisms of angiotensin-converting enzyme, angiotensin-II receptor, angiotensinogen or heme oxygenase-1. *Pharmacogenet Genomics*. May 2006;16:331-337.

[153] Tiroch K, Koch W, von Beckerath N, Kastrati A, Schomig A. Heme oxygenase-1 gene promoter polymorphism and restenosis following coronary stenting. *European heart journal*. Apr 2007;28:968-973.

[154] Holweg CT, Balk AH, Uitterlinden AG, Niesters HG, Maat LP, Weimar W, Baan CC. Functional heme oxygenase-1 promoter polymorphism in relation to heart failure and cardiac transplantation. *J. Heart Lung Transplant*. Apr 2005;24:493-497.

[155] Ullrich R, Exner M, Schillinger M, Zuckermann A, Raith M, Dunkler D, Horvat R, Grimm M, Wagner O. Microsatellite polymorphism in the heme oxygenase-1 gene promoter and cardiac allograft vasculopathy. *J. Heart Lung Transplant*. Oct 2005;24: 1600-1605.

[156] Lin R, Fu W, Zhou W, Wang Y, Wang X, Huang W, Jin L. Association of heme oxygenase-1 gene polymorphisms with essential hypertension and blood pressure in the Chinese Han population. *Genetic testing and molecular biomarkers*. Jan-Feb 2011;15:23-28.

[157] Yun L, Xiaoli L, Qi Z, Laiyuan W, Xiangfeng L, Chong S, Jianfeng H, Shufeng C, Hongfan L, Gu D. Association of an intronic variant of the heme oxygenase-1 gene with hypertension in northern Chinese Han population. *Clin. Exp. Hypertens*. Oct 2009;31:534-543.

[158] Dick P, Schillinger M, Minar E, Mlekusch W, Amighi J, Sabeti S, Schlager O, Raith M, Endler G, Mannhalter C, Wagner O, Exner M. Haem oxygenase-1 genotype and cardiovascular adverse events in patients with peripheral artery disease. *European journal of clinical investigation*. Dec 2005;35:731-737.

[159] Funk M, Endler G, Schillinger M, Mustafa S, Hsieh K, Exner M, Lalouschek W, Mannhalter C, Wagner O. The effect of a promoter polymorphism in the heme oxygenase-1 gene on the risk of ischaemic cerebrovascular events: the influence of other vascular risk factors. *Thrombosis research*. 2004;113:217-223.

[160] Bai CH, Chen JR, Chiu HC, Chou CC, Chau LY, Pan WH. Shorter GT repeat polymorphism in the heme oxygenase-1 gene promoter has protective effect on ischemic stroke in dyslipidemia patients. *Journal of biomedical science*. 2010;17:12.

[161] Wu MM, Chiou HY, Lee TC, Chen CL, Hsu LI, Wang YH, Huang WL, Hsieh YC, Yang TY, Lee CY, Yip PK, Wang CH, Hsueh YM, Chen CJ. GT-repeat polymorphism in the heme oxygenase-1 gene promoter and the risk of carotid atherosclerosis related to arsenic exposure. *Journal of biomedical science*. 2010;17:70.

[162] Wu MM, Chiou HY, Chen CL, Wang YH, Hsieh YC, Lien LM, Lee TC, Chen CJ. GT-repeat polymorphism in the heme oxygenase-1 gene promoter is associated with cardiovascular mortality risk in an arsenic-exposed population in northeastern Taiwan. *Toxicology and applied pharmacology*. Nov 1 2010;248:226-233.

[163] Kanai M, Tanabe S, Okada M, Suzuki H, Niki T, Katsuura M, Akiba T, Hayasaka K. Polymorphisms of heme oxygenase-1 and bilirubin UDP-glucuronosyltransferase genes are not associated with Kawasaki disease susceptibility. *The Tohoku journal of experimental medicine*. Jul 2003;200:155-159.

[164] Abe T, Kakyo M, Tokui T, Nakagomi R, Nishio T, Nakai D, Nomura H, Unno M, Suzuki M, Naitoh T, Matsuno S, Yawo H. Identification of a novel gene family encoding human liver-specific organic anion transporter LST-1. *J. Biol. Chem.* Jun 11 1999;274:17159-17163.

[165] Lin R, Wang X, Zhou W, Fu W, Wang Y, Huang W, Jin L. Association of polymorphisms in the solute carrier organic anion transporter family member 1B1 gene with essential hypertension in the Uyghur population. *Annals of human genetics.* Mar 2011;75:305-311.

[166] Kivisto KT, Niemi M, Schaeffeler E, Pitkala K, Tilvis R, Fromm MF, Schwab M, Lang F, Eichelbaum M, Strandberg T. CYP3A5 genotype is associated with diagnosis of hypertension in elderly patients: data from the DEBATE Study. *Am. J. Pharmacogenomics.* 2005;5:191-195.

[167] Link E, Parish S, Armitage J, Bowman L, Heath S, Matsuda F, Gut I, Lathrop M, Collins R. SLCO1B1 variants and statin-induced myopathy--a genomewide study. *N. Engl. J. Med.* Aug 21 2008;359:789-799.

[168] Zhang W, Chen BL, Ozdemir V, He YJ, Zhou G, Peng DD, Deng S, Xie QY, Xie W, Xu LY, Wang LC, Fan L, Wang A, Zhou HH. SLCO1B1 521T-->C functional genetic polymorphism and lipid-lowering efficacy of multiple-dose pravastatin in Chinese coronary heart disease patients. *Br. J. Clin. Pharmacol.* Sep 2007;64:346-352.

[169] Gajdos V, Petit FM, Perret C, Mollet-Boudjemline A, Colin P, Capel L, Nicaud V, Evans A, Arveiler D, Parisot F, Francoual J, Genin E, Cambien F, Labrune P. Further evidence that the UGT1A1*28 allele is not associated with coronary heart disease: The ECTIM Study. *Clin. Chem.* Dec 2006;52:2313-2314.

[170] Hsieh CJ, Chen MJ, Liao YL, Liao TN. Polymorphisms of the uridine-diphosphoglucuronosyltransferase 1A1 gene and coronary artery disease. *Cellular and molecular biology letters.* 2008;13:1-10.

[171] Kubo M, Hata J, Ninomiya T, Matsuda K, Yonemoto K, Nakano T, Matsushita T, Yamazaki K, Ohnishi Y, Saito S, Kitazono T, Ibayashi S, Sueishi K, Iida M, Nakamura Y, Kiyohara Y. A nonsynonymous SNP in PRKCH (protein kinase C eta) increases the risk of cerebral infarction. *Nature genetics.* Feb 2007;39:212-217.

[172] Yamada Y, Fuku N, Tanaka M, Aoyagi Y, Sawabe M, Metoki N, Yoshida H, Satoh K, Kato K, Watanabe S, Nozawa Y, Hasegawa A, Kojima T. Identification of CELSR1 as a susceptibility gene for ischemic stroke in Japanese individuals by a genome-wide association study. *Atherosclerosis.* Nov 2009;207:144-149.

[173] Turner ST, Peyser PA, Kardia SL, Bielak LF, Sheedy PF, 3rd, Boerwinkle E, de Andrade M. Genomic loci with pleiotropic effects on coronary artery calcification. *Atherosclerosis.* Apr 2006;185:340-346.

[174] Ikram MA, Seshadri S, Bis JC, Fornage M, DeStefano AL, Aulchenko YS, Debette S, Lumley T, Folsom AR, van den Herik EG, Bos MJ, Beiser A, Cushman M, Launer LJ, Shahar E, Struchalin M, Du Y, Glazer NL, Rosamond WD, Rivadeneira F, Kelly-Hayes M, Lopez OL, Coresh J, Hofman A, DeCarli C, Heckbert SR, Koudstaal PJ, Yang Q, Smith NL, Kase CS, Rice K, Haritunians T, Roks G, de Kort PL, Taylor KD, de Lau LM, Oostra BA, Uitterlinden AG, Rotter JI, Boerwinkle E, Psaty BM, Mosley TH, van Duijn CM, Breteler MM, Longstreth WT, Jr., Wolf PA. Genomewide association studies of stroke. *N. Engl. J. Med.* Apr 23 2009;360:1718-1728.

[175] Hata J, Matsuda K, Ninomiya T, Yonemoto K, Matsushita T, Ohnishi Y, Saito S, Kitazono T, Ibayashi S, Iida M, Kiyohara Y, Nakamura Y, Kubo M. Functional SNP in an Sp1-binding site of AGTRL1 gene is associated with susceptibility to brain infarction. *Human molecular genetics*. Mar 15 2007;16:630-639.

[176] Pruissen DM, Kappelle LJ, Rosendaal FR, Algra A. Genetic association studies in ischaemic stroke: replication failure and prospects. *Cerebrovascular diseases (Basel, Switzerland)*. 2009;27:290-294.

[177] Lanktree MB, Dichgans M, Hegele RA. Advances in genomic analysis of stroke: what have we learned and where are we headed? *Stroke*. Apr 2010;41:825-832.

[178] Genome-wide association study of 14,000 cases of seven common diseases and 3,000 shared controls. *Nature*. Jun 7 2007;447:661-678.

[179] Erdmann J, Grosshennig A, Braund PS, Konig IR, Hengstenberg C, Hall AS, Linsel-Nitschke P, Kathiresan S, Wright B, Tregouet DA, Cambien F, Bruse P, Aherrahrou Z, Wagner AK, Stark K, Schwartz SM, Salomaa V, Elosua R, Melander O, Voight BF, O'Donnell CJ, Peltonen L, Siscovick DS, Altshuler D, Merlini PA, Peyvandi F, Bernardinelli L, Ardissino D, Schillert A, Blankenberg S, Zeller T, Wild P, Schwarz DF, Tiret L, Perret C, Schreiber S, El Mokhtari NE, Schafer A, Marz W, Renner W, Bugert P, Kluter H, Schrezenmeir J, Rubin D, Ball SG, Balmforth AJ, Wichmann HE, Meitinger T, Fischer M, Meisinger C, Baumert J, Peters A, Ouwehand WH, Deloukas P, Thompson JR, Ziegler A, Samani NJ, Schunkert H. New susceptibility locus for coronary artery disease on chromosome 3q22.3. *Nature genetics*. Mar 2009;41:280-282.

[180] Kathiresan S, Voight BF, Purcell S, Musunuru K, Ardissino D, Mannucci PM, Anand S, Engert JC, Samani NJ, Schunkert H, Erdmann J, Reilly MP, Rader DJ, Morgan T, Spertus JA, Stoll M, Girelli D, McKeown PP, Patterson CC, Siscovick DS, O'Donnell CJ, Elosua R, Peltonen L, Salomaa V, Schwartz SM, Melander O, Altshuler D, Ardissino D, Merlini PA, Berzuini C, Bernardinelli L, Peyvandi F, Tubaro M, Celli P, Ferrario M, Fetiveau R, Marziliano N, Casari G, Galli M, Ribichini F, Rossi M, Bernardi F, Zonzin P, Piazza A, Mannucci PM, Schwartz SM, Siscovick DS, Yee J, Friedlander Y, Elosua R, Marrugat J, Lucas G, Subirana I, Sala J, Ramos R, Kathiresan S, Meigs JB, Williams G, Nathan DM, MacRae CA, O'Donnell CJ, Salomaa V, Havulinna AS, Peltonen L, Melander O, Berglund G, Voight BF, Kathiresan S, Hirschhorn JN, Asselta R, Duga S, Spreafico M, Musunuru K, Daly MJ, Purcell S, Voight BF, Purcell S, Nemesh J, Korn JM, McCarroll SA, Schwartz SM, Yee J, Kathiresan S, Lucas G, Subirana I, Elosua R, Surti A, Guiducci C, Gianniny L, Mirel D, Parkin M, Burtt N, Gabriel SB, Samani NJ, Thompson JR, Braund PS, Wright BJ, Balmforth AJ, Ball SG, Hall AS, Schunkert H, Erdmann J, Linsel-Nitschke P, Lieb W, Ziegler A, Konig I, Hengstenberg C, Fischer M, Stark K, Grosshennig A, Preuss M, Wichmann HE, Schreiber S, Schunkert H, Samani NJ, Erdmann J, Ouwehand W, Hengstenberg C, Deloukas P, Scholz M, Cambien F, Reilly MP, Li M, Chen Z, Wilensky R, Matthai W, Qasim A, Hakonarson HH, Devaney J, Burnett MS, Pichard AD, Kent KM, Satler L, Lindsay JM, Waksman R, Knouff CW, Waterworth DM, Walker MC, Mooser V, Epstein SE, Rader DJ, Scheffold T, Berger K, Stoll M, Huge A, Girelli D, Martinelli N, Olivieri O, Corrocher R, Morgan T, Spertus JA, McKeown P, Patterson CC, Schunkert H, Erdmann E, Linsel-Nitschke P, Lieb W, Ziegler A, Konig IR, Hengstenberg C, Fischer M, Stark K, Grosshennig A, Preuss M, Wichmann HE, Schreiber S, Holm H, Thorleifsson G, Thorsteinsdottir U, Stefansson K, Engert JC, Do

R, Xie C, Anand S, Kathiresan S, Ardissino D, Mannucci PM, Siscovick D, O'Donnell CJ, Samani NJ, Melander O, Elosua R, Peltonen L, Salomaa V, Schwartz SM, Altshuler D. Genome-wide association of early-onset myocardial infarction with single nucleotide polymorphisms and copy number variants. *Nature genetics*. Mar 2009;41:334-341.

[181] Karvanen J, Silander K, Kee F, Tiret L, Salomaa V, Kuulasmaa K, Wiklund PG, Virtamo J, Saarela O, Perret C, Perola M, Peltonen L, Cambien F, Erdmann J, Samani NJ, Schunkert H, Evans A. The impact of newly identified loci on coronary heart disease, stroke and total mortality in the MORGAM prospective cohorts. *Genetic epidemiology*. Apr 2009;33:237-246.

[182] Samani NJ, Erdmann J, Hall AS, Hengstenberg C, Mangino M, Mayer B, Dixon RJ, Meitinger T, Braund P, Wichmann HE, Barrett JH, Konig IR, Stevens SE, Szymczak S, Tregouet DA, Iles MM, Pahlke F, Pollard H, Lieb W, Cambien F, Fischer M, Ouwehand W, Blankenberg S, Balmforth AJ, Baessler A, Ball SG, Strom TM, Braenne I, Gieger C, Deloukas P, Tobin MD, Ziegler A, Thompson JR, Schunkert H. Genomewide association analysis of coronary artery disease. *N. Engl. J. Med*. Aug 2 2007;357:443-453.

[183] Samani NJ, Deloukas P, Erdmann J, Hengstenberg C, Kuulasmaa K, McGinnis R, Schunkert H, Soranzo N, Thompson J, Tiret L, Ziegler A. Large scale association analysis of novel genetic loci for coronary artery disease. *Arterioscler. Thromb. Vasc. Biol*. May 2009;29:774-780.

[184] Helgadottir A, Thorleifsson G, Magnusson KP, Gretarsdottir S, Steinthorsdottir V, Manolescu A, Jones GT, Rinkel GJ, Blankensteijn JD, Ronkainen A, Jaaskelainen JE, Kyo Y, Lenk GM, Sakalihasan N, Kostulas K, Gottsater A, Flex A, Stefansson H, Hansen T, Andersen G, Weinsheimer S, Borch-Johnsen K, Jorgensen T, Shah SH, Quyyumi AA, Granger CB, Reilly MP, Austin H, Levey AI, Vaccarino V, Palsdottir E, Walters GB, Jonsdottir T, Snorradottir S, Magnusdottir D, Gudmundsson G, Ferrell RE, Sveinbjornsdottir S, Hernesniemi J, Niemela M, Limet R, Andersen K, Sigurdsson G, Benediktsson R, Verhoeven EL, Teijink JA, Grobbee DE, Rader DJ, Collier DA, Pedersen O, Pola R, Hillert J, Lindblad B, Valdimarsson EM, Magnadottir HB, Wijmenga C, Tromp G, Baas AF, Ruigrok YM, van Rij AM, Kuivaniemi H, Powell JT, Matthiasson SE, Gulcher JR, Thorgeirsson G, Kong A, Thorsteinsdottir U, Stefansson K. The same sequence variant on 9p21 associates with myocardial infarction, abdominal aortic aneurysm and intracranial aneurysm. *Nature genetics*. Feb 2008;40:217-224.

[185] Tregouet DA, Konig IR, Erdmann J, Munteanu A, Braund PS, Hall AS, Grosshennig A, Linsel-Nitschke P, Perret C, DeSuremain M, Meitinger T, Wright BJ, Preuss M, Balmforth AJ, Ball SG, Meisinger C, Germain C, Evans A, Arveiler D, Luc G, Ruidavets JB, Morrison C, van der Harst P, Schreiber S, Neureuther K, Schafer A, Bugert P, El Mokhtari NE, Schrezenmeir J, Stark K, Rubin D, Wichmann HE, Hengstenberg C, Ouwehand W, Ziegler A, Tiret L, Thompson JR, Cambien F, Schunkert H, Samani NJ. Genome-wide haplotype association study identifies the SLC22A3-LPAL2-LPA gene cluster as a risk locus for coronary artery disease. *Nature genetics*. Mar 2009;41:283-285.

In: Bilirubin: Chemistry, Regulation and Disorder
Editors: J. F. Novotny and F. Sedlacek
ISBN: 978-1-62100-911-5
© 2012 Nova Science Publishers, Inc.

Chapter III

Molecular Simulation of Bilirubin Transport and Its pK_a Values

Rok Borštnar[*]

National Institute of Chemistry, Slovenia, Hajdrihova, Ljubljana, Slovenia

Abstract

Nowadays there is a large increase of interest in interdisciplinary sciences that have the ability to interconnect and integrate many different fields of science which have in recent years become very specific, oriented, and focused only on a certain problem without the introduction and consideration of the accompanying features, and therefore they often seem to yield unreliable results, without any applicable value. At first sight this might be true, but when we take a closer look, and in conjunction with the results of other sciences, these results can make a lot of sense, which is why we have to take them into consideration when we want to study chemistry with molecular simulations. It is also very important in the design of novel drugs, because we have to understand and to know how to design and later to synthesize a drug in a way that it will be able to reach and have favorable interaction energies with the desired target. Since there are many "borders" (membranes) a drug has to cross on its journey to the final destination, if the transport mechanism through a certain biological membrane is understood, it is then much easier to design a new drug. With molecular simulations some new insights and useful directions can be given in order to simplify the synthesis of a novel drug, consequently resulting in the lab process being much easier and more efficiently planned, and in results being much more reliable at the end.

Hence we would like to present a brief review of what has already been done in the field of bilirubin research (transport, etc.) and give some new insights into the future of the molecular simulation of bilirubin transport and studies of its chemistry.

---

[*] E-mail: rok.borstnar@guest.arnes.si.

Preface

In our study we decided [1] that we would critically evaluate and calculate pK_a values with quantum mechanical methods and solvation models (Langevin Dipoles (LD) model [2-4] and Solvent Reaction Field (SCRF) of Tomasi and co-workers [5]) to calculate the free energies of solvation. At first we did the calculations for a few carboxylic acids, and then for the bilirubin molecule, which is one of the end products of the heme group metabolism of hemoglobin. In solution bilirubin co-exists in an equilibrium of three forms: diacid form, monoanionic and dianionic form. Bilitranslocase as an organic anion transporter (its 3-D structure is not known) participates in the transport of bilirubin through the hepatocytic cell membrane. However, the hepatic uptake mechanism is not yet completely understood. Thus, with calculations of pK_a values we contribute some new insights towards understanding the role of the protonation states of bilirubin, which are most probably changed during the transport process. Our calculations implied reasonable agreement with the experimental value of 4.4 for the first protonation step (5.5 (HF/6-31G(d)) and 6.5 (B3LYP/6-31*G(d,p)) in conjunction with the LD solvation model, which peformed better than SCRF. There were severe disagreements with the experimental value of 5.5 in the case of the second protonation step. The reason for that could lie in the fact that we have to compensate the bond energy of almost 400 kcal mol^{-1}. It was also evident that results very much depended on the flexibility of the basis set used in the calculations. This study is preliminary to all atom simulation of bilirubin transport by bilitranslocase, where protonation states are most probably changed.

Introduction

Bilirubin is the yellow or brownish yellow water insoluble bile pigment, one of the end products of the heme group catabolism of hemoglobin and other hematoproteins. The yellow color of bruises, yellowish discoloration of jaundice, and brownish color of feces are all "consequences" of bilirubin "activity". Moreover, bilirubin is produced in the process where the liver breaks down old red blood cells (source of hemoglobin), and is afterwards removed from the body through the stool (feces), which explains its brownish color.

The liver, besides its other functions, is also responsible for the removal of a variety of organic anions (including products of metabolism like bilirubin, bile salts etc.) from plasma, and their disposal into the bile. One of the most important hepatic functions for the digestive tract is bile secretion, which helps in lipid digestion and absorption, and is also a significant mechanism for the excretion of cholesterol and other organic pigments like bilirubin. Bile acids are steroid acids which make up about 65% of the dry weight of bile and are normally conjugated with glycine or taurine. Conjugated bile acids are amphipathic in nature. The entire bile acid pool is recirculated two or more times between the liver (where it is synthesized) and the small intestine (where it helps with fat digestion) in response to a typical meal, and this recirculation of bile is known as enterohepatic circulation. Besides bile acids, the bile contains phospholipids, cholesterol, proteins, bilirubin and related pigments [6].

In humans, bilirubin is produced at a rate of 300 mg/day, 80% of which comes from the degradation of the porphyrin moiety of hemoglobin of degrading erythrocytes. Bilirubin is a tetrapyrrole dicarboxylic acid consisting of two dipyrrole units with propionic side chains connected by a tetrahedral carbon atom. During catabolism of heme, heme oxygenase produces iron, carbon monoxide and biliverdine – a water-soluble compound. Biliverdin is a tetrapyrrole with a double bond between the second and the third pyrrole ring. The cytosolic enzyme NAD(P)H-biliverdin reductase reduces this double bond to a single bond, thus converting biliverdin to bilirubin. Although bilirubin is a tetrapyrrole like biliverdin, the conformations of the four pyrrole rings are different. Lately it has been recognized as a potentially cytotoxic agent, as a consequence of being a water-insoluble waste product. In the liver, uridine diphosphate (UDP)-glucuronyl transferase converts bilirubin to a mixture of monoglucuronides and diglucuronides, which are then secreted into the bile for excretion. When oxidized, bilirubin reverts back to biliverdin and is believed to act as a cellular antioxidant of physiological importance [7].

The normal total bilirubin concentration in blood serum is less than 1 mg/dL (17 µmol/liter). An elevated concentration of serum bilirubin results in a condition called Hyperbilirubinemia, which may be caused by the overproduction of bilirubin, impaired bilirubin uptake by the liver, abnormalities of bilirubin conjugation, hepatocellular diseases, impaired canalicular excretion, or biliary obstruction. It can also be caused by inherited conditions such as the Criglar-Najar, Gilbert's, Dubin-Johnson, or Rotor syndromes. With blood tests that determine the concentrations of total and direct bilirubin we are able to recognize patients who have Hyperbilirubinemia. Hyperbilirubinemia results in pathological conditions such as different kinds of jaundice [8,9].

If one takes a closer look at the pathway of bilirubin production, it can be seen that biliribun is a product of biliverdin reductase activity on biliverdin, the latter being the precursor of bilirubin constituted as a conjugate system (single-double-single-double-... bond), meaning that bilirubin is the reduced form of biliverdin. Bilirubin can be oxidized back into biliverdin, which is claimed to be the argument for the potent antioxidant activity of bilirubin, hence its proposed function is to be a cellular antioxidant, as well as a scavenger of the ultimate carcinogens [7,10].

In the beginning it was stated that bilirubin is insoluble in water, which is true for its unconjugated form called indirect or unconjugated bilirubin (UCB), circulating in the blood stream before entering the liver, where it is then transformed into the direct or conjugated bilirubin, which is soluble in water. When UCB is in solution it can co-exist as three forms in equilibrium – diacid (fully protonated), monoanionic and dianionic (deprotonated in both carboxylic groups) [11]. There are many speculations on which of these three forms bilirubin is transported through through the membrane of the liver. One of these speculations claims that it is transported through the membrane in free form [12], nevertheless the mechanism of bilirubin hepatic uptake is not yet completely understood despite the great effort that has been put into finding an answer to this question.

There is a wide spectrum of options regarding the hepatic uptake of bilirubin, however lately two general directions concerning the bilirubin transport mechanism have been proposed. One states that bilirubin can spontaneously diffuse through the phospholipids bilayer by the diffusion mechanism, which is an energy independent process [12]. The other leans towards the assumption that bilirubin hepatic uptake occurs against the concentration gradient, supported by the protein mediated active transport. It involves a transport protein which is energy dependent, meaning that an energy source is required, and in biochemical processes that is usually ATP [13,14].

In plasma, bilirubin is found in a complex with human serum albumin. The formation of the bilirubin-albumin complex is a reversible process with low dissociation constant, thus allowing the concentration of free bilirubin in plasma to be within the solubility limit of bilirubin in water at normal pH 7.4 [14]. The free form of bilirubin is believed to be the transported species and it is thus expected that any mechanism involved in its transport would tend to decrease the actual concentration of bilirubin in the vicinity of the hepatocytic membrane, favoring the dissociation of the bilirubin-albumin complex. Results from several studies claim that unconjugated bilirubin is able to diffuse spontaneously across the phospholipids bilayer via a "flip-flop" passive diffusion mechanism [15]. On the other hand, various *in vivo* and *in vitro* studies suggest hepatic bilirubin uptake is a saturable process occurring against the concentration gradient, and thus supporting the protein mediated active transport system [15]. Other analyses suggest that the uptake of bilirubin is concentrative, implying participation of the liver cell plasma membrane in the establishment of the concentration gradient and therefore suggest facilitated transport [16].

One of the carriers of organic anions isolated from the sinusoidal domain of liver cells is *bilitranslocase*. As the name already suggests, it functions as a bilirubin carrier across the hepatocytic membrane. The primary structure of bilitranslocase reveals no sequence homology with the known protein sequences. It was found that only the amino acid motif, identified to be the bilirubin-binding motif, is conserved in a number of α-phycocyanins. The 3-D structure of the protein is not known but it is proposed that it exists in two metastable forms with different affinities for the substrates that include bilirubin, nicotinic acid, sulfobromopthalein, a number of anthocyanins and other organic anions. The transport mechanism of bilirubin through bilitranslocase is not known in detail, hence for now it is only believed that the transport is electrogenic and requires a quasi-planar or planar structure of the substrate. It is suggested that the intramolecular hydrogen bond of bilirubin can be at least partially broken in order to attain a stretched conformation while it is transported by bilitranslocase. The hydroxyl groups provided by the serine residues in the bilitranslocase bilirubin-binding sequence may successfully compete for the establishment of a protein-pigment hydrogen bond network forcing bilirubin to loosen its ridge-tile conformation and attain an extended interaction with a carrier [14].

The presumed transport mechanism of bilirubin through bilitranslocase suggests that it is transported in anionic form with co-transportation of sodium cation. In this chapter, we decided to study and determine the pK_a values of bilirubin (in various protonation states) in order to be able to predict and to make a few reasonable assumptions about which form it is transported in and thus to shed some light on its transport mechanism. The pK_a values of the different forms of bilirubin (diacidic, monoanionic, and dianionic) can provide some information about the stability of the protonation states. If the pK_a value is low, suggesting the anionic form is stable enough, one could say that there is a great possibility for a bilirubin molecule to pass through the membrane in anionic form, and that proton transfer has occurred. However, when one makes such an assumption, one neglects many other factors involved in and contributing to the transport of bilirubin. Hence one has to be aware and take into consideration the fact that pK_a values are a very strong indicator of protonation states stability, which nevertheless can not explain the whole transport process of bilirubin on their own.

The pK_a of the two carboxylic acid groups of bilirubin is a debated issue, since a wide range of values from 4.4 to 9.3 measured with various techniques is reported. The value also depends on the type of solvent used. Many contradictory ideas are proposed in order to explain the experimental values. A few researchers believe that pK_a values are attributed to the effects of the intramolecular hydrogen bonds, such as the donation of a hydrogen bond from the –OH moiety of the –COOH group, hindered solvation of –COO⁻ group, and restricted rotation of –COOH and –COO⁻ groups [17]. These three factors arise due to the complicated hydrogen bonded structure of UCB that was argued to act in concerted way, causing the pK_a values of UCB to be much higher than those expected for propionic acid, for which none of the three factors is operative. On the other hand, most experiments have indicated the range of pK_a values between 4-5.5, which is in accordance with the expected values for aliphatic propionic acids. Contrary to the previous explanation for high pK_a values, the intramolecular hydrogen bonding is considered to have no influence on the pK_a values of UCB to explain the values close to those expected for propionic acid [15,16]. Here we tried to determine the pK_a values of UCB with computational methods, using water as a solvent.

The question here is why is there such great ambiguity regarding the mechanism of bilirubin hepatic uptake (transport mechanism through the liver membrane). One of the main reasons is that the 3-D structure of the transmembrane protein bilitranslocase, an organic anion transporter in the hepatocytic cell membrane, is not known. Actually it is the structure of its transmembrane region that is not known, since it is a very demanding task to isolate a transmembrane protein in its pure form, without causing any changes or damage to its transmembrane region during the isolation and crystallization process. Until the 3-D structure of bilirubin is clearly mapped out, there will be more or less only assumptions, speculations, and random guesses made about the possible transport mechanism. By knowing the chemistry of bilirubin we can gain insight and develop new directions towards resolving the problem. Molecular simulations are useful and helpful tools when dealing with such an issue, as they can save us a lot of money and energy. They design the process by calculating out the parameters for the ensuing experimental research and provide useful information beforehand.

Overview of the Bilirubin Production Process

If we want to understand the process of hepatic uptake of bilirubin, we should also dedicate a few lines to the description and discussion of its production from hemoglobin. As it has already been mentioned, bilirubin is one of the end products of the hemoglobin or more exactly heme catabolism. This might sound very straightforward, but in reality this process is much more complicated. Why? Because different types of enzymes are involved in this process, and when there are "systems" like this in action, one needs to know and to understand :

   a. how they work (the mechanism of enzyme catalytic reaction);
   b. what they are there for;
   c. what they actually do, what their mission is (are they selective or non-selective, specific, etc.).

As soon as the enzymes come onto the scene, their co-factors, which are usually situated at the enzyme active site where the enzyme catalytic reaction takes place, also need to be taken into consideration. They have one of the crucial roles for the enzyme catalytic step to occur, hence they require special attention. We can divide enzyme co-factors into two main groups or we can differentiate between two main groups of enzyme co-factors:

(a) inorganic (metals, etc.);
(b) organic (flavin, etc.).

It is hard to say which type of co-factors is predominant, since that depends on the environment in which the enzyme is situated. It is also important how a co-factor is connected to the rest of the enzyme: covalently (strong bonding) or non-covalently (weak bonding), since this has an influence on the flexibility of a co-factor during the catalytic reaction in the active site of an enzyme.

Now if we take a closer look at which enzymes and other compounds are involved in bilirubin production, we can see that:

1) In order to get biliverdin IX-α from heme, heme oxygenase (HO), NADPH and oxygen are required. Heme oxygenase catalyzes the degradation of heme, and besides biliverdin produces carbon monoxide (CO) and iron (Fe) with help from NADPH and oxygen [18].

The following equation represents the heme oxygenase catalytic reaction

$$\text{Heme} + \text{NADPH} + \text{H}^+ + 3\ O_2 \rightarrow \text{biliverdin} + \text{Fe}^{2+} + \text{CO} + \text{NADP}^+ + \text{H}_2\text{O} \quad [19]$$

This reaction actually represents the formation of a bruise, which in the macro world can be seen as changing the color of the skin during the gradual healing: red heme converts to green biliverdin, and the latter further on to yellow bilirubin. The highest activity of the heme monoxygenase was observed in the spleen, which is usually the source of old erytrocytes which are then sequestrated and "destroyed".

NADPH, actually NADPH-cytochrome P450 reductase serves as a source of electrons for the catalytic turnover of the heme monoxygenase in mammalian systems [20].

At this point it has to be emphasized that two isoforms of HO have been characterized so far:

a. HO-2 (constitutive form) – present under physiological conditions;
b. HO-1 (stress-induced form).

HO is a potential target for the control of hyperbilirubinemia in newborns and in a number of hereditary disorders [21].

2) In the next step biliverdin is converted to bilirubin with the enzyme biliverdin reductase (BVR), which reduces the biliverdin double-bond between the second and third pyrrol ring to a single bond, by using NADPH + H+ as an electron donor (Figure 1, Figure 2, Figure 3) [22]. Biliverdin dissociates from heme oxygenase (HO) by overlapping lysines and arginines as key residues of BVR [23].

Therefore, in order to understand the disorders involving changed concentration levels of bilirubin, one has to deal with the mechanisms of catalytic reactions in enzymes, since this is the only way a reasonable explanation can be given as to why there are different levels (concentrations) of bilirubin in different environments.

Baranano and co-workers showed with their experiments that bilirubin protects cells from a large excess of H2O2, which is known to be a strong oxidant responsible for the degradation of many cells, and whose harmful activity has to be stopped. Bilirubin could be one of the major physiologic antioxidant cytoprotectants [24]. Understanding the role of biliverdin reductase in the process of bilirubin production is paramount, just as it is also important to consider the possibility of reoxidation of bilirubin back to biliverdin (reverse reaction).

In the publication of Baranano and co-workers one very interesting question is raised: why is biliverdin reduced to bilirubin, which is known to be toxic and insoluble in water, while the former is water-soluble [24]? Answering this question requires an investigation of the differences in the chemical structures of biliverdin and bilirubin.

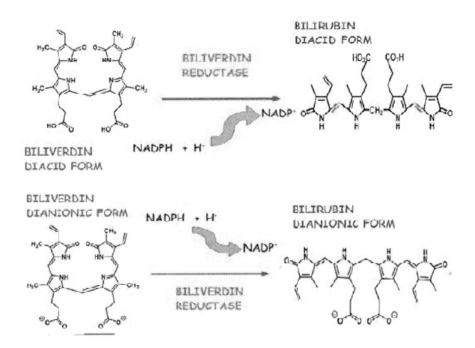

Figure 1. Conversion of biliverdin to bilirubin with biliverdin reductase. Up – diacid, down – dianionic.

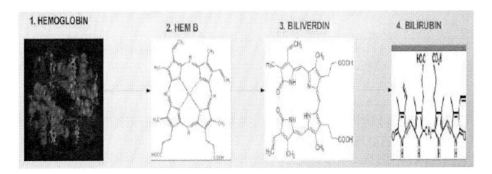

Figure 2. Scheme of hemoglobin degradation to bilirubin.

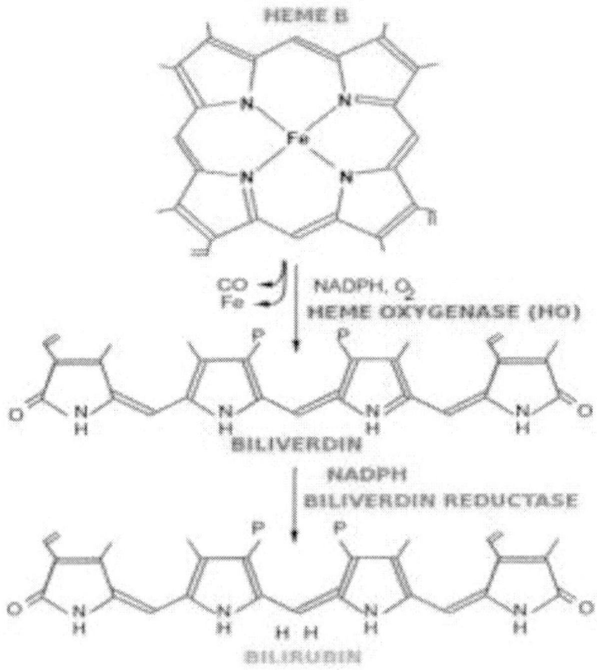

Figure 3. Scheme of Heme B catabolism process.

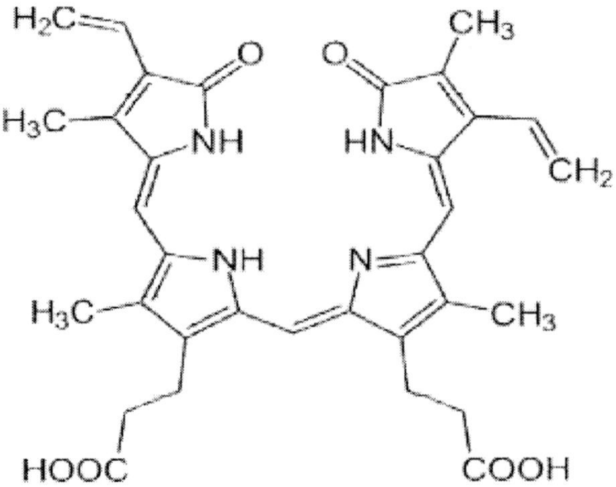

Figure 4. chemical structure/representation of biliverdin molecule.

The biliverdin molecule (Figure 4) has a type of conjugated system and is therefore very attractive to water molecules, while bilirubin loses its conjugated double bond between the second and third pyrrole rings (Figure 5), turning into a methylen group. This is much less reactive than a conjugated double bond, but on the other hand much more flexible than the double bond and more acceptable for rotations. Because bilirubin is insoluble in water when it is in its unconjugated form, it has to find a partner to turn into conjugated form. At this point it has to be clearly emphasized that here the term conjugated does not mean a system of single-double-single-double-..... bonds, but making a connection with a suitable partner.

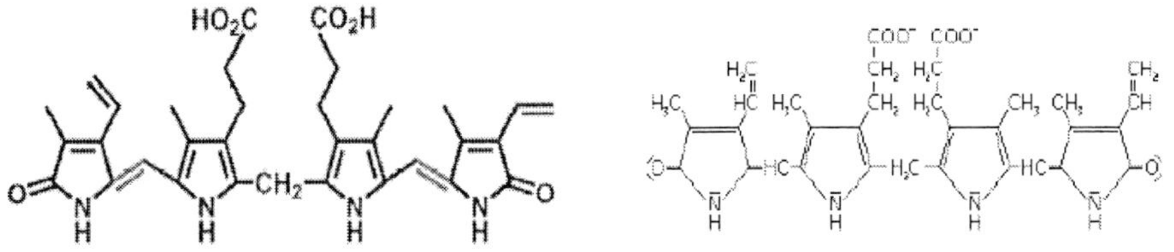

Figure 5. chemical structure/representation of bilirubin molecule. Left – fully protonated (diacid), right – deprotonated (dianionic form).

Again, there is a palette of questions arising, and some answers need to be given before moving on to the next step.

1) Who/What is a suitable partner for bilirubin to get conjugated with?
2) Where can a partner like that be found?
3) How much does bilirubin have to pay the new partner?
4) What are the costs of this partnership, how much does bilirubin have to invest in such a partnership?
5) Is one of the partners dominant in this relationship?
6) What is the payment currency?

Bilirubin is conjugated in the liver with glucuronic acid by involving the enzyme glucuronyltransferase, the latter catalyzing the glucuronidation reaction, where the glycosyl group from a UTP-sugar is added to a hydrophobic molecule [25].

There are many factors that determine the characteristics of bilirubin in certain environments. One of the key factors are the protonation states of bilirubin, which can be characterized by pK_a values, and critically depend on the environment. Our motivation was to study the protonation states of bilirubin in different environments. However, before we could do that, we had to critically examine and evaluate how different computational methods perform in aqueous solution for which experimental pK_a values are known.

If we look back at what has been said at the beginning and what has been written in the Introduction part of this review, we become aware of the fact that in solution unconjugated bilirubin (UCB) co-exists as three forms in equilibrium – diacidic, monoanionic and dianionic form [11]. These forms or the protonation states of bilirubin are defined by its two carboxylic groups. The pK_a value of these two carboxylic groups in water is still a debated issue, because several different experimental values are reported. Why is that? As it has been repeatedly emphasized, the pK_a value critically depends on the environment. Thus by changing the pH of the environment, the pK_a value consequently changes. From the results obtained it can also be concluded that the process of bilirubin deprotonation is a stepwise rather than a concerted procedure. Moreover, since we modeled the reaction of bilirubin deprotonation as a stepwise process, our calculations indicate the same thing. We assumed there is a first (abstract of the first proton from bilirubin) and a second (abstract of the second proton from bilirubin) protonation step. The experimental pK_a value of the first carboxylic group of bilirubin is reported in the 4.4 range, and for the second carboxylic group about 5.0 [15-17]. So far, these have been the most accurate measurements, carried out with a variety of NMR methods and techniques.

pH and pK$_a$

Since the terms pH and pK$_a$ often give rise to confusion (one is often mistaken for the other), the following lines are going to be dedicated to explaining what pH and pK$_a$ actually are. pH is defined as a measure of the acidity or basicity of an aqueous solution. In more detail, it is a measure of the concentration of free hydrogen ions (protons) in an aqueous solution. It was first introduced by Danish chemist Søren Peder Lauritz Sørensen in 1909 as pcH, where c represents the concentration of hydrogen ions in an aqueous solution. The negative exponent in the expression shows how the concentration of hydrogen ions in an aqueous solution correlates with the pcH which later became known as

$$\text{pH}: c_H = 10^{-pcH} \tag{1}$$

If this expression is then mathematically modified, pcH becomes the decadic logarithm of hydrogen ion concentration:

$$pcH = -\log c_H = -\log 10^{-pcH} \quad [26] \tag{2}$$

The actual meaning of "p" in pH is still a matter of debate, but one of the most common explanations is that it stands for power, whereas H represents hydrogen.

Later in history, pH became defined as the negative decadic logarithm of the activity of the hydrogen ion or the hydrogen ion activity in a solution:

$$pH = -\log_{10}(a_H+) \tag{3}$$

Activity for hydrogen ions can also be determined as the product of the activity coefficient (γ) and concentration of the hydrogen ions in the solution:

$$(a_H+) = \gamma \cdot (c_H+) \tag{4}$$

The activity coefficient is usually estimated by numerous theoretical approaches, and the expression most commonly used is the Debye-Huckel limiting law:

$$\log \gamma_+ = -A \mid (z_+)(z_-) \mid (I^{0.5})(1 + BI^{0.5})^{-1} \tag{5}$$

where A is a constant that depends on the properties of the solvent (0.5085 for water at 25°C), z_+ and z_- are the ionic charges, I is the ionic strength, and B is an empirical constant. Here, the activity coefficient is molal-based and is not a measurable quantity. Instead it can only be calculated by a well-defined and designed system of mathematical equations. Therefore, molecular simulations might be very attractive to calculate the activity coefficient, although an adequate and efficient ensemble must be chosen, meaning that the proper variables need to be kept constant. This is a very demanding task, hence the pH is not very suitable for molecular simulation. The activity of ions can be measured by electrochemical techniques, and pH is then expressed as the difference of the electrochemical potentials between an electrode where the change of activity of ions is measured, and a reference electrode.

On the other hand, pK_a values are not measured, but calculated as the negative decadic logarithm of an acid dissociation constant

$$(K_a): pK_a = -\log_{10}(K_a) \tag{6}$$

and that provides information about the strength of an acid. In comparison to pH, where the change in concentration of hydrogen ions in the solution is actually measured, in pK_a the equilibrium of the deprotonation or protonation reaction of an acid is studied, and this is how one can see whether the acid can dissociate easily (strong acid – release or abstraction of the proton is easy, almost without an energy barrier), or the dissociation of the acid has an energy barrier of a certain height, which prevents "free abstraction/release" of the proton (weak acid). Another noticeable difference between pH and pK_a values should also be pointed out: while pH values are divided into two classes on a scale from 0 to 7 (acids) and 7 to 14 (bases), pK_a values have no limitations, and can even be negative..

If pH values are not so attractive to be studied by molecular simulations, pK_a values are much more. They represent and offer a wide spectrum of studies that can be done in future with molecular simulations including various methods of computational chemistry.

Calculation of the pK_a Values

The experimental measurements of pK_a values are a demanding and difficult task, whereas calculating them is an even more challenging assignment, as the energy of the OH bond has to be traded for the favorable hydration energy of an anion and a hydronium ion. Protons act as payment currency here, and the question is how many protons bilirubin has to give to environmental partners if it wants to change its protonation states. Just as in economics we have models and equations for calculating how much money we have to pay someone, in the field of computational chemistry we have solvation models and many other different methods for the calculation of the pK_a values. Quantum chemistry offers a wide range of tools to build a proper computational model. The most important are of course the different levels of quantum chemical theories in conjunction with numerous different basis sets and solvation models. There are two well established procedures for the theoretical calculation of pK_a values:

1) calculation of absolute pK_a values;
2) calculation of relative pK_a values.

Although it is difficult to make a proper prediction of absolute pK_a values, we decided to calculate them for bilirubin. For calculating relative pK_a values an adequate reference is required, and one might argue that the reference was not chosen properly. Usually calculations of the relative pK_a values are performed relative to the acetate ion. We assumed that the reaction of deprotonation of bilirubin occurs in two protonation steps as follows:

I. abstraction of the first proton from the carboxylic group (Figure 6);
II. abstraction of the second proton from the other carboxylic group (Figure 6).

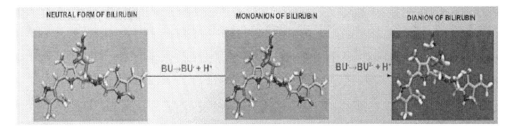

Figure 6. Reaction scheme of bilirubin deprotonation. All of the depicted geometries in the scheme were optimized at B3LYP/6-31+G(d,p) level of theory.

How did we design the model for the calculation of absolute pK_a values?

First we calculated *in vacuo* energies (full geometry optimizations) with inclusion of zero-point corrections that we obtained from the vibrational analysis, and then used the solvation model to calculate the free energy of solvation, which is a construct of the free energy of hydration contributions of a single species (Figure 7).

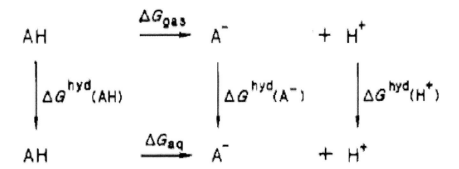

Figure 7. Model for calculating absolute pK_a values. ΔG_{gas} represents free energy *in vacuo*, ΔG_{aq} is free energy of solvation (free energy in the presence of solvent), and ΔG^{hydr} is free energy of hydration of a certain species. The values of free energies are commonly given in kcal mol^{-1}.

In the "world of experiments" *in vacuo* represents a blank sample (sample without the environment), and the solvation model represents the environment (when we put the sample into the reaction solution). Evaluating the hydration free energies calculated with different solvation models is just as important as *in vacuo* energetics, so we decided to use two well established solvation models: the Langevin dipoles (LD) model by Florian and Warshel [2-4] and the solvent reaction field (SCRF) of Tomasi and co-workers [5].

Langevin Dipoles Model (LD)

In the LD procedure, hydration free energy is calculated as a reversible work necessary to build a solute charge distribution in the field of dipoles on a cubic grid (Figure 8). The charge distribution is represented by Merz–Kollman atomic charges and thermal averaging is performed by changing the coordinate origin.

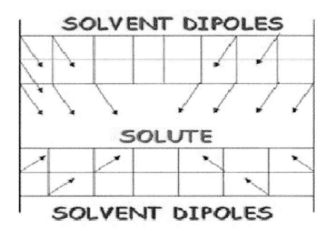

Figure 8. Representation of the Langevin dipoles model for the ethanol molecule.

Solvent Reaction Field (SCRF)

The SCRF method treats the solute embedded in a cavity of interlocking spheres and the solvent is treated as a dielectric continuum (Figure 9). In the applied version of SCRF the reaction field polarizes the electron wave function and the procedure is iterative.

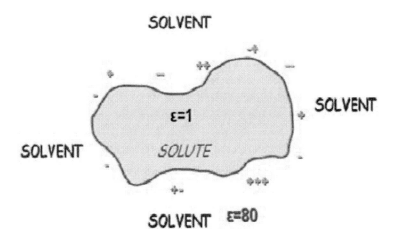

Figure 9. Solvent reaction field model representation.

Cui and Li presented a new approach for predicting the pK_a value for a specific residue in complex environments based on a combination of hybrid quantum mechanical/molecular mechanical (QM/MM) potential and the free energy perturbation (FEP) technique [27]. Cui and Riccardi used the SCC-DFTB/MM-GSBP for the pK_a analysis for the zinc-bound water in Human Carbonic Anhydrase II (CAII) based on the QM/MM approach, where SCC-DFTB (the standard second-order self-consistent charge density functional tight-binding approach) represents the QM region and GSBP is generalized solvent boundary potential. They were exploring applicability of this QM/MM approach as a framework for studying chemical reactions in biomolecules. They calculated the pK_a of the zinc-bound water and compared it with the experiments for the wild type CAII [28]. Besides the GSBP for the combined QM/MM simulations they also implemented the Ewald summation electrostatic treatments [29]. Combined QM/MM potential function was used to calculate the pK_a of residue 66 in *Staphylococal nuclease* in a thermodynamic integration approach in order to see how the behaviour in two mutants V66E and V66D of this bacteria varies [30].

Extended Description of Langevin Dipoles and Solvent Reaction Field Procedure

In the Langevin dipoles solvation model the solvent is approximated by polarizable dipoles fixed on a cubic grid [2-4], while in the SCRF method of Tomasi and co-workers we have a solute placed in a cavity (spherical), which is formed by a union of spheres centered on each atom [5]. This is one of the continuum methods called Polarizable Continuum Model (PCM), which includes the electrostatic interaction with the medium in the vicinity. The electro potential of the solute generates an "apparent surface charge" on the surface of the cavity in which the solute is placed [2-4]. In our calculations this surface is then divided into small tesserae, on which the charge and contributions to the gradient are evaluated. It is known that the solute charge density can penetrate outside the cavity and presents certain technical difficulty, which is here solved by renormalization, with the loss of the specific interaction between the solute and the solvent. There is no special attention dedicated to the entropy effect in this method. On the other hand, the LD model deals with the entropy as one of the main contributions to the ΔG_{hydr}, especially when we have charged solutes. It takes into consideration the possibility of entropy loss, when dipolar solvent molecules and charged solutes get close to each other [2-4]. From the tables we can see big differences between the SCRF method of Tomasi and co-workers and the LD model in ΔE. The LD model also includes the thermal average, while the SCRF does not.

As we have stated before, we used Merz-Kollman partial atomic charges encoded in the Gaussian03 [Gaussian] as an input for the LD model. In the Merz-Singh-Kollman (MK) scheme by U. C. Singh and P. A. Kollman [31], atomic charges are fitted to reproduce the molecular electrostatic potential (MEP) in a number of points around the molecule. As a first step of the fitting procedure, the MEP is calculated in a number of grid points located on several layers around the molecule. The layers are constructed as an overlay of van der Waals spheres around each atom. All points located inside the van der Waals volume are discarded. Best results are achieved by sampling points not too close to the van der Waals surface and the van der Waals radii are therefore modified through scaling factors. The smallest layer is obtained by scaling all radii with a factor of 1.4. The default MK scheme then adds three more layers constructed with scaling factors of 1.6, 1.8, and 2.0.

After the evaluation of the MEP in all valid grid points located on all four layers, atomic charges are derived reproducing the MEP as closely as possible. We can say that the MK procedure fits the ESP to the atomic charges.

The major difference is that once we have a solute in the cavity (SCRF), and a solute surrounded by envelopes forming a grid (MK) with the distances between the envelopes of 3 Å corresponding to the O-O distance, it results in more significant differences in ΔE.

As previously notified, the critical component is the hydration free energy of an anion and a hydronium ion, if looking at pK_a values relative to other Brønsted acids, since the error of hydration free energy for the hydronium ion is the same for all acids. Free energy of hydration can be evaluated by microscopic description of the solvent in conjunction with thermal averaging and one of the established methods for free energy calculations (thermodynamic perturbation, thermodynamic integration, particle insertion method). The drawback of microscopic description are the high computational costs.

In this chapter we applied the solvent reactive field (SCRF) of Tomasi and co-workers [5] and the Langevin dipoles (LD) of Florian and Warshel [2-4].

Since the free energy of proton hydration is difficult to calculate, consideration of its experimental value seems to be a reasonable strategy to evaluate the pK_a values.

Experimental data concerning the transfer of proton from the gas-phase to the aqueous solution very much depend on the method applied and the laboratory in which the experiment was conducted. The textbook of Marcus reports a value of 252.4 kcal mol^{-1} [32]. Karplus and co-workers suggested a value of 259.2 kcal mol^{-1} as an average of values obtained from measurements of standard hydrogen potentials in the range between 253.6 and 260.8 kcal mol^{-1} [33]. The most recent cluster-ion solvation data give a value of 264 kcal mol^{-1}, which is the value we adopted [34] when we were calculating pK_a values.

Now if we transform the quantum chemical terms into economic terms, we can conclude that the proton or protons are the legal payment currency in the world of bilirubin interactions with other biological molecules. The question which still remains is: how many protons does bilirubin have to pay a suitable partner for conjugation (glucuronic acid), for companionship (serum albumin), to enter the liver (lipid bilayer)? The answer certainly depends on the specific environment and its characteristics. Bilirubin is like a traveling salesman: it travels to and from the liver and on its way it is "forced" to make deals with "business partners". Of course bilirubin looks for business partners that would bring it certain advantages and benefits at a reasonable price. Changing protonation states in quantum chemistry seems to be an efficient way of doing business that way. Without a system of equations to evaluate the expected costs, not much can be done nowadays, because one would easily get embroiled in a business fraud. Even when a system like that exists it has to be presented and set up in a way that everyone can understand it, and consequently can not be manipulated. Therefore, we believe that our system of equation fulfills all of the conditions listed above.

For the calculation of free energies of solvation we used the following equation:

$$\Delta G_{solv} = -264 \text{ kcal mol}^{-1} + \Delta G^{hydr}(RCOO^-) - \Delta G^{hydr}(RCOOH) \qquad (7)$$

where -264 kcal mol^{-1} is the experimental value of the correction for proton hydration, which includes formation of the hydronium ion. $\Delta G^{hydr}(RCOO^-)$ is the free energy of hydration for anionic species and $\Delta G^{hydr}(RCOOH)$ for acidic species, respectively.

The free energy of the process ΔG was then calculated as the sum of contributions from the gas phase ΔE_{dry} and solution ΔG_{solv}:

$$\Delta G = \Delta E_{dry} + \Delta G_{solv} \qquad (8)$$

Finally we were able to calculate the pK_a of the proton transfer process:

$$pK_a = \frac{(\Delta G)}{(2.303\,RT)} \qquad (9)$$

where R denotes gas or the Raoult constant of 8.314 J mol^{-1} K^{-1} and T temperature in Kelvins, respectively.

This way we procured ourselves the proper tools to calculate pK_a values. We then carried out a test to establish whether this model was suitable for the calculation of the pK_a values, and we used it to calculate the pK_a values of a few carboxylic acids: formic, acetic, and nicotinic acids, for which the experimental pK_a values and free energies of hydration are well known. It revealed what we had suspected: the calculated hydration free energy of the hydronium ion is clearly not favorable enough, and therefore application of its experimental value for proton hydration free energy (corresponding to the formation of H_3O^+) of -264 kcal mol^{-1} [34] is fully justified.

The pK$_a$ Values Calculation Results

The results for formic (Figure 10), acetic (Figure 11) and nicotinic acid (Figure 12) were in quite good agreement with the experimental ones (results can be obtained from reference [1]). However, for bilirubin, being a much larger system than all of the former ones (Table 1, Figure 13), only the calculated pK_a values of the first protonation step were in reasonable agreement with the experimental value (Table 2, Figure 13), while for the second protonation step they were in severe disagreement with the experimental value (Table 3, Figure 13).

Because bilirubin is a large system, we limited our study to the Hartree-Fock level in conjunction with 6-31G(d) and B3LYP in conjunction with 6-31+G(d,p) basis sets. We considered both the LD and the SCRF solvation models. We are aware that application of more flexible basis sets or alternative accurate post-HF methods would be desirable but the size of the system did not allow for more accurate treatment.

Free energies of hydration for neutral bilirubin and both deprotonated forms are collected in Table 1. It is evident that the results significantly depend on the flexibility of the basis set. Addition of diffuse functions to heavy atoms and simultaneous addition of polarization functions to hydrogen atoms lowers hydration free energies of neutral bilirubin and of its singly charged anion. It is interesting that the opposite is observed for the doubly charged anion. In Table 2 are collected pK_a values for bilirubin of 5.5 (HF/6-31G(d)) and 6.5 (B3LYP/6-31+G(d,p)) in conjunction with the LD solvation model, while with the SCRF these values are 13.0 (HF/6-31G(d)) and 10.3 (B3LYP/6-31+G(d,p)), respectively.

However, the calculated pK_a values for the first protonation step are in reasonable agreement with the experimental value of 4.4 [15]. The calculated pK_a values for the second protonation step are in severe disagreement with the experimental value of 5.0 [15]. Namely, the HF and B3LYP methods do not even agree on the sign for the pK_a value, the differences between the LD and SCRF methodologies are large, and the absolute values are far from the experimental one. The reason might be that calculating the second pK_a value is much more demanding since we are compensating for the bond energy of almost 400 kcal mol^{-1} for hydration. There is still scope for improvement in the model.

Nevertheless, we obtained some valuable data and information about the protonation states of bilirubin which can help us explain its transport mechanism.

Figure 10. formic acid (left) and formiate (right). The atoms are colored as follows: red-oxygen, orange-carbon, white-hydrogen.

Figure 11. acetic acid (left) and acetate anion (right). For atom coloring see Figure 10.

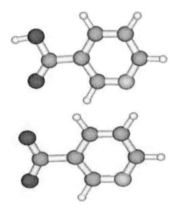

Figure 12. Structure of the nicotinic acid (top), nicotinate anion (bottom) optimized at the B3LYP/631+G(d,p) level of theory. The atoms are colored as follows: blue-nitrogen, red-oxygen, orange-carbon, white-hydrogen.

Table 1. Free Energies of Hydration in kcal mol^{-1} for Bilirubin (diacid (BU), monoanionic (BU$^-$) and dianionic (BU^{2-}) form)

Species	BU		BU$^-$		BU^{2-}	
Method	LD	SCRF	LD	SCRF	LD	SCRF
HF/6-31G(d)	-29.07	-36.52	-100.14	-97.31	-232.30	-247.03
B3LYP/6-31+G(d,p)	-34.81	-36.91	-106.82	-103.81	-208.97	-209.54

Table 2. pK_a calculation for bilirubin BU →BU⁻ +H⁺

Method	ΔEdry [kcal mol⁻¹][a]	ΔGsolv [kcal mol⁻¹][b]		ΔG [kcal mol⁻¹][c]		pK_a [d]	
Method		LD	SCRF	LD	SCRF	LD	SCRF
HF/6-31G(d)	342.59	-335.07	-324.79	7.52	17.8	5.5	13.0
B3LYP/6-31+G(d,p)	344.90	-336.01	-330.90	8.9	14.0	6.5	10.3

Table 3. BU⁻ → BU²⁻+H⁺

Method	ΔEdry [kcal mol⁻¹][a]	ΔGsolv [kcal mol⁻¹][b]		ΔG [kcal mol⁻¹][c]		pK_a [d]	
Method		LD	SCRF	LD	SCRF	LD	SCRF
HF/6-31G(d)	370.69	-396.16	-413.72	-25.47	-43.03	-18.7	-31.5
B3LYP/6-31+G(d,p)	397.24	-366.15	-369.73	31.09	27.51	22.8	20.2

[a] Dry energy difference with zero-point energy correction.
[b] Free energy of solvation provided by the Langevin dipoles (LD) method and solvent reaction field (SCRF) [41].
ΔG_{solv}= -264 kcal mol⁻¹ + ΔG^{hydr}(RCOO⁻) - ΔG^{hydr}(RCOOH).
[c] $\Delta G = \Delta E_{dry} + \Delta G_{solv}$.
[d] pK_a value.
LD- Langevin dipoles.

Experimentally obtained results suggest there is a co-transport with sodium [35], meaning that bilirubin passes through the membrane in anionic form along with the sodium cations. If this is true the question then is which of the two anionic forms, co-existing in an equilibrium along with the diacid form, dominates the other and which is transported through a membrane in the company of sodium cations?

Experimental pK_a Measurements

In the work of Boiadjiev and co-workers are presented results of measurements they carried out with the ^{1}H, ^{13}C NMR and absorbance spectroscopy in CDCl$_3$ and DMSO-d$_6$ with the water soluble conjugate of bilirubin to which poly(ethylene glycol) thiol (MPEG-SH) was added regiospecifically to its *exo* vinyl group to determine the pK_a values of bilirubin. Their experiments confirmed what had been previously implied, that the pK_a values of bilirubin are in the range of pK_a values for carboxylic acids [15]

The cause for the occurrence of high pK_a values is suggested to be the artifacts of aggregation or inaccuracy of a technique.

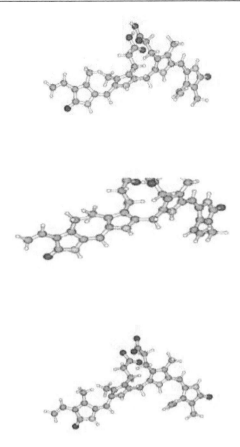

Figure 13. Structures of bilirubin: neutral form (top), monoanion (middle), dianion (bottom) optimized at the B3LYP/6-31+G(d,p) level of theory. The atoms are colored as follows: blue-nitrogen, red-oxygen, orange-carbon, white-hydrogen.

With the ^{13}C NMR spectroscopy measurements it was shown that the signals belonging to carboxylic acids respond differently to the change of the protonation state (ionization), accompanied by high sensitivity. The chemical shifts in the ^{13}C NMR can be thus measured much more accurately [36], excluding the possibility that the pK_a values of two carboxylic groups in bilirubin have a high value. In literature there are still reports of high pK_a values without considering them as a consequence or artifacts obtained from aggregation, or inaccuracy of the technique used [37-39]. Bilirubin is usually represented as a linear tetrapyrrole [40], however Boiadjiev and co-wokers suggested that bilirubin adopts the structure of a half-opened book, being common for crystals of bilirubin [41,42] stabilized by an intramolecular hydrogen bond. The latter finding is very interesting: in our calculations the bilirubin molecule was not rigid, it slightly folded, while still no intramolecular hydrogen bonds were observed for the fully protonated (diacid) form of bilirubin; this was the case even when the molecule was surrounded by the solvent molecules, while in the dianionic form each of the two deprotonated carboxylic groups formed one intramolecular hydrogen bond with the hydrogen attached to the nitrogen of the neighbouring pyrrole ring (Figure 14).

Nevertheless, the phenomenon of intramolecular hydrogen bonding in bilirubin is not an entirely new idea, and was usually used as an argument to explain why the measured pK_a values of bilirubin are as high or higher than expected [43-44].

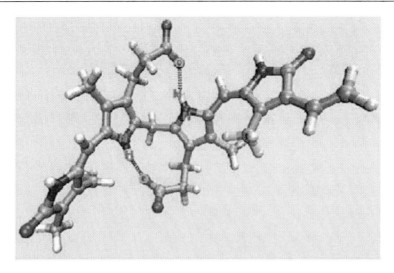

Figure 14. Representation of the intramolecular hydrogen bonds formed during the dianion optimization.

In order to obtain more accurate results from the pK_a calculations, membranes and protein environment have to be included. Experiments have shown that they have a great influence on the pK_a values, since they depend on the effects of aggregation and accompanying processes taking place in the different parts of the human body. The effects of aggregation should also be very carefully investigated and evaluated through molecular simulations, not only the aggregation effects alone, but their influence on the pK_a values as well in order to see whether they cause the increase or decrease of the pK_a values.

Experiments suggest that the monomeric bilirubin has normal mean pK_a values, while the aggregation causes the pK_a values of bilirubin to increase [15,16]. Moreover, it is necessary to point out that the existence of intramolecular hydrogen bonding would not have the same effect on all of the bilirubin protonation states, because it strongly depends on the level of the ionization, which is another reason why the determination and detection of the protonation states of bilirubin is so important. Furthermore, the effects of different solvents on the protonation states of bilirubin should be more carefully studied, and a complete computational study has to be carried out considering and including all of the relevant experimental data and facts known so far.

For instance, the structure of bilirubin in the dimethyl formamide (DMF) [45] is still not completely understood, and molecular simulations have proved to be a great supportive tool in cases like this. They offer almost cost free help to provide answers to such questions or uncertainties arising during experimental studies. When bilirubin is in the presence of the DMSO, which acts as a solvent, it would be interesting to calculate how much of the intramolecular hydrogen bond in bilirubin, expressed as a percentage, is conserved in comparison to the intermolecular hydrogen bonds formed between the bilirubin and DMSO. These might also be formed, as the DMSO is known as a polar aprotic solvent that can be a hydrogen acceptor or in the presence of really strong bases (pK_a > 35) as a hydrogen donor. Its methyl groups can then be deprotonated (pK_a ~ 35). Moreover, three different laboratories reported similar pK_a values in the DMF, DMSO and H_2O/DMSO solvent systems, suggesting that the interference of an organic solvent does not play a major role in the intramolecular hydrogen bonding in bilirubin [16,46,47].

Bilirubin Transport – Experimental and Molecular Simulations Aspects

Results of pK_a calculations for bilirubin obtained with the Density Functional Theory (DFT) tend more towards the dianionic form, as it resembles the α-chain more than it does the monoanionic form, whereas with the Hartree-Fock (HF) method nothing can be said for certain because of the H-bond formation in the dianion. If it is true that bilirubin passes through the membrane along with the sodium cations, that would mean that the dielectric constant ε inside the membrane region (intermembrane space) of the transporter protein (bilitranslocase) should be smaller than the one in the extracellular environment, which is basically an aqueous medium with ε≈80. That allows the existence of ions, leading to the conditions of electro-neutrality being fulfilled. The concentration of sodium cations in the extracellular environment is 140mM, in cytoplasm 12mM, whereas the potassium cations concentrations are 4mM (extracellular) and 139mM (cytoplasm), respectively. Therefore, one could suggest that bilirubin is co-transported with sodium cations from an extracellular environment into the cell, working as a Lewis base, which can donate electron pairs to a Lewis acid (sodium and potassium cations). However, at this point one may argue about the former statement, because when the discussion is about Lewis acids and Lewis bases it needs to be known whether one has to deal with a hard or soft Lewis acid and Lewis base. Sodium cations are known to be hard Lewis acids with stable low energetic HOMO orbitals and big energy difference between HOMO and LUMO orbitals.

Therefore, for an effective interaction, a hard Lewis base is needed, while on the other hand very weak covalent and ionic bonds are formed when a hard Lewis acid and a soft Lewis base interact. If conjugated bonds are present in the molecule, it comes to the phenomenon of perturbation of HOMO and LUMO orbitals, meaning that when one has a free electron pair in the p-orbital (electron donor) new HOMO and LUMO orbitals can be formed, being energetically higher than the ones in an ordinary carbon-carbon double bond (C=C) without the single electron pair. In this case the system becomes more nucleophilic and more reactive. When a conjugated double bond with an atom having an empty p-orbital (electron acceptor) is present, HOMO and LUMO orbitals have lower energies than the HOMO and LUMO orbitals of the C=C bond. The system becomes more electrophilic, and consequently more reactive. However, in the bilirubin there are many conjugated double bonds present outside and inside the pyrrole rings along with all of the accompanying ionic charges that can be formed as a response to environmental influences and changes. The pK_a values calculated [in this work] and experimentally determined [15-17] have led to the assumption that bilirubin is, as a weak acid, somewhere in the range of the acetic acid.

When bilirubin is deprotonated a negative charge is formed on the carboxylic group, which is then distributed between both oxygens (resonance - mesomeric effect) and it behaves as a Lewis base. What kind of Lewis base is it, hard or soft? One could immediately say it reacts as a hard Lewis base, because sodium cations are hard Lewis acids, but is it really so simple? It is rather dangerous to draw such conclusions. If we take a closer look at the whole bilirubin molecule, we can see there are methyl groups attached to the pyrrole rings, known in organic chemistry as electron donating groups, further on accompanied by the system of conjugated C=C bonds, acting as nucleophils when interacting with cations. Hence we should ask ourselves what kind of effects the conjugated system of double bonds has on the deprotonated carboxylic groups. Further experimental and molecular simulation studies that could answer this question must be carried out.

The UCB diacid isomer with minimum energy in nature is the IX$_\alpha$-Z,Z isomer which is a bis-lactum. The molecule is folded with an angle of about 100° with the two dipyrrinone units in two planes and the angle at the C$_{10}$ tetrahedral carbon atom [16] taking a ridge-tile conformation. Each carboxyl group forms three hydrogen bonds with lactam oxygen and ring nitrogen but the conformation is not dependent on the internal hydrogen bonds, it is only stabilized by them. Computational analysis has revealed that the non-bonded steric repulsion between the propionic acids and lactame rings favors the ridge-tile conformation even in the absence of hydrogen bonding. Involvement and inclusion of polar groups in hydrogen bonds account for the water insolubility of UCB. Two enantiomers of this conformation exist, and are interconvertible at room temperature.

In plasma, bilirubin is found in a complex with human serum albumin. The formation of the bilirubin-albumin complex is a reversible process with a low dissociation constant thus allowing the concentration of free bilirubin in plasma to be within the limit of solubility of bilirubin in water at normal pH 7.4 [16]. The free form of bilirubin is believed to be the transported species, and thus it is expected that any mechanism involved in its transport would tend to decrease the actual concentration of bilirubin in the vicinity of the hepatocyte membrane, favoring the dissociation of the bilirubin-albumin complex. However, the mechanism of hepatic uptake of bilirubin is still not entirely understood. Results of several studies suggest that unconjugated bilirubin has the ability to diffuse spontaneously across the phospholipids bilayer through a "flip-flop" passive diffusion mechanism [15]. On other hand, various *in vivo* and *in vitro* studies suggest that hepatic bilirubin uptake may be a saturable process occurring against the concentration gradient, thus supporting the protein mediated active transport system [15]. Other analyses suggest that the uptake of bilirubin is concentrative, implying the participation of the liver cell plasma membrane in the establishment of the concentration gradient and the facilitated transport [15].

Most authors claim that bilitranslocase works by co-transport with Na$^+$ ions and that nicotinic acid is the smallest anion which binds to it so far, suggesting there is a bilirubin-nicotinate binding site [48]. Its physiological function is assumed to mediate the facilitated diffusion of an organic anion from the blood into the liver since the negative membrane potential opposition bromosulphthalein anion is taken up from the blood into the cell [49]. Nevertheless, *in vivo* experiments using rats have shown that the purified bilitranslocase reconstitutes the electrogenic sulphobromophthalein (BSP) transport as a uniport type without being stimulated by ATP [50,51]. It is also assumed that bilitranslocase can occur in low-affinity and high-affinity state, these two forms co-existing in a sort of equilibrium modulated by the substrate concentration. The substrates may accelerate conversion from low-affinity to high-affinity state [50]. At low substrate concentrations a fast, exponential step, typical of the carrier-mediated process was evident, where at the higher concentrations of substrate it resembled a slow, linear, passive diffusion. With the proteins isolated from the basolateral plasma-membrane domain of the hepatocyte it has been shown that bilitranslocase is a higher affinity electrogenic transporting system of organic ions, operating at low substrate concentrations, whereas the bilirubin-binding protein is a lower affinity electroneutral transporter, taking part in the higher substrate concentration [52,53]. Here it is important to emphasize that electroneutral transport was observed in the absence of the transmembrane potassium gradient [52]. It was suspected that the overall activity of bilitranslocase is regulated by allosteric effectors, as a result of intracellular metabolism. The allosteric equilibrium of bilitranslocase was shown to be modulated by the redox equilibrium of the nicotinamide nucleotides. Under physiological conditions a low NADH/NAD$^+$ ratio favors the low-affinity state of bilitranslocase, thus increasing the NADH concentration in the cytoplasm and activating bilitranslocase. When lowering the NAD$^+$/NADH ratio, the bilirubin concentration in the cell medium decreases by about 40% at 37 °C, but not at 0 °C, which could be a possible effect of temperature dependent uptake into the cell monolayer [54].

It was also suggested that bilitranslocase is a carrier which mediates the bilirubin transport from blood to the liver in humans and rats [53,54] and its homolog mediating anthocyanin transport from cytoplasm into the central vacuole of the carnation petal [52]. As the bilirubin uptake was completely blocked when 1μM of nicotinic acid was added to HepG2 cells, it was assumed that bilirubin and nicotinic acid share a common binding site on the carrier. It was assumed this is a competitive process going on between them, meaning that if the addition of 1μM of nicotinic acid blocked the bilirubin uptake, then this is a saturating concentration of nicotinic acid for bilitranslocase [54].

Several experiments have been conducted, showing that the bilitranslocase can exist in two metastable forms which differ in affinities for different substrates whether or not they are electrogenic [55]. This means that one form has high, whereas the other one has low affinity for the substrate. When we speak about allostery, we have to consider and think about cooperativity and non-cooperativity. However, as long as we do not know what kind of subunits the bilitranslocase has, we can not talk about that in details. Therefore, the presence of the allosteric activators is possible. Allosteric activators are effectors that enhance the protein's activity, whereas those that decrease the activity of the protein are known as allosteric inhibitors. Neither of the allosteric effectors binds to the active site of the protein, but to a site other than the active site. Currently in pharmacology allosteric modulators are widely used as drugs for specific protein receptors interactions. Allosteric modulators generally bind to the protein in the cell membrane through the non-competitive mechanism, affecting the signal transduction after ligand binds to the active site. Two types of allosteric modulators are known on the basis of how they affect the transduction signal. Negative allosteric modulators (NAM) decrease the signal, and positive allosteric modulators (PAM) increase it. The future perspective in the studies of the allostery phenomenon is to have an allosteric agonist and an allosteric antagonist. However, as long as we do not know the exact structure of the bilitranslocase, we can not say anything about which region substrates like bilirubin bind to and consequently about how the conformation of the bilitranslocase changes during the bilirubin uptake.

It is worth dedicating a few lines to discussing transport numbers, ion mobility, and conductivity of strong and weak electrolytes. It is important how many water molecules can surround and organize themselves around a certain neutral molecule, ion, etc., and to take into consideration the size of the formed hydration shell, as it has a notable effect on the hydration free energies. In the field of computational chemistry a lot of time has been used to explain how many hydration shells water molecules can form, and still no conclusive answer has been provided. However, when one plans to talk about which of the species present in a certain environment, for instance ions, travels faster through the ionic channel, etc., special attention needs to be paid before drawing any conclusion. It can not and should not be logically assumed that smaller ions pass through the channels faster than bigger ones. It is the radius size of a certain ion that plays a major role here, as this is one of the decisive factors regarding ion mobility. It might happen that a smaller ion has a bigger hydration shell or even more hydration shells than a bigger ion, because more water molecules can form a stronger net around a smaller ion than around a bigger ion. When this is true, the bigger ion moves faster than the smaller one. In order to be sure which of the effects dominates, it is good to measure the conductivity of the species in water solution and the transport numbers to see whether we have a weak or strong (a) electrolyte, and to obtain information about the electric current density. The transport number provides information about how much of the total charge is transported or transfered from point A to point B by a certain ionic species. Ionic species with higher mobility can transfer more electricity through a solution than the ones with lower mobility, because mobility is in correlation with velocity (v_i) and the electric field potential (E) of ionic species:

$$v_i = u_i \frac{E}{l} \tag{10}$$

where l is the distance between two electrodes. What is usually measured is how much of an electric (ionic) current is transferred between two electrodes (cathode and anode):

$$i_i = \frac{EF}{l} z_i c_i u_i \tag{11}$$

where F is the Faraday constant of 96487 As mol^{-1}, z_i is the charge of an ionic species, c_i is the concentration of ionic species in a solution, and finally i_i is the amount of (electric) current (the electric current density) transferred between the cathode and anode or vice versa, depending on the type of ionic species (anion or cation). The total amount of current transferred between the electrodes is:

$$i = \sum_i i_i = \frac{EF}{l} \sum_i z_i c_i u_i \tag{12}$$

In general, the mobilities of cations and anions are not the same, hence the following needs to be considered,

$$i_+ \neq i_- \tag{13}$$

For strong electrolytes, e.g. strong electrolyte of type z-z (1:1) the following simplifications can be made: $z_+ = z_- = z$, and $c_+ = c_- = c$. Therefore, when one wants to calculate the transport number of a species, the following procedure should be considered:

a) calculation of the total transport number (transport number of all ionic species present in the solution), which is equal to 1:

$$\sum_i t_i = \sum_i \frac{i_i}{\sum_i i_i} \tag{14}$$

b) the transport number of a certain species is:

$$\sum_i t_i = \frac{i_i}{\sum_i i_i} \qquad (15)$$

c) when we have strong electrolytes:

$z_+c_+ = z_-c_-$, and

$t_+ = i_+/i = u_+/(u_+ + u_-)$, for cations,

$t_- = i_-/i = u_-/(u_+ + u_-)$, for anions.

This is valid only in the presence of strong electrolytes that are fully dissociated and do not form complex ions.

One of the easiest techniques for the determination of a transport number is the moving border method (a colorized border between two liquids or solutions shows the movement of ionic species), where the resistance and the time of border movement is measured. At the end one can calculate the transport number for cation and anion, while their velocity, and consequently the time and direction of border movement between two solutions changes during the process. If for instance a small positive ion diffuses behind the border, it comes to an environment with higher electrical field potential, whereas if a larger negative ion diffuses in front of the border, it is slowed down, because it comes into an area of lower electrical field potential. This process is called and known as the "regulatory effect". Hence solutions to carry out such an experiment should be carefully chosen and prepared with sufficient accuracy, so that the border moves in the direction of a cation with higher mobility.

It is thus much more convenient to measure the ionic conductivity of a solution, because again two different electrodes are used, only this time the specific conductivity or more exactly the specific resistance is measured. This is very important when the specific conductivity of a weak electrolyte has to be measured, because bilirubin most likely belongs to the group of weak electrolytes, while its pK_a values are in the range of carboxylic acids, which are known as weak electrolytes.

When weak electrolytes such as bilirubin are present in a solution, things can get much more complicated. Since they do not dissociate completely, transport numbers can not be determined so easily, because several mathematical approximations have to be made. Nevertheless, some information can be obtained by measuring the ionic conductivity of weak electrolytes, where the Ostwald law is valid as long as the concentration is so low that the interionic forces can be neglected. In that point the activities of certain components are equal to concentrations, and mobilities of the ions are equal to mobilities at infinite dilution.

How could this be used in the case of bilirubin? As it is already well known, bilirubin has two carboxylic groups with pK_a values in the range of common carboxylic acids, which are known as weak electrolytes. Bilirubin might be included in the same group, hence the best solution would be to first measure the ionic conductivity, and afterwards transport numbers, ion mobility, etc. can be determined. This way some insight could be given into how fast bilirubin moves (mobility of bilirubin). Moreover, its radius could be calculated, and some useful information could be obtained by combining them with pK_a values, blood test results, etc. One would be able to tell something about the protonation states of bilirubin, since the mobility changes along with the protonation states, and the hydration shell size also adjusts to circumstances. Overall, ionic conductivity measurements are very useful for defining the stability of ionic species, especially in the aqueous solution, and in the example given for bilirubin they could be a very useful supplement in determining the protonation states of bilirubin in an aqueous solution.

We will therefore stop at this point and save any further discussions for future articles on this theme, which shows a great perspective to work on.

Function of Na/K-ATPase and Possible Involvement in the Bilirubin Transport

Usually when Na^+ transport through the membrane is studied, one must also have in mind and consider the existence of Na/K-ATPase as an ion-translocating ATPase. This is a transmembrane protein with two subunits, α and β, forming a dimer that translocates 3 Na^+ ions from the cytoplasm to the cell exterior, and 2 K^+ ions from the cell exterior to the cytoplasm. It cleaves one ATP to ADP and Pi during the transport process. α and β subunits have hydrophobic sequences embedded in the lipid bilayer and hydrophilic sequences on both sides of the membrane.

There are two more types of Na/K-ATPase: the H^+-ATPase and Calcium-ATPase [55]. Although one could say that H^+-ATPase plays a major role in the bilirubin transport, one should also pay great attention to the location where H^+-ATPase is usually situated (endosomes, lysosomes, mitochondria, plasma membranes of certain epithelial cells).

The ion-translocating ATPase pumps perform active transport, meaning that they "offer" help to solute molecules so that they are able to to move against a concentration gradient by an energy-dependent mechanism. There are two kinds of active transport performed by these types of ion pumps: primary and secondary active transport. The former moves a solute against a concentration gradient, being directly powered by the hydrolysis of ATP, while the latter has no directly hydrolyzed ATP in the process. Instead, it uses the potential energy produced by the ion-translocating ATPases in the way that one solute moves down a gradient, while the other moves up a gradient [55].

This seems to fit very well with what we have written about the bilirubin transport. The experimental data [48,51-54] suggest co-transport, which is actually a secondary active transport that does not receive energy from the ATP hydrolysis, but directly from the Na^+ gradient. The question arising here is what causes the donation of and what donates the protons to bilirubin once it arrives into the cell if the assumptions about how it is co-transported with sodium cations are valid. Perhaps this is the moment when the H^+-ATPase comes into play and pumps the protons up into the cell (H^+ : 40nM extracellular, 100nM cytoplasm) during the ATP hydrolysis.

ATP is recognized here as a mediator between the transmembrane protein and the molecule that wants to go through the membrane. Just as a towboat, it signals to the protein when is the right time to open the door on the other side, and close it on the incoming side (entrance side) in order to maintain a certain level of membrane potential. If this level of membrane potential is broken down, meaning that the transmembrane protein can no longer distinguish among the different signals it receives, in the macro world this can cause severe disorders manifested as a disease, which is a response to unfavorable conditions. However, if one takes a closer look at the figure below (Figure 15) showing how Na/K-ATPase works we can see that it predicts the closure of the cytoplasm site of the transporter with the phosphate molecule formed during the process of hydrolysis of the ATP molecule (ATP + $H_2O \rightarrow$ ADP + P_i), consequently opening the extracellular side.

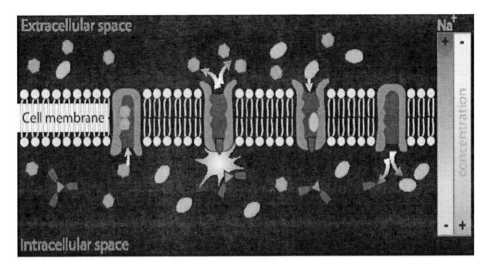

Figure 15. Na+/K+ pump, sodium-potassium pump [56].

The missing part of this figure is the one that would show whether or not the protein turns around or if there is any conformational change in it. It seems that when potassium cations want to enter the cytoplasmic side, they have to push away the phosphate molecule and they would need a certain amount of energy to do that. Therefore, one could ask where this energy came or was taken from. Yet another problem that arises here is the one associated with hydrophilicity and hydrophobicity. Since transmembrane proteins are known to have their hydrophilic part facing the extracellular environment and cytoplasm, whereas the hydrophobic part can be found inside the membrane, the turnaround process would be very unfavorable given that the hydrophilic part of the protein would have to pass through the hydrophobic part.

It is more likely for a transmembrane protein, bilitranslocase in our case, to be acting as a transporter that receives a signal from the molecule which shows an interest in going through the membrane, or from another molecule acting as a mediator (ATP,...) which needs help to make this possible. So it opens the door on the side where it receives the signal and helps the molecule to come through with the assistance of ions (sodium, potassium cations).

The complete structure of bilitranslocase is still not known, therefore we can only assume how transport processes assisted by it really occur. Nowadays it is suggested that this is more a carrier-mediated transport than the ATP driven process. For any further discussion one would need to know the complete 3-D structure of this transmembrane protein, so that the complete comparison between experimental results and molecular simulation results would be possible. Here it should be pointed out that molecular simulations can not exist and survive without the real experiment, conducted in a trustworthy laboratory.

The following sections will be dedicated to the disorders connected with the bilirubin and its regulation.

Bilirubin Disorders and Its Regulation: Experimental Facts and Perspectives Introducing Molecular Simulations

Hyperbilirubinemia

I. Conjugated

When accumulation of bilirubin and/or its conjugated forms (conjugates) occurs, the consequences are high levels of bilirubin in plasma, which results in the appearance of jaundice (also known as icterus). This can be seen as the yellowish pigmentation of the skin, sclerae, mucous membranes, and other less visible tissues [57-59].

Conjugated bilirubin, as it has already been discussed, is a mixture of monoglucuronides and diglucuronides produced in the liver. It is converted to this mixture by uridine diphosphate (UDP)-glucuronyl transferase, and later secreted into the bile by an ATP-dependent transporter. The process is highly sufficient when the body conditions are normal, keeping concentrations of unconjugated bilirubin in plasma low. As soon as there are any "deviant actions", the levels of unconjugated bilirubin in plasma and of conjugated bilirubin in the liver change. Bilirubin is then distracted or disordered from the liver into the bile, resulting in a disease, and a proper diagnosis has to be made for the treatment of the symptoms.

What actually happens in the case of "conjugated hyperbilirubinemia" is that the secretion of the conjugated bilirubin from the liver into the bile, and consequently of the flow of bile into the intestine is reduced, which causes reflux of conjugates back into the plasma, in most cases pointing to hepatobiliary disease.

II. Unconjugated

The undesirable increase in the rate of bilirubin formation is usually caused by hemolysis. In this case we have accumulation of unconjugated bilirubin in plasma.

Measurements of Bilirubin Levels in Blood

In order to differentiate between conjugated and unconjugated bilirubin certain blood tests are required. Laboratory measurements are usually done spectrophotometrically, as the bilirubin is a system of pyrrole rings which are excellent "chromofors". Therefore, it is broken down by light in the presence of diazotized sulfanilic acid, which generates a colored azodipyrrole that can be assayed spectrophotometrically. As the conjugated bilirubin is water soluble, it quickly reacts with the dye (diazotized sulfanilic acid), whereas unconjugated bilirubin, being poorly soluble in water, reacts more slowly even in the presence of accelerators such as ethanol. The conjugated bilirubin is measured without an accelerator.

Although the unconjugated (indirect) bilirubin produces the azobilirubin, it still needs to be calculated from the equation:

Indirect bilirubin = Total bilirubin - Direct bilirubin,

where the direct bilirubin is the conjugated bilirubin, and total bilirubin is the measure for both indirect and direct bilirubin in the blood. Several authors claim that the terms conjugated and direct, and unconjugated and indirect bilirubin are not the same, and the similarity of these terms is still a matter of debate. Since the kidneys are not capable of filtering the unconjugated bilirubin because it is attached to albumin, the presence of bilirubin in urine indicates the presence of conjugated hyperbilirubinemia.

Measurement Methods of Bilirubin Levels in Blood in the Past

The Van den Bergh reaction was originally used for the qualitative estimation of bilirubin [61], where all of the bilirubin in plasma reacts in less than an hour with diazonium salts in acid, and by adding methyl alcohol to obtain a red coloration. It reacts with varying velocities in the aqueous solution of an acid, and consequently different fractions are collected [62].

In the past, there were two main theories or two main classes of theory attempting to explain the differences between the percentages of the total bilirubin (collected from different plasmas of different patients) and the reaction with the aqueous acid solution. The first one upholds the thesis that bilirubin does not react in aqueous acid solution, or as was later revealed it can react very slowly. The plasma proteins fraction originating from hemoglobin prevents bilirubin from reacting with other species by forming a valence bond with it [62]. The second theory was leaning towards the presence of catalytic or inhibiting substances which accelerate, slow down or prevent the direct reaction. Malloy and Evelyn conducted a few experiments pointing in the direction of the first theory, which speaks about the fraction of plasma proteins making a valence linkage, rather than about the presence of catalytic and inhibitor substances [62]. Cataphoresis was the method of choice to quantitatively show how the bilirubin is attached to the plasma albumin in the case of jaundice plasma. This complex can move in the electric field [62-64]. However, some doubts were expressed about the accuracy of the method because it was noticed that accuracy changes in the presence of various substances in plasma. Since with the ultrafiltration through the cellophane membranes it was shown that no bilirubin passes through, this led to the conclusion that all of the bilirubin is attached to the plasma albumin [65]. Mixing two equal volumes of plasmas, and later adding diazotized sulfanilic acid was another indicator implying that differences in the reactivity of the direct and indirect bilirubin are more the consequence of structural rather than catalytical effects or causes [62]. The reaction rate of the plasma mixture was equal to the sum of the reaction rate of bilirubin in single plasma. It was assumed that in the presence of the catalyst the rate of the overall reaction of the mixture would have been altered in comparison with the reaction rate in single plasma [62].

Measurement Methods of Bilirubin Levels in Blood in the Present

Blood Tests

Nowadays the direct bilirubin is often measured by the Jendrassik and Grof method [66], while the amount of total bilirubin is determined by the reaction with 2,5-dichlorophenyldiazonium (DPD).

Rolinski and co-workers used and evaluated a new method for measuring total bilirubin. They used a blood gas analyzer with multiple-wavelength photometry to measure concentrations of total bilirubin in blood taken from newborns with and without jaundice. All the results were then compared with the results that had been obtained from standard clinical procedures for measuring total bilirubin in plasma. The advantages of the blood gas analyzer with multiple-wavelength photometry include the small volumes of whole blood needed for sampling and the fast turnaround time [66].

Urine Test

So far all of the measurement methods have dealt with bilirubin levels in blood, therefore blood tests have been described. Certainly the bilirubin concentrations can also be measured in the urine since bilirubin is secreted from the liver into the bile, and from there partially into the urine. That means that in the urine we can expect the presence of conjugated or direct bilirubin, which reacts quickly with an azodye. However, before carrying out a bilirubin urine test, the person who will do the test and examine the samples must be aware of the patient's medical record. That will help to know whether the patient is taking any drugs which may have a severe effect on the results of this test, and to later decide whether the patient is eligible to take such a urine test [67].

Natural Regulation of Bilirubin Levels

Measurement methods are designed in order to obtain information about the natural substances created by nature in the human body and involved in the appearance of the disorder. After obtaining this information one knows more about how the system is regulated. However, one must be aware of the fact that we are talking about nature, which has surely thought of everything when constructing a system like the human body. The major regulators and some of the most potent ones, although so small (invisible to the naked eye) are the enzymes, which we can compare in the macro world with the stock brokers, factories or governments in different countries speaking different languages. They have to make smart and right decisions at the right time if they want to keep a normal state in their country.

In the human body there are different environments with their own characteristics which must be taken into consideration. What is more, they must be unconditionally respected, otherwise their defense system responds by not allowing any room for compromise and that can cause incurable disorders. These environments are separated by borders (membranes), and if a substance wants to pass a border like this, it has to fulfill all of the eligibility conditions. All of the criteria required by a certain customer must be met or it must meet all the conditions of a particular vendor in order to be allowed (to get permission) to do that. Eligibility of the substance that wants to enter is determined by the environment (federal state/country) and its characteristics. At some borders the substance must pay a certain fee to enter the state/country (environment), and that is exactly what we were discussing about, when the discussion was about pK_a values, which tell us how much we have to pay and what is the official payment currency. Most of the time the enzymes work in our favor, but when it comes to disorders, meaning that things have gone wrong, they can become destructive, hence their activity needs to be stopped or inhibited. This is like in the case when a certain company (enzyme), for instance now when there is so much discussion about the global crisis spends money irrationally, thus putting the whole system (human body) in danger. Therefore, destructive action like this must be prevented until the "country" establishes a normal state again. Afterwards companies' activities can be resumed if they were stopped (inhibited) reversibly (reversible inhibition). However, if they were irreversibly inhibited, the resumption of activity is hard to be achieved.

The most important enzymes involved in bilirubin regulation are those that participate in bilirubin production, homoxygenase and biliverdin reductase. The latter is perhaps slightly more involved in the bilirubin regulation than the former, since it more directly participates in the bilirubin production. In order to regulate disorders such as hyperbilirubinemia or jaundice, one has to understand the catalytic mechanism of biliverdin reductase. The biliverdin reductase exists in more than one isomer. The fetus express billiverdin reductase as biliverdin reductase IXβ (BVRB), while adults later express it as biliverdin reductase IXα (BVRA) [68-70]. Human BVR is suggested to be a zinc binding protein with one zinc atom per protein molecule [71].

Introduction of Molecular Simulations into Investigation of Bilirubin Disorders

How can molecular simulation help in setting up and planning the treatment of hyperbilirubinemia?

With molecular simulations we can mimic a real experimental system, but its accuracy depends on the reliability, stability and efficient design of the model system. The background of every molecular simulation is represented by different theories, which have not yet been tested for the real system. This way one can save a lot of money, since it is much cheaper to do molecular simulations than to carry out an experiment that could additionally be very dangerous, especially when it is not known how chemicals interact with one another and the proper conditions of the chemical reaction are not known. Problems appear when experimentalists ask theoreticians how to test the theory or how to make a junction between theory and real experiment.

Nowadays, molecular dynamics are probably one the most established theoretical techniques or procedures for solving equations of motion. With their help we get information about how fast a chemical process will be, and what the values of the parameters involved in the equations of motion (pressure, temperature, etc.) are. The problem with molecular dynamics is the limitation of sampling to a subset of all possible states of the system i.e. non-ergodicity, which is strongly noticeable at low temperatures. Too high temperatures are also a problem, so this is solved with the parallel simulation or simulated annealing to locate a good approximation of a given function in a large space to the global optimum. This is very important in the case of membranes, proteins, where the initial temperature has to be carefully chosen: it should not be too high, since this can cause undesired effects. When the initial temperature is properly set (usually we start with a high value), the annealing procedure or annealing schedule should be followed so that at each step of the simulation the temperature is decreased (the temperature is gradually decreased during the simulated annealing process). At the end the system is at the minimum possible energy level [72-74].

Statistical mechanics, especially classical and statistical thermodynamics, which (can) provide information about the particular configuration, conformation, and other properties that can be expressed in terms of averages, represent the "alma mater" of molecular simulations. The most common term used for the description of the thermodynamic properties of a system in molecular simulations are the free energies, which in our case were used to calculate and evaluate the pK_a values. Molecular simulations actually provide microscopic information about the macroscopic properties of a real system, obtained from the molecular simulation. It is therefore important to know how to interpret microscopic information. If the interpretation is wrong, the results of a certain simulation could lead to severe misunderstandings and misjudgments, which would consequently represent a serious danger when designing the real experiment in the next phase. If one knows the position, velocity and interaction between particles in the study system, assumptions about the future behaviour can be made. People ask why molecular simulations were not used in the past. The reason is that a real system contains so many molecules that it would have taken a very long time to do molecular simulations or calculations manually. Therefore, the advent of the computer is best summarized in the following quotation: "That's one small step for a man, a giant leap for mankind "(Neil Armstrong's words on first setting foot on the moon in 1969). Up until that time it had been practically impossible to do or to run a molecular simulation and predict the properties of a real system supported by reliable calculation results. On the other hand, the theoretical part was much stronger at that time, as scientists were forced to study mathematical equations in detail. Consequently, more analytical solutions were obtained, as opposed to the current situation, where more or less numerical solutions are provided, with computers doing the rest of the job much faster than humans. However, this demands the junction of different scientific disciplines in order to provide, ensure and offer the best possible and reliable results.

Applicability of Molecular Simulation Methods in Bilirubin Studies

Molecular simulations can play an important role in the design of a drug used for the treatment of a certain disease, in this case hyperbilirubinemia. One of the most popular methods in drug design these days is virtual screening and QSAR. In the former we have a sort of griddle drawn up on researchers' assumptions, through which a number of compounds try to pass. Compounds suitable for usage are extracted and used in the laboratory assays, the others are discarded. Quantitative structure-activity relationship (QSAR) is more about the characteristics of the chemical structure of a compound, and is a drug candidate (clinical use, etc.) in quantitative correlation with the chemical reactivity of a compound or system or any other biological activity of a substance. Chemical reactivity or biological activity is expressed as a quantity, most often this is a chemical quantity (concentration, etc.) whose value is a number, hence a mathematical relationship can be formed between the properties and structure of a substance. Moreover, in the QSAR process one wants to express activity as a function of the physical-chemical properties and structural characteristics of a compound. In the case of the bilirubin this would mean that one wants to know the concentration levels of bilirubin in blood and the liver by correlating them to the physical-chemical properties and structural characteristics of the bilirubin in relation to plasma albumin and other compounds in plasma (blood), and in relation to UDP in the liver. The pK_a values appear very interesting for application in the QSAR procedure. However, pK_a probably plays a major role in Lipinski's Rule of Five, as the first two criteria of this rule deal with hydrogen bond acceptors and donors [75] which can be characterized by pK_a values. Hence calculating them with computational chemistry methods is fully justified when the methods are accurate enough with a strong and stable mathematical model.

Quantum chemistry methods are very helpful in the molecular modeling of chemical reactions, especially in constructing what is now called an active site model. Following analogy of this model we construct the active site of the enzyme, which usually contains co-factor and nearby amino acid residues that have great influence on the catalytic step of the reaction [76]. How can one imagine the modeling of the active site of an enzyme? It is like building a port (Figure 16), where the docks have to be carefully designed and then built, so that ships can come in and put down anchor. It is also important to know that not all types of ships can come in. In the human body the ships are represented by substrates and inhibitors (Figure 17). The question then is in which dock the ship will lay anchor and what type of bond it will form (or not form) with the port berth. It is also important how the berth looks, therefore its construction with all its accessories (amino acid residues and accompanying waters near the co-factor and residues participating in the catalytic reaction) is perhaps of crucial importance in the construction of the port - in our case of the active site model.

Where to put or how to apply the idea of the active site model to the story of bilirubin? This idea can be used when studying biliverdin reductase, where one is eager to know where the catalytic step in the enzyme of interest occurs. Are there any co-factors, amino acid residues without which the catalytic step can not be possible (Figure 18)? This can be explained by point mutations, where the amino acid residues from the wild-type enzyme are substituted by other residues.

The main disadvantage of the active site model is the size of the system that can still be included in calculations. As the system grows, the time needed for calculations increases even faster, especially with the more flexible basis sets and modern quantum chemical methods. In the calculation itself this can reflect as a problem with convergence, since the calculation can not converge, and the molecular simulation is stopped at a certain point without reaching the end.

Calculations could be performed in the following way: we could take the NADPH and an overlapping binding site including Lys^{18}, Lys^{22}, Lys^{179}, Arg^{183}, and Arg^{185} as key residues (Figure 17). At each step corresponding to the fixed distance we would optimize the geometry and calculate the corresponding zero point energy. From this data we would construct the reaction diagram, and we would be able to see how high the energy barrier for this process is. Finally, we would take the optimized structures corresponding to the reactant complex, perform a full non–constrained geometry optimization and calculate the vibrational zero point energies in the harmonic approximation. For the transition state, we would take the highest energy geometry, perform a non–constrained optimization towards the transition state, and calculate the corresponding zero point energy and vibrational frequencies.

Ab initio and Density Functional Theory (DFT) calculations would be performed on the Hartree–Fock (HF) and B3LYP [77,78] levels of theory in conjunction with the basis sets that are flexible enough to allow for reasonably accurate calculations.

As a good compromise between accuracy and practicality of the model, all molecular geometries could be optimized by the very efficient B3LYP/6–31+G(d) method. Zero-point vibrational energies (ZPVEs) and thermal corrections would be then extracted from frequency calculations, obtained at the same level of theory, where the scaling of the calculated harmonic vibrational frequencies would not be applied.

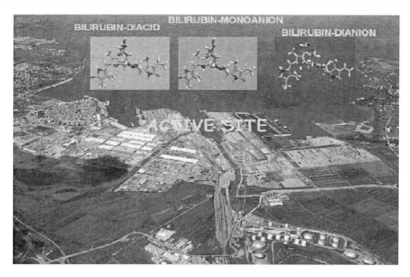

Figure 16. Example of an active site model presentation in a real world perspective for biliverdin reductase.

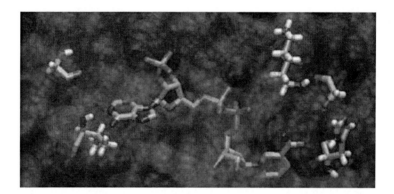

Figure 17. Active site model design with NADP, and residues 153, 155, and 178 for biliverdin reductase.

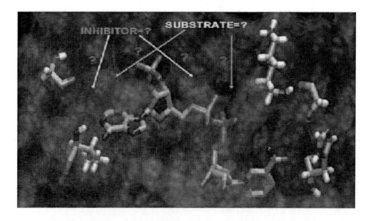

Figure 18. Active side model design with NADP, and residues 153, 155, and 178 for biliverdin reductase.

Analysis of all normal vibrational modes would be used to verify that all stationary structures correspond to true minima on the electronic potential energy surfaces by checking the absence of imaginary vibrational frequencies. All transition state structures are then examined to possess one imaginary frequency, and the visualization of the corresponding eigenvector should reveal promotion of the chemical reaction in a desired direction. All atomic charges should be also obtained at the same level of theory employing Mulliken's electron density partitioning scheme [79]. The final single-point energy calculations could be attained using a highly flexible 6–311++G(2d,2p) basis set employing B3LYP functional and Boese–Martin for kinetics (BMK) method [80]. The first functional has frequently been shown in the literature to provide very accurate thermodynamic parameters for simple organic molecules, whereas BMK approach has been particularly parameterized and demonstrated to offer excellent accuracy for the kinetic data. To account for the polarization effects caused by the rest of the enzyme, which is not explicitly included in our model, we would perform single-point calculations on optimized geometries with the conductor-like polarisable continuum model (CPCM) [81,82], as employed by Himo and co-workers [76,83]. This approximation assumes that the enzyme surrounding is a homogeneous polarisable medium with some conventionally defined dielectric constant. Calculations could be carried out at B3LYP/6–311++G(2d,2p) and BMK/6–311++G(2d,2p) levels of theory employing the usual dielectric constant of $\varepsilon = 4$ and taking the rest of parameters for pure water [76].

With a proper design of the active site model one can calculate the activation free energy for the rate-limiting step at biliverdin reductase. However, we should not forget the convergence problems that might occur during the calculations, hence we must keep this in mind while running the simulations.

The QM/MM methods become key players in later phases, when the system is too large to be still considered only with quantum mechanical methods. With quantum-chemical (QM) calculations one gets an impression of how the system works, as the QM part usually describes a part of the system where a chemical reaction takes place (bond breaking and bond making). Particles are treated as quantum atoms, which are any atoms that undergo a change in their bonding pattern as the reaction proceeds, while the MM part serves for the description of the environment around the particle of interest, being well suited for the representation of molecules that do not undergo chemical transformations (no bond breaking or making). However, problems will arise if one wants to include or use the MM part. The most important issue here is which classical force field should be used, since there is still no unique one that could be used in all cases.

By the introduction of thermal averaging and of established methods for free energy calculations, activation free energy can be calculated and critically compared with the experiment. Moreover, the introduction of an explicit protein and aqueous environment allows for studies of point mutations. As a critical test of the reaction mechanism one can address nuclear quantum effects giving rise to the H/D kinetic isotope effect if a proton transfer occurs.

Additional spectroscopic studies (NMR, EPR, ESR) would be of major benefit in elucidating other potential mechanisms, in proving or disproving the mechanism considered in a certain study. Performing additional cross–over isotope experiments ($^{13}C, ^{15}N$) with isotope labelled inhibitors in order to gain additional insight into the molecular mechanism of biliverdin reductase and mono oxygenase inhibition remains a challenge. From the computational point of view, besides the use of QM/MM methodology on the EVB level, the challenge is to calculate the H/D nuclear quantum effects by the quantization of the nuclear motion [84]. All these efforts may result in a deeper understanding of the chemistry of bilirubin, bilirubin transport, regulation, and disorders.

Conclusion

We have calculated pK_a values for several carboxylic acids, including bilirubin by using solvent reaction field (SCRF) and Langevin dipole hydration models in order to provide information about the nature of bilirubin transport, which could later be applied in the explanation of bilirubin regulation and disorders. Calculations of pK_a are demanding since an essential feature of pK_a is the compensation of the energy of covalent bonds by hydration, and consequently hydration free energies play one of the major roles in the calculation of the pK_a values.

We have demonstrated that a high level of quantum-chemical theory is required if one wants to do calculations that are in good agreement with the experiment. We reproduced bilirubin pK_a values with B3LYP DFT level in conjunction with flexible basis sets, starting with the 6- 31+G(d,p) basis set and Langevin dipoles solvation model. We demonstrated that the Langevin dipoles solvation model performs better than the solvent reaction field model.

If we now summarise, we can conclude that the transport of bilirubin is a complex process, considering the fact that under physiological conditions the Fick law is not obeyed, and the simple diffusion process is suddenly not the only possible one. There is growing evidence that besides bilitranslocase the bilirubin-binding protein acts as a lower affinity electroneutral transporter taking part in the higher substrate concentration [51-53]. Several studies have shown that the transport of bilirubin is also potassium ion mediated [49]. Moreover, redox equilibrium of nicotinamide nucleotides (NADH/NAD+) modulates bilitranslocase transport properties, which are in addition highly temperature sensitive [85]. Current opinion leans towards the idea that bilirubin is transported by bilitranslocase with co-transport of sodium [10, 48-51]. It is possible that other ions, particularly K^+, are involved in the co-transport and the process is mediated by Na/K-ATPase [52,53]. However, detailed studies of the effect of the ion radius on the free energy of hydration, and later on the bilirubin transport, have to be carried out. Experimental determination of bilirubin protonation states during the protonation remains a challenge, while it is very difficult to isolate bilirubin in a complex with bilitranslocase. We are aware that the protonation states will most probably change during the transport since we are dealing with a proton rich environment. Proton transport across the membrane is driven by H^+-ATPase and is believed to be much more evolved and faster than bilitranslocase, still the latter is believed to be specific for bilirubin, while H^+-ATPase only transports protons. Because bilirubin protonation states in the interior of the transport protein are difficult to determine experimentally, we believe that molecular simulation will contribute to clarifying this issue.We have performed the first step towards molecular simulation of bilirubin transport by the critical examination of the methodology for pK_a values calculation. We have demonstrated that relatively flexible basis sets starting with 6-31+G(d,p) on the Hartree–Fock and B3LYP levels of theory and the Langevin dipoles solvation model are able to predict the pK_a values for small carboxylic acids. Unfortunately, the size of bilirubin precludes application of these levels and especially the second pK_a value was not reproduced. We expect that the future of the molecular simulation of bilirubin transport will consider the available experimental data for protonation states. The methodology is developed and readily available [85-87]. However, another necessary requirement are the structures of the transport proteins, which are presently not known.

This study is preliminary to all atom simulation of bilirubin transport by bilitranslocase when most probably the protonation states of bilirubin are changed during the transport process. A very recent review paper by Lee and Crippen demonstrates that prediction of pK_a values by different methods remains a rapidly evolving field that is of enormous relevance for (bio) chemistry, physiology and drug design [88].

References

[1] Borštnar, R., Roy Choudhury, A., Stare, J., Novic, M., Mavri, J. (2010). Calculation of pK_a values of carboxylic acids: Application to bilirubin. *Journal of molecular structure*: *THEOCHEM, 947*, 76-82.

[2] Florian, J. and Warshel, A. (1999). Calculations of hydration entropies of hydrophobic, polar, and ionic solutes in the framework of the Langevin dipoles solvation model, *J. Phys. Chem. B, 103*, 10282-10288.

[3] Florian, J. and Warshel, A. (1997). Langevin dipoles model for ab initio calculations of chemical processes in solution: Parametrization and application to hydration free energies of neutral and ionic solutes and conformational analysis in aqueous solution, *J. Phys. Chem. B, 101*, 5583-5595.

[4] Strajbl, M., Florian, J., and Warshel, A. (2000). *Ab initio* evaluation of the potential surface for general base-catalyzed methanolysis of formamide: a reference solution reaction for studies of serine proteases, *J. Am. Chem. Soc., 122*, 5354-5366.

[5] Miertus, S., Scrocco, E., and Tomasi, J. (1981). Electrostatic interaction of a solute with a continuum. A direct utilization of *ab initio* molecular potentials for the prevision of solvent effects, *Chem. Phys., 55*, 117-129.

[6] Berne, R. M., Levy, M.N. Physiology. 4th ed. New York: Mosby, Inc.; 1998.

[7] Stocker, R, Yamamoto, Y, McDonagh, AF, Glazer, AN, Ames, BN. (1987) Bilirubin is an antioxidant of possible physiological importance. *Science, 235*, 1043-1046.

[8] Bhutani, V. K., Johnson, L., Sivieri, E. M. (1999). Predictive ability of a predischarge hour-specific serum bilirubin for subsequent significant hyperbilirubinemia in healthy term and near-term newborns. *Pediatrics, 103*, 6-14.

[9] O'Sullivan, A. J., Kennedy, M.C., Casey, J.H., Day, R. O., Corrigan, B., Wodak, A.D. (2000). Anabolic-androgenic steroids: medical assessment of present, past and potential users, *MJA, 173*, 323-327.

[10] Passamonti, S., Vanzo, A., Vrhovšek, U., Terdoslavich, M., Cocolo, A., Decorti, G., Mattivi, F. (2005). Hepatic uptake of grape anthocyanins and the role of bilitranslocase. *Food Res. Int., 38*, 953-966.

[11] Ostrow, J. D., Mukerjee, P., and Tiribelli, C. (1994). Structure and binding of unconjugated bilirubin: relevance for physiological and pathophysiological function. *J. Lipid Res., 35*, 1715-1737.

[12] Zucker, S.D, Goessling W., Gollan J.L. (1995) Kinetics of bilirubin transfer between serum albumin and membrane vesicles: insight into the mechanisms of organic anion delivery to the hepatocyte plasma membrane. *J. Biol. Chem.*, 270, 1074–1081.

[13] Scharschmidt, B.F, Waggoner, J.G, Berk, P.D. (1975). Hepatic organic anion uptake in the rat. *J. Clin. Invest.*, 56, 1280–1292.

[14] Paumgartner, G. and Reichen, J. (1976). Kinetics of hepatic uptake of unconjugated bilirubin. *Clin. Sci. Mol. Med.*, 51, 169-176.

[15] Boiadjiev, S.E, Watters, K., Wolf, S., Lai, B.N., Welch, W.H., McDonagh, A.F., Lightner, D.A. (2004). pKa and aggregation of bilirubin: titrimetric and ultracentrifugation studies on water-soluble pegylated conjugates of bilirubin and fatty acids. *Biochemistry, 43,* 15617-15632.

[16] Trull, F.R., Boiadjiev, S., Lightner, D.A., McDonagh, A.F (1997). Aqueous dissociation constants of bile pigments and sparingly soluble carboxylic acids by 13C NMR in aqueous dimethyl sulfoxide: effects of hydrogen bond. *J. Lipid Res.* 38, 1178-1188.

[17] Lightner, D.A., Holmes, D.L., McDonagh, A.F. (1996). On the acid dissociation constants of bilirubin and biliverdin. *J. Biol. Chem., 271,* 2397-2405.

[18] Kikuchi, G., Yoshida, T., Noguchi, M. (2005). Heme oxygenase and heme degradation. *Biochem. Biophys. Res. Commun., 338*, 558–567.

[19] Evans, J.P, Niemevz, F., Buldain G., de Montellano, P.O. (2008). Isoporphyrin intermediate in heme oxygenase catalysis. Oxidation of alpha-meso-phenylheme. *J. Biol. Chem., 283*, 19530–19539.

[20] Wang, M., Roberts, D. L., Paschke, R., Shea, T. M., Siler Masters, B. S., Kim, J. J. P. (1997). Three-dimensional structure of NADPH-cytochrome P450 reductase: Prototype for FMN- and FAD-containing enzymes. *PNAS, 94,* 8411-8416.

[21] Schuller, D. J., Wilks, A., Ortiz de Montellano, P. R., Poulos, T. L. (1999). Crystal structure of human heme oxygenase-1. Nat. Struct. Biol., 6, 860–867.

[22] Rigney, E., Mantle, T.J. (1988). The reaction mechanism of bovine kidney biliverdin reductase. *Biochim. Biophys. Acta, 957*, 237–242.

[23] Wang, J., de Montellano, P. R. (2003). The binding sites on human heme oxygenase-1 for cytochrome p450 reductase and biliverdin reductase. *J. Biol. Chem., 278,* 20069–20076.

[24] Baranano, D. E., Rao, M., Ferris, C. D., Snyder, S. H. (2002). Biliverdin Reductase: A Major Physiologic Cytoprotectant. *PNAS, 99,* 16093–16098.

[25] King, C., Rios, G., Green, M., Tephly, T. (2000). UDP-glucuronosyltransferases. *Curr. Drug. Metab., 1*, 143–161.

[26] Sørensen, S. P. L. (1909). Enzymstudien. II, Über die Messung und die Bedeutung der Wasserstoffionenkonzentration bei enzymatischen Prozessen, *Biochem. Zeitschr., 21,* 131–304.

[27] Li, G. and Cui, Q. (2003). pK_a Calculations with QM/MM Free Energy Perturbations. *J. Phys. Chem. B, 107,* 14521-14528.

[28] Riccardi, D. and Cui, Q. (2007). pK_a Analysis for the Zinc-Bound Water in Human Carbonic Anhydrase II: Benchmark for "Multiscale" QM/MM Simulations and Mechanistic Implications. *J. Phys. Chem. A, 111,* 5703-5711.

[29] Riccardi, D., Schaefer, P., Cui, Q. (2005). pK_a calculations in solution and proteins with QM/MM free energy perturbation simulations: a quantitative test of QM/MM protocols. *J. Phys. Chem. B, 109*, 17715-17733.

[30] Ghosh, N. and Cui, Q. (2008). pK_a of residue 66 in *Staphylococal nuclease*. I. Insights from QM/MM simulations with conventional sampling. *J Phys Chem B., 112*, 8387–8397.

[31] *Singh, U. C. and Kollman, P. A. (1984).* An approach to computing electrostatic charges for molecules. *Journal of Computational Chemistry, 5, 129–145.*

[32] Marcus, Y. Ion Solvation. New York: Willey; 1985.

[33] Lim, C., Bashford, D., Karplus, M. (1991). Absolute pKa calculations with continuum dielectric methods. *J. Phys. Chem., 95*, 5610-5620.

[34] Tissandier, M. D., Cowen, K.A., Feng, W. J. (1998). The Proton's Absolute Aqueous Enthalpy and Gibbs Free Energy of Solvation from Cluster-Ion Solvation Data. *J. Phys. Chem., 102*, 7787-7794.

[35] Passamonti, S., Battiston, S., Sottocasa, G. L., *Mol. Membr. Biol.*, 1999, 16, 167-172

[36] Cistola, D. P., Small, D. M., Hamilton, J. A. (1982). Ionization behavior of aqueous short-chain carboxylic acids: a carbon-13NMR study. *J. Lipid Res., 23*, 795-799.

[37] Ostrow, J. D., Pascolo, L., Shapiro, S. M., Tiribelli, C. (2003). New concepts in bilirubin encephalopathy, *Eur. J. Clin. InVest., 33*, 988-997.

[38] Ostrow, J. D. and Tiribelli, C. (2001). New concepts in bilirubin neurotoxicity and the need for studies at clinically relevant bilirubin concentrations. *J. Hepatol., 34*, 467-470.

[39] Brito, M. A., Brondino, C. D., Moura, J. J., Brites, D. (2001). Effects of bilirubin molecular species on membrane dynamic properties of human erythrocyte membranes: a spin label electron paramagnetic resonance spectroscopy study. *Arch. Biochem. Biophys., 387*, 57-65.

[40] Falk, H. The Chemistry of Linear Oligopyrroles and Bile Pigments. Wien, New York: Springer-Verlag; 1978.

[41] Bonnett, R., Davies, J. E., Hursthouse, M. B., Sheldrick, G. M. (1978). The structure of bilirubin, Proc. R. Soc. London, *Ser. B, 202*, 249-268.

[42] Le Bas, G., Allegret, A., Mauguen, Y., De Rango, C., Bailly, M. (1980). The structure of triclinic bilirubin chloroform-methanol solvate. *Acta Crystallogr., Sect. B, 36*, 3007-3011.

[43] Ostrow, J. D., Celic, L., Mukerjee, P. (1988). Molecular and micellar associations in the pH-dependent stable and metastable dissolution of unconjugated bilirubin by bile salts. *J. Lipid Res., 29*, 335-348.

[44] Hahm, J.-S., Ostrow, J. D., Mukerjee, P., Celic, L. (1992). Ionization and self-association of unconjugated bilirubin, determined by rapid solvent partition from chloroform, with further studies of bilirubin solubility. *J. Lipid Res., 33*, 1123-1137,

[45] McDonagh, A. F. (1979) The Porphyrins (Dolphin, D., Ed.) Vol. VI, pp 293-491, Academic Press, New York.

[46] Hansen, P. E., Thiessen, H., Brodersen, R. (1979). Bilirubin acidity. Titrimetric and 13C NMR studies. *Acta Chem. Scand. B, 33*, 281-293.

[47] Lee, J. J., Daly, L. H., Cowger, M. L. (1974). Bilirubin ionic equilibria; their effects on spectra and on conformation. *Res. Commun. Chem. Pathol. Pharmacol., 9*, 763-770.

[48] Passamonti, S. and Sottocasa, G.L. (2002). Bilitranslocase: structural and functional aspects of an organic anion carrier. *Recent research developments in biochemistry, 3,* 371-391.

[49] Novič, M., Passamonti, S., Karawajczyk, A., Drgan, V., Medič, N., Oboh, G. (2007). Properties of flavonoids influencing the binding to bilitranslocase investigated by neural network modelling. *Biochemical Pharmacology, 73,* 308-320.

[50] Passamonti, S., Battiston, L., Sottocasa, G. L. (1998). Bilitranslocase can exist in two metastable forms with different affinities for the substrates-evidence from cysteine and arginine modification. *Eur. J. Biochem, 253,* 84-90.

[51] *Passamonti, S., Battiston, L., Sottocasa, G. L. (2000).* Gastric uptake of nicotinic acid by bilitranslocase. *FEBS Lett., 482, 167-168.*

[52] Tiribelli, C., Torres, A. M, Lunazzi, G. C., Stremmel, W. (1993). Bilitranslocase and sulfobromophthalein/bilirubin-binding protein are both involved in the hepatic uptake of organic anions. *Proc. Natl. Acad. Sci., 90,* 8136-8139.

[53] Passamonti, S., Cocolo, A., Braidot, E., Petrussa, E., Peresson, C., Medič, N., Macri, F., Vianello, A. (2005). Characterization of electrogenic bromosulfophthalein transport in carnation petal microsomes and its inhibition by antibodies against bilitranslocase. *FEBS Journal, 272,* 3282-3296.

[54] Passamonti, S., Terdoslavich, M., Margon, A., Cocolo, A., Medič, N., Micali, F., Decorti, G., Franko, M. (2005). Uptake of bilirubin into HepG2 cells assayed by thermal lens spectroscopy. Function of bilitranslocase. *FEBS Journal,* 272, 5522-5535.

[55] John B. West (Ed.), Best and Taylor's: Physiological Basis of Medical Practice. 12th ed. Baltimore: Williams and Wilkins; 1990.

[56] http://en.wikipedia.org/wiki/File:Scheme_sodium-potassium_pump-en.svg;   Author LadyofHats Mariana Ruiz Villarreal.

[57] Lathe, G.H. (1972). The degradation of haem by mammals and its excretion as conjugated bilirubin. *Essays Biochem., 8,* 107-148.

[58] Muraca, M., Fevery, J., Blanckaert, N. (1988). Analytic aspects and clinical interpretation of serum bilirubins. *Semin Liver Dis., 8,* 137-147.

[59] Westwood, A. (1991). The analysis of bilirubin in serum. *Ann. Clin. Biochem., 28,* 119-130.

[60] Iyanagi, T., Emi, Y., Ikushiro, S. (1998). Biochemical and molecular aspects of genetic disorders of bilirubin metabolism. *Biochim. Biophys. Acta., 1407,* 173-184.

[61] Coolidge T. (1939). Chemistry of the van den Bergh reaction. *J. Biol. Chem., 132,* 119-128.

[62] Malloy, H.T. and Evelyn, K.A. (1937). The determination of bilirubin with the photoelectric colorimeter. *J. Biol. Chem., 119,* 481-490.

[63] Bennhold, H. (1932). Uber die Vehikelfunktion der Serumeiweiss- korper. *Ergebn. inn. Med. u. Kinderh., 42,* 273-375.

[64] Pedersen, K. O. and Waldenström, J. (1937). Studien iiber das Bilirubin in Blut und Galle mit Hilfe von Elektrophorese und Ultrazentrifugierung. *Z. physiol. Chem., 245,* 152.

[65] Gregory, R. L. and Anderson, M. (1937). The filtrability of bilirubin in obstructive jaundice. *J. Lab. And Clin., 22,* 1111.

[66] Rolinski, B., Küster, H., Ugele, B., Gruber, R., Horn, K. (2001). Total bilirubin measurement by photometry on a blood gas analyzer: potential use in neonatal testing at the point of care. *Clin. Chem. 47*, 1845–1847.

[67] Berk PD, Korenblat KM. Approach to the patient with jaundice or abnormal liver test results. In: Goldman L, Ausiello D, eds. *Cecil Medicine*. 23rd ed. Philadelphia, Pa: Saunders Elsevier; 2007:chap 150.

[68] Blumenthal S. G., *Stucker, T., Rasmussen, R. D., Ikeda, R. M., Ruebner, B. H., Bergstrom, D. E., Hanson, F. W. (1980). Changes in bilirubins in human prenatal development. Biochem. J., 186*, 693-700.

[69] Yamaguchi T. *and Nakajima, H. (1995). Changes in the Composition of Bilirubin-IX Isomers During Human Prenatal Development. Eur. J. Biochem., 233*, 467-472.

[70] Yamaguchi T., *Komoda, Y., Nakajima, H. (1994). Biliverdin-IX alpha reductase and biliverdin-IX beta reductase from human liver. Purification and characterization. J. Biol. Chem., 269*, 24343-24348.

[71] Maines, M.D., Polevoda, B.V., Huang, T.J. and McCoubreyJr. *Eur. J. Biochem. 235*, 372−381 (1996).

[72] Kirkpatrick, S., Gelatt, C. D., Vecchi, M. P. (1983). Optimization by Simulated Annealing. *Science New Series, 220*, 671–680.

[73] Cerny, V. (1985). A thermodynamical approach to the travelling salesman problem: an efficient simulation algorithm. *JOTA, 45*, 41-51.

[74] Granville, V., Krivanek, M., Rasson, J.P. (1994). Simulated annealing: A proof of convergence. *IEEE Transactions on Pattern Analysis and Machine Intelligence, 16*, 652–656.

[75] Lipinski, C.A., Lombardo, F., Dominy, B.W., Feeney, P.J. (2001). Experimental and computational approaches to estimate solubility and permeability in drug discovery and development settings. *Adv. Drug Del. Rev., 46*, 3–26.

[76] Georgieva, P. and Himo, F. (2010). Quantum chemical modeling of enzymatic reactions: the case of histone lysine methyltransferase. *J. Comput. Chem., 31*, 1707–1714.

[77] Becke, A. D. (1992). Density functional thermochemistry. I. The effect of the exchange only gradient correction. *J. Chem. Phys., 96*, 2155–2160.

[78] Lee, C. T., Yang, W. T., Parr, R. G. (1988). Development of the Colle-Salvetti correlation-energy formula into a functional of the electron density. *Phys. Rev. B, 37*, 785–789.

[79] Mulliken, R. S. (1995) Electronic Population Analysis on LCAO-MO Molecular Wave Functions. *J. Chem. Phys., 23*, 1833-1841.

[80] Boese, A. D., Martin, J. M. L. (2004) Development of density functionals for thermochemical kinetics . *J. Chem. Phys., 121*, 3405–3416.

[81] Barone, V., Cossi, M. (1998) Quantum Calculation of Molecular Energies and Energy Gradients in Solution by a Conductor Solvent Mode. *J. Phys. Chem. A, 102*, 1995-2001.

[82] Cossi, M., Rega, N., Scalmani, G., Barone, V. (2003) Energies, Structures, and Electronic Properties of Molecules in Solution with the C-PCM Solvation Model. *J. Comput. Chem., 24*, 669-681.

[83] Himo, F., Siegbahn, P. E. M. (2003) Quantum Chemical Studies of Radical-Containing Enzymes. *Chem. Rev., 103*, 2421–2456.

[84] Kamerlin, S. C. L., Mavri, J., Warshel, A. (2010). Examining the Case for the Effect of Barrier Compression on Tunneling, Vibrationally Enhanced Catalysis, Catalytic Entropy and Related Issues (Review). Febs Letters, 584, 2759–2766.
[85] Mavri, J., Gaberšček, M. (1999). Phenol forms complexes with tetramethylammonium ions in aqueous solution. *Chem. Phys. Lett., 308*, 421-427.
[86] Pitera, J. W., van Gunsteren, W. F. (2001). One-Step Perturbation Methods for Solvation Free Energies of Polar Solutes. *J. Phys. Chem. B, 105*, 11264-1274.
[87] Billeter, S. R., van Gunsteren, W. F. (2000). Computer Simulation of Proton Transfers of Small Acids in Water. *J. Phys. Chem. A, 10*, 3276-3286.
[88] Lee, A. C., Crippen, G. M.J. (2009). Predicting pK_a. *Chem. Inf. Model., 49*, 2013-2033.

In: Bilirubin: Chemistry, Regulation and Disorder
Editors: J. F. Novotny and F. Sedlacek
ISBN: 978-1-62100-911-5
© 2012 Nova Science Publishers, Inc.

Chapter IV

Hyperbilirubinemia-Associated Renal Disorders

Khajohn Tiranathanagul[1], Asada Leelahavanichkul[2], Somchit Eiam-Ong[*3] *and Somchai Eiam-Ong[1]*

[1]Division of Nephrology, Department of Medicine,
[2]Immunology Unit, Department of Microbiology,
[3]Department of Physiology,
Faculty of Medicine, Chulalongkorn University and King Chulalongkorn Memorial Hospital, Bangkok, Thailand

Abstract

Renal dysfunction in patients with extensive jaundice has been firstly described a century ago. The renal disorders after jaundice vary from tubular injury-induced electrolyte imbalances (eg. hyponatremia, hypokalemia, hypouricemia, normal gap metabolic acidosis, and Fanconi syndrome) to acute kidney injury (AKI). Hyponatremia is commonly found in more than 70% of reversible obstructive jaundice (OJ) patients. Moreover, intestinal malabsorption also causes hypocalcemia, hypophosphatemia, and hypomagnesemia. Billirubin and bile acid have been hypothesized to be the major renal toxic components in jaundice patients. From the OJ animal model studies, 2 phases of renal response have been purposed, early natriuresis (diuresis) and late renal sodium absorption phase. The natriuresis phase is the result of hepatic metabolite excretion from the kidney, instead of the liver, and/or bile acid-induced direct tubular toxicity due to detergent effect. Sulfated bile acid also neutralizes amiloride-sensitive Na^+/H^+ antiporter in the proximal tubule, leading to more distal tubular Na^+ delivery which could accelerate cortical collecting duct function and kaliuresis. Moreover, bilirubin inhibits uric acid absorption at S1 and S3 segments of the proximal tubule, causing hyperuricosuria. In the

---

[*]Please send all correspondence to Professor Somchai Eiam-Ong, MD. Division of Nephrology, Department of Medicine, Faculty of Medicine, Chulalongkorn University and King Chulalongkorn Memorial Hospital, Bangkok, Thailand 10330 Fax & Phone (662) – 2526920. Email address : Somchai80754@yahoo.com

late phase of OJ (3-7 days after OJ), renal adaptation leads to more salt and water retention as measured by reduced fractional excretion of Na^+ and solute free-water whereas the urine osmolality is still persistently low. In comparison, the natriuresis phase even during sodium restriction had been reported in cholangiocarcinoma patients with severe hyperbilirubinemia (up to 40 mg/dL). Then, the liver parenchymal damage occurs at 10-20 years later and leads to hypoalbuminemia, ascites, and sodium conservation.

In OJ patients, AKI develops in approximately 10-30% and contributes to more than 70% mortality. The pathophysiology of AKI after OJ, is more complicated and might consist of 1). renal macrocirculatory changes from natriuresis with negative cardiac inotropic and defective vasoactive vascular responses 2). renal microcirculatory alterations from endotoxinemia, nitric oxide, prostaglandin, or other mediators3). direct toxicity of the biliary products. Nevertheless, the pre-existing deleterious cardiovascular function and hypovolemia are the main predisposing factors to AKI. In addition, the prognosis of AKI in OJ patients seems to relate to the jaundice severity. The mortality rates of postoperative AKI in OJ are 33 and 85% in patients with serum bilirubin< 10 mg/dL and > 20 mg/dL, respectively. Thus, several modalities of extracorporeal liver support systems have been shown to reduce jaundice and biochemical parameter severity although the survival benefit has not yet been demonstrated. This article outlines the scope of various important topics and studies that contribute to the current knowledge regarding hyperbilirubinemia-associated renal disorders, including electrolyte imbalances and AKI.

Keywords: hyperbilirubinemia, obstructive jaundice, electrolyte imbalances, acute kidney injury, extracorporeal liver support systems

Introduction

Liver, the largest internal organ, is an important vital organ responsible for several essential metabolism control systems such as nutritional metabolism, detoxification, and protein synthesis. Both liver and kidney are the important sites for neutralizing toxic substances and excreting by-product metabolites. Normal hepatic function requires normal hepatocyte, Kuffer cells, and intact biliary tree. The hepatic failure is either a result of hepatocyte injury (parenchymal liver disease) or biliary tract abnormality (obstructive jaundice: OJ). Biliary tract obstruction or bile excretion abnormality causesaccumulationof bile and toxic substances in vascular system. Bile, the dark green yellowish fluid produced by hepatocyte, consists of water (85%), bile salts (10%), mucus and pigments (3%), fats (1%), inorganic salts (0.7%), and cholesterin (0.3%) [1]. Bile acts as a surfactant to emulsify fat by attachingthe bile salt hydrophobic side with fat droplet and exposure ofthe hydrophilic side to form water soluble micelle. Additionally, bile is also the excretion route of bilirubins, the erythrocyte recycling by products [1]. The bile excretion abnormality results in hyperbilirubinemia, which might be caused by either biliary tract obstruction or parenchymal liver injury. Then bile, especially the water soluble part, is excreted through the kidney. Bile salts (e.g. deoxycholate, taurocholate, etc.), the oxidative products of cholesterol via hepatic cytochrome P450, are the anionic compounds used as lipid surfactant [2]. High bile salt concentration in urine may be mediated by the organic anion-transporting polypeptide transporter and could act as nephrotoxic substance [3]. Rats with massive oral administration of sodium cholate/ deoxycholate died within 8-24 h [4]. The renal injuries after hyperbilirubinemia can vary from tubular dysfuction-induced electrolyte imbalances to acute kidney injury (AKI). In this chapter, the evidence of correlation between hyperbilirubinemia and renal dysfunction and the current bilirubin-lowering procedures will be reviewed.

The Observation of Renal Dysfunction in Jaundice Patients

The first report of renal dysfunction in jaundice patients was in 1911. Haberer and Clairmont described AKI, previously called "acute renal failure (ARF)", in 5 patients with OJ from common bile duct lithiasis [5]. Helwing and Schultz introduced the name "hepatorenal syndrome" or "liver kidney syndrome" in 1932 to describe patients with renal failure after biliary tract surgery [5]. At present, hepatorenal syndrome refers to the condition which has deteriorated kidney function in patients with cirrhosis or fulminantliver failure[6]. Several patient series of AKI in OJ have been reported [7-13]. Jaundice became one of the risk factors for postoperative AKI[11, 13]. A following study found that renal dysfunction in OJ patients was associated with the degree of biliary obstruction, patient age, and urinary sodium excretion [14].Nevertheless, the spectra of renal dysfunction in hyperbilirubinemia patients do not only limit to AKI but also include tubular dysfunctionssuch as electrolyte imbalances. Several series of preoperative OJ patients revealed that mild hyponatremia was common [15]. The association between hypouricemia and OJ had been reported in earlier studies [16, 17]. Bairaktari et al reported 35 cases of OJ patients with hypouricemia and hypophosphatemia with glucosuria and phosphaturia diagnosed as Fanconi syndrome [18]. Interestingly, 12 patients in this case series showed improvement of proximal renal dysfunction after partial or total correction of hyperbilirubinemia.

In summary, the association between hyperbilirubinemia and renal dysfunction has been observed around a century ago, starting from AKI, the most obvious renal dysfunction,to electrolyte disturbances. Several studies have been performed to find the mechanisms to explain these disorders.

Level of Serum Bilirubin and Renal Dysfunction

After bile duct obstruction, only conjugated bilirubin which is water soluble and not bound to the protein is filtered through the renal glomerulus with no tubular secretion [19]. Unconjugated bilirubin, which is strongly protein bound, is not filtered except when extremely high plasma concentration is present [20]. The human hepatic bilirubin production is approximately 250 mg/day while the maximum renal bilirubin excretion capacity is 220 mg/day [21, 22]. As such, the plateau of human serum total bilirubin (TB) is 25-30 mg/dL but can be higher in case of AKI [23, 24]. However, the plateau of TB varies between animal species due to the differences in the ratio of conjugated/ unconjugated bilirubin and liver production rate. With more percentage of conjugated bilirubin in dog and rat, the less TB level was found after bile duct ligation in these animals [24]. Concerning the association between the severity of hyperbilirubinemia and renal dysfunction, Sitprija et al reported that there was no change in renal function in the patients with mild to moderate jaundice,TB ranging 8-15 mg/dL, while increased urinary sodium excretion and negative water clearances were observed in the patients with severe jaundice,TB ranging 27-40 mg/dL[25]. In addition, AKI was identified in cholangiocarcinoma patients with TB more than 26 mg/dL [26].However, Bairaktari et al found that hypokalemia and renal tubular acidosis were noted in the patients with TBranging 8-23 mg/dL[18]. Padillo et al reported mild azotemia in the patients with TB ranging 9-21mg/dL [14]. On the other hand, cardiac dysfunction in hyperbilirubinemia can be found at higher TB level (TB > 15-18 mg/dL) and the cardiac function was improved after bile drainage operation (TB < 14 mg/dL) [27]. Thus, it seems that mild renal tubular dysfunction can be observed in OJ patients with TB level more than 8 mg/dL while more obvious renal tubular dysfunction or AKI should be concerned in patients with TB over 26 mg/dL.

Animal Models of Hyperbilirubinemia

To understand the pathophysiology of hyperbilirubinemia and renal dysfunction observed in the real patient condition, the animal models are crucially needed. The most widely used model is bile duct ligation (BDL) which has been performed in the rat, rabbit, guinea pig, cat, dog, and baboon [28-31]. Although bile productionsare quite different among species[30], highest production in rabbit and guinea pig and less indog or rat (8-12 times), the early hyperbilirubinemia uaually develops within few days with the complete stagnant of bile flow in around 1 week [31]. After several weeks, the level of hyperbilirubinemia is less due to lower hepatocyte bile production from liver disease progression. The variation in BDL model depends on the surgical technique and animal species used. In BDL without duct resection or division, the peak level hyperbilirubinemia occurs by the 7th and 2nd -5th postoperative days in dog and rat, respectively with the slightly increased serum level of TB which is rarely more than 5 mg/dL.Then,the serum TB reduces to near baseline level at the 3rd and 2nd post-operative weeks in dog and rat, respectively [28, 32, 33]. On the other hand, in BDL with complete ligation by duct resection or division, the peak hyperbilirubinemia occurs at the same time as BDL without duct resection but the serum TB level istremendously escalated, possibly as high as 15 mg/dL, in either dog or rat andis consistently high even by the 8th and 5th postoperative weeks in dog and rat, respectively [34-36]. Thus, the natural course of the BDL model without fortified procedures is comparable to the human condition only in the 1st week, but BDL with fortified procedure, the acute early cholemia with subsequently hepatocellular damage,could mimic the natural course of cholangiocarcinoma patients [25]. Choledochocaval anastomosis (CDCA), the surgical anastomosis of the bile duct and the venous system, is another model that has been developed to study the isolated effect of cholemia [37, 38]. In CDCA model, the serum TB level is more severe which might elevate as high as 60 mg/dL within few days [38]. The hepatic function remains minimally altered and the serum TB levelis persistently high. The CDCA model is a useful model for the isolated high hyperbilirubinemia studies. In a comparison study between CDCA and BDL dog model, the TBlevel in CDCA (16.5 ± 3.67 mg/dL) was significantly higher than BDL (3.4±1.75 mg/dL) with mostly conjugated bilirubin in both models (CDCA, 13.4±2.24 mg/dL vs. BDL, 3.2±1.7 mg/dL) [39]. Nevertheless, the rapidly and enormously high serum bilirubin in the CDCA model is not as physiologic as the BDL model and might not represent the actual patient condition. The last model that has been used is continuous bile acid infusion [40, 41]. This model is less mimic to the real patient condition and needs continuous anesthesia which might affect the study outcome. The continuous bile infusion model mainly aims to study the bile acid effect more than the jaundice consequence. However, all of the animal models mentioned above have been used to explore the renal dysfunction due to hyperbilirubinemia in different aspects.

Electrolyte Disturbancesin Hyperbilirubinemia

1. Hyponatremia

Hyponatremia is a result of water regulation abnormality [42]. Several animal model studies demonstrated mild hyponatremia in chronic bile duct obstruction [43, 44]. The prevalence of hyponatremia in OJ patients is not properly mentioned in the literatures. However, several series demonstrated that OJ is frequently associated with mild hyponatremia. Padillo et al. found hyponatremia in all of 63 cases of preoperative OJ patients either from malignant or benign causes[14]. In another study, more than half of the cholangiocarcinoma patients showed serum Na levelbelow 135mEq/L [26].

Pathophysiology of Hyponatremia

Due to the clinical variation of jaundice patients such as obscure disease onset, biliary obstruction severity, and several underlying diseases, renal pathophysiology study in human is difficult. Thus, most of the hyponatremia studies in hyperbilirubinemia were based on animal models.

Biphasic Renal Responsesafter BDL

In the early phase after BDL (1-3 days), preserved plasma Na level was observed due to the inappropriate diuresis/natriuresis with preserved glomerular filtration rate [43, 44](Figure 1). The natriuresis was observed in few hours after BDL which might be a result of hepatic metabolite excretion from the kidney instead of the liver. At 24 h, the natriuresis was obviously demonstrated in BDL; urine Na excretions in BDL and sham-operated rabbits were 4.8±4.2 and 1.7±1.8 mEq/24h, respectively with lower Na fractional excretion (FENa) in the sham-operated rabbits(0.61±0.3% in BDL vs. 0.27±0.31% in sham, respectively). During the first few days, the frail BDL-rabbits drank 60% and ate only 8% when compared with the sham-operated rabbits. Of interest, hypodipsia occurred as early as 24h after BDL [45]. The natriuresis together with hypodipsia, poor intake and mild to moderate negative water balance stabilized serum Na level (145±4 mEq/L) at 24 h after BDL in rabbit [45]. Subsequently, the renal adaptation to reduce natriuresis and water loss were observed at the 4th day after BDL; urine Na excretion in BDL and sham-operated rabbits were 1.9±1.5 and 4.7±1.7 mEq/24h, respectively with no significantly different FENa [44]. Despite diuresis, hypodipsia, and rapid weight loss in the early phase of bile duct obstruction, hypernatremia was not apparent. Instead, the normal plasma Na level, 139±1 mEq/L, was observed [44].

In the late phase of BDL (3-7 days), the renal compensation for water loss and natriuresis continued. At the 10th day after BDL, urine Na excretion in BDL and sham-operated rabbits were 1.1±0.7 and 8±4.4 mEq/24h, respectively. Even with renal compensation, mild hyponatremia occurred at 10th days after BDL [44]. However, it is important to mention that the cut point between acute and late phase biliary tract obstruction in human condition is difficult to determine due to the vague onset of the complete obstruction. Indeed, some patients might be in the incomplete obstruction for some periods then the disease progresses into the complete obstruction late after the onset of disease. Of note, Sitprija et al reported natriuresis in cholangiocarcinoma patients even after 5-7 weeks from the jaundice onset [25].

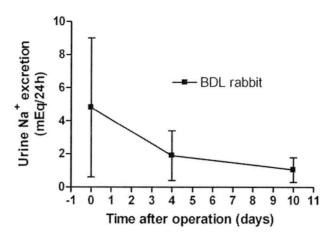

Adapted from reference [44].

Figure1. The natural course of natriuresis after BDL in rabbit. The 24 h urine Na collection from the BDL rabbits at 1st, 4th, and 10th day after BDL surgery demonstrated early rapid natriuresis (1st day) and more Na reservation in the later phase (10th day).

Hyperbilirubin Induced Natriuresis, the Direct Bile Toxicity toRenal Tubule

Topuzlu and Stahl intravenously infused 15 mL of bile in pentobarbital-anesthetic dogs, weighing 20-25kg, and found that natriuresis with increased urine output can be immediately detected[41]. Bile acid is the natural detergent which directly affects the proximal tubule. The underlying mechanism was proposed to be, at least in part, the neutralization of amiloride-sensitive Na^+/H^+ antiporter by sulfated bile acid in the *in vitro* study on human membrane brush border kidney cortex (Figure 2) [46].

Additionally, in situ microperfusion of taurocholate to rat proximal tubular cell caused 30% less fluid absorption [47]. It seems likely that the natriuresis from the bile effect triggers the initial loss of sodium and water in urine. The volume loss in the early phase of OJ referred to as "dry jaundice" was opposite to the volume retention in liver cirrhosis recognized as "wet jaundice" [48].

Cholangiocarcinoma jaundice patients with significant liver damage, determined from low serum albumin, showed sodium and water retention compatible with "wet jaundice". In contrast, the cholangiocarcinoma jaundice patients with less liver damage, reflected by normal serum albumin, demonstrated natriuresis and clinically hypovolemia comparable to "dry jaundice" [25]. Thus, the sodium retention after natriuresis in OJ patients might be an indicator of hepatocellular damage.

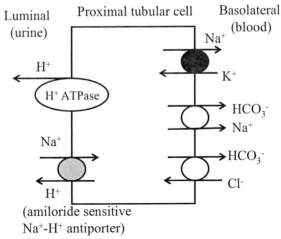

Figure 2. Bile acid inhibits amiloride sensitive Na^+/H^+ antiporter and causes natriuresis.

Neurohormonal Alteration-Induced Hyponatremia

The neurohormonal response to control body volume status is essential for sodium homeostasis. Hypovolemia in jaundice had been observed nearly a century ago. An earlier experiment on intravenous injection of bile in anesthetic dogs revealed diuresis, natriuresis, and hypovolemia [41]. The blood volume measurement with radioiodinated albumin showed 15% reduction in BDL rats [49]. In parallel, the body composition measurement with multi-isotope tracer dilution also found 15% and 30% reduction of total body water and extracellular fluid compartment, respectively, at 6 days after BDL in rabbit [43]. A previous study in jaundice patients with this tracer dilution technique was comparable with the BDL rabbit study [50]. The reduction in the total body water (41.8 vs. 46.2% body weight; $p = 0.02$) was caused by decreases in both interstitial fluid and plasma volume of the extracellular compartment. Additionally, bioelectrical impedance study in 63 OJ patients also demonstrated lesser extracellular fluid volume[27]. The total body water was 10% less in OJ patients. Despite volume depletion, atrial natriuretic peptide (ANP), a hormone excreted from myocyte to reduce total body water through natriuresis, was increased in OJ. Under normal physiology, ANP is released from atrium in response to the mechanical stretching in the heart due to hypervolemia or high blood pressure (Figure 3) [51, 52]. However, in OJ, ANP seems to be a primary response directly from bile duct obstruction. Within 24 h after BDL in the rabbit, ANP surge (41±7 vs 10.7±2 fmol/mL; $p = 0.02$ in OJ and control, respectively) had been demonstrated, then the ANP level reduced but was still in a higher value at 72 h after BDL (28.4±5 vs 11.6±3 fmol/mL; $p = 0.02$ in OJ and control, respectively) [45]. The heart immunohistochemistry of BDL rabbit confirmed that ANP was identified in both atriums as early as 24 h after obstruction in BDL rabbit and can be detected at 72 h after obstruction [53].

To determine if ANP response was stimulated by the mechanical stretching at biliary system or the absence of bile in duodenum or the direct effect of bile components on the atrium, ANP levels from rabbits with BDL and CDCA were measured[54]. The ANP levels increased from baseline, 10 to140 fmol/L as early as 1 h after CDCA and raised to 350 fmol/L at 4 h. On the other hand, the ANP level in BDL was 40 fmol/L at 4h.

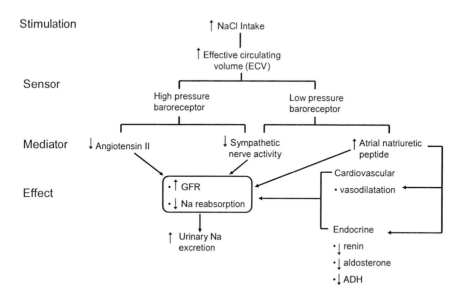

Figure 3. The normal neurohormonal responses of atrial natriuretic peptide (ANP), renin, aldosterone, and antidiuretic hormone (ADH). ANP is normally stimulated in response to increased effective circulatory volume (hypervolemia). Hyperbilirubinemia directly activates ANP despite lower sodium intake and hypovolemia.

From the rapidly onset of ANP raising at 1h and the ninth fold higher ANP at 4h in CDCA compared with BDL, the increased ANP in OJ should be the direct chemical effect of bile component on atrial myocyte but not biliary tract stretching or bile absence from the duodenum. It was hypothesized that cholemia might cause cardiopressor effect, leading to cardiac distension and ANP hypersecretion [48]. After ANP induction of natriuresis and volume depletion, other hormones, antidiuretic hormone; (ADH), aldosterone, and renin, thatregulate water homeostasis increase as a secondary phase response. In BDL rabbit, aldosterone and ADH increased as early as 24h after obstruction while plasma renin activity (PRA) elevated at 72h [45]. In OJ patients which were generally comparable with chronic BDL model, the higher plasma ANP, aldosterone, PRA, and ADH were observed in several studies [14, 25, 55].

Taken together, the neurohormonal responses after OJ seem to begin with the rapid primary ANP surge which has been shown in early phase of the animal models (Figure 3). Then, other hormonal responses secondary to water and Na depletion are provoked to maintain body homeostasis.

"Jaundiced Heart": Another Hyponatremia Contributing Factor

Low effective circulatory volume due to hypotension from vasodilatation/cardiopressor is one of the most important factors contributing to the heightened ADH and hyponatremia. The effect of hyperbilirubinemia on cardiac function had been reported since 1932. It was noted that the blood pressure in essential hypertension patients with severe jaundice returned to normal. Animal studies in parallel to this observation also revealed the reduction in blood pressure of renal artery stenosis dog model after BDL surgery [5, 56]. In CDCA dogs, hypotension with vasodilatation and blunted response to vasoactive drugs had been demonstrated [57, 58]. On the other hand, the values of systolic blood pressure in BDL rat and baboon were normal but more susceptible to hemorrhagic shock and had blunt response to vasoactive drugs [5]. In the *ex vivo* study, the bile acid derivatives suppressed the contractile response of isolated rat portal vein to norepinephrine [59, 60]. The cellular mechanism responsible for the impaired response to norepinephrine was shown to be the defect in the expression of α-1 adrenergic receptor. Evaluation of the ∝-adrenergic receptor in 3-day BDL rat demonstrated blunted pressor responsiveness to norepinephrine, electrical stimulation, and the ∝1-adrenergic receptor agonists, such as methoxamine and phenylephrine. However, normal responsiveness to BHT-933 and clonidine, the ∝2-adrenergic receptor agonists, was noted in these animals [61].

In addition to the low effective circulatory volume from vasodilation-induced hypotension, the negative inotropic and chronotropic effects were observed. OJ-induced bradycardia causing negative chronotropic effect had been mentioned for over a century [5]. The bradycardia had been demonstrated in BDL rat during bile acid injection [62]. Interestingly, the negative chronotropic effect was dose dependent and vagally mediated. Atropine (anticholinergic drug) and vagotomy attenuated the negative chronotropic effect whereas reserpine (sympatholytic drug) showed a lesser effect [62]. It was hypothesized that the bile acid-induced negative chronotropic effect through the formation of a monolayer on the surface of cell membrane, thereby mechanically interfering with the membrane function[63]. Regarding the bilirubin-induced negative inotropic effect, the dose dependent negative inotropic effect on the rat papillary muscle and isolated ventricular myocytes had been demonstrated in an *ex vivo* study [64]. However, the results of the inotropic effect of bile were conflicting. In BDL model, the cardiac output results were inconsistent which, in part, might be due to different cardiac function assessment methods [65-67]. However, the noninvasive measurement of cardiac function in conscious CDCA dogs showed negative inotropic effect [48]. A prospective study in 13 OJ patients by a Swan-Ganz catheter insertion to determine the hemodynamic parameters before and after biliary drainage showed subclinical myocardial dysfunction that was improved after relieving the obstruction; cardiac index 3.4±0.6 vs. 3.8±0.7 L/min/m^2(p = 0.03); systolic index 48±6 vs. 52±7 systolic volume/m^2 (p = 0.02) in before and after biliary drainage condition, respectively [27]. In addition, the volume required to reach pulmonary capillary pressure of 14 mmHg was 427±355 vs 703±40 mL (p = 0.01) in before and after biliary drainage condition, respectively.

Therefore, hyperbilirubinemia-induced hyponatremia seems to be a result of bile-induced natriuresis and water loss from either direct toxic effect or increased ANP. The rapid ANP response might be one of the hyperbilirubinemia-induced negative cardiopressor effects. The lower effective circulatory volume from the combination of fluid loss and negative cadiopressor trigger the secondary hormonal responses which include aldosterone, ADH, and PRA, leading to hyponatremia.

2. Other Electrolyte Abnormalities

Hypokalemia

The prevalence of hypokalemia was not properly collected but was reported only as patient series or case report. Theoretically, the effect of high PRA and aldosterone in hyperbilirubinemia might cause kaliuresis by activation of Na^+-K^+ ATPase at basolateral membrane of the principal cell, mostly in cortical collecting duct (CCD) and by stimulation of epithelial sodium channel (ENaC) and renal outer medullary potassium channel (ROMK) at luminal side. The Na^+-K^+ exchanges aims to decrease intracellular Na^+ and increase intracellular K^+. Then, the increased intracellular K^+ will be excreted through the ROMK (Figure 4). Additionally, the hyperbilirubin-induced natriuresis results in continuously high sodium concentration at CCD, and, increased Na^+ absorption through the ENaC. The choride (Cl^-) ion outside the lumen acts as negative charges (potential difference), leading to more K^+ excretion through the ROMK (Figure 4). Nevertheless, there were only few reports of hypokalemia in jaundice patients [26, 68-70]. This might be explained from an easier reversibility of hypokalemia than hyponatremia. Sitprija et al reported hypokalemia and hyponatremia in cholelithiasis-induced jaundice patients with total bilirubin of 15 mg%[70]. At 11th postoperative day, hypokalemia was improved but not hyponatremia (Table 1) [70]. Other reports on hypokalemia were associated with other proximal tubular defects which will be mentioned in the part of Fanconi syndrome.

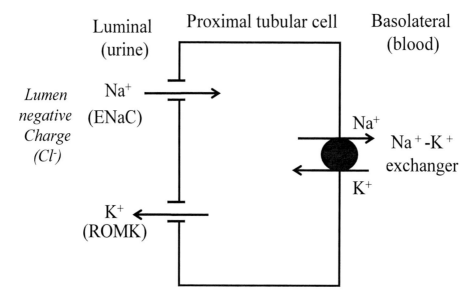

Figure 4. Kaliuresis is activated by high aldosterone (and PRA) and increased potential difference from natriuresis. Aldosterone activates Na^+-K^+ exchanger, resulting in more intracellular K^+. High NaCl from natriuresis induces Na^+ absorption through ENaC. More negative charge in the luminal side from retained Cl^- with increased intracellular K^+ causes more K^+ excretion through ROMK.

Table 1. OJ-induced hypokalemia patient parameters

	Postoperative day		
Laboratory parameters	7	9	11
Plasma Na (mEq/L)	125	127	133
Plasma K (mEq/L)	2.8	2.8	3.9
Plasma osmolarity (mOsm/kgH$_2$O)	235	-	270
BUN (mg/dL)	19	-	17
Creatinine (mg/dL)	0.8	-	0.6
Urine Na$^+$ (mEq/day)	230	215	190
Urine K$^+$ (mEq/day)	130	78	70

Adapted from reference [70].

Hypouricemia

The data on the incidence and prevalence of hypouricemia with hyperbilirubinemia are limited. Bairaktari et al. found 90 hypouricemic patients (1.24%) from 7,250 patients at the admission, of which more than 90% were caused by inappropriately increased renal excretion [68].

Interestingly, the most common causes were OJ (18 from 90 patients). Schlosstein et al. reported 15 hypouricemic patients from over 4,000 screenings of admitted patients [17]. More than half of these patients had hyperbilirubinemia. The mechanism of hyperbilirubinemia-induced hyperuricemia in cholangiocarcinoma patients was hypothesized to be a result of hyperuricosuria due to a presecretory reabsorptive defect associated with a relatively high urate secretion at the proximal tubule [16]. Uricosuria was slightly suppressed by pyrazinamide (strongly uricosuric inhibitor) and was surprisingly enhanced by probenecid (uricosuric enhancer).

Thus, the pyrazinamide nonsuppressible urate excretion with a post-probenecid huge uricosuria suggests that bile impairs uric acid absorption at S1 and S3 segments of the proximal tubule.

Malnutrition-Induced Electrolyte Imbalance (Calcium, Phosphate, and Magnesium)

The decreased intestinal bile from prolonged obstruction leads to impaired absorption of fat and results in steatorrhea. Vitamin D, one of the fat soluble vitamins, might become deficiency in chronic bile duct obstruction and could be responsible for hypocalcemia and hypophosphatemia [71, 72].

Moreover, OJ patients might have severe weight loss from their primary disease, usually malignancy or chronic intermittent infection, leading to hypomagnesemia [73, 74]. Although occasionally found in clinical practice, there were very limited case reports in adult with these electrolyte imbalances.

Renal Tubular Acidosis and Fanconi Syndrome

There was only a single report case of OJ from choledocholithiasiswith serum total bilirubin of 32 mg/dL and diagnosed as Fanconi syndrome [69].

The authors showed pan proximal tubular dysfunction; renal potassium, calcium, phosphate, magnesium, urate, and glucose losses in urine with mild hyponatremia (136 mEg/L), hypokalemia, hypocalcemia, hypophosphatemia, hypomagnesemia, and hypouricemia. Although OJ with Fanconi syndrome seems to be uncommon, it might be concluded that hyperbilirubinemia could affect the proximal tubular function.

The concise pathophysiology of hyperbilirubinemia-induced electrolyte disturbances can be concluded in Figure 5.

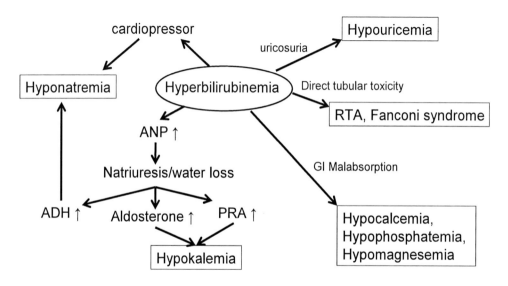

ADH, Antidiuretic hormone; ANP, Atrial natriuretic peptide; PRA, Plasma renin activity; RTA, renal tubular acidosis

Figure 5. Pathophysiology of hyperbilirubinemia-induced electrolyte imbalances.

Acute Kidney Injury in Hyperbilirubinemia

As previously mentioned, AKI was the first renal dysfunction that was recognized to be associated with hyperbilirubinemia. Several patient series described jaundice as a preoperative risk factor for postoperative AKI in the surgical patients. Although different causes of jaundice were reported, the overall mortality and the incidence of AKI were consistently high [24]. From the data in perioperative OJ patients, AKI develops in approximately 10-30% and contributes to more than 70% mortality [7, 75-78].The prevalence of AKI in preoperative patients seems to be not correlated with the etiologies of hyperbilirubinemia or the duration of jaundice. The relationship between AKI and jaundice was more prominent in postoperative OJ patients. The rapid oliguria/anuria in the postoperative period and hypotension due to septic, hypovolemic, or cardiogenic shock had been reported [79, 80]. The degree of biliary obstruction might relate to the incidence of postoperative AKI. Dawson et al reported 28 postoperative AKI patients out of 103 laparotomy-treated jaundice patients; 6 AKI patients had preoperative bilirubin above 20 mg/dL [81]. In the series of pre-postoperative creatinine clearance study of 15 OJ patients, a drop in creatinine clearance was greater with higher serum bilirubin level [82, 83]. When 168 AKI patients without jaundice were compared with 67 AKI patients with jaundice, the mortality rate was 85% in patients with serum TB above 20 mg/dL and only 33% in patients with serum TB below 10 mg/dL ($p < 0.01$) [84].

Regarding the human kidney histopathology in OJ patients, only few reports had been mentioned in the literatures. The term "chloremic nephrosis" had been introduced from the patient autopsy with bile stain kidney [85, 86]. An autopsy case report of cholangitis patient who died from sepsis with the total serum bilirubin of 15 mg/dL showed macroscopically deeply bile stain kidney [70]. The histopathology revealed bile casts in the lumen of the distal convoluted and collecting tubules with occasionally noted vacuolization of the proximal convoluted tubular epithelial cell without any changes in the glomeruli and interstitium [70]. However, the vacuolization of the proximal tubular cell was nonspecific and could be found not only from bile acid toxicity but also in other conditions such as sepsis and ischemic reperfusion injury [87, 88]. Kawaguchi et al reported an immunohistochemical and clinicopathologic analyses of glomerular alterations in 20 autopsy cases with OJ [89]. The authors found IgA trapping in some patients which was not influenced by the duration or intensity of jaundice. However, the renal function was not mentioned in the study. Recently, Uslu et al. studied kidney histology from 20 OJ patients with TB levels ranged 4.5-21 mg/dL and serum creatinine concentration ranged 0.6-2.7 mg/dL [90]. The most prominent renal histology findings were dilatation of peritubular venule and ATN in more than 60% of the patients. The dilated peritubular vessels werehypothesized to be the effect of endotoxinemia-induced inducible nitric oxide synthase (iNOS) production [90].

Pathogenesis of Hyperbilirubinemia-Induced AKI

The pathophysiology of AKI after OJ is still unestablished and complicated, but might consist of 1). renal macrocirculatory change from natriuresis with negative cardiac inotropic and defective vasoactive vascular responses 2). renal microcirculatory alteration from endotoxinemia, nitric oxide, prostaglandin, or other mediators 3). direct toxicity of biliary products. The renal macrocirculatory change was the most important contributing factor of AKI and was already mentioned in the hyponatremia section. The other two topics will be discussed in this section.

Hyperbilirubinemia-Induced RenalMicrocirculatory Alteration

The renal microcirculatory alteration might be caused by endotoxinemia and other mediators. Endotoxin, the gram negative surface component from intestinal organisms, had been detected in 25-85% of OJ patients [91, 92]. In normal patients, endotoxin is normally released from gut into the portal circulation and eliminated through the liver. The Kuffer cells trap and change endotoxin while hepatocytes pinocytose and eliminate it [93]. The absorption of endotoxinis inhibited, in part, by bile salts [91]. Endotoxin was identified in either liver parenchymal disease [94-96] or extrahepatic biliary obstruction [97-99]. Bailey et al reported that all of the patients with preoperative endotoxemia showed higher mortality rate [91]. Endotoxin-induced vasoconstriction in the microcirculatory vessels had been shown in endotoxin injection model [100]. The BDL model in rodents and dogs demonstrated a decrease in hepatic microcirculation [101-104]. The OJ rats were highly susceptible to endotoxin injection [98, 105, 106]. However, the association between endotoxemia and AKI in human patients was still controversial [107]. Administration of lactulose or bile acid was used to prevent endotoxin absorption in OJ in either animal models or human patients. The results were inconsistent among various studies [108-113]. Several vasoactive mediators associated with endotoxemia such as nitric oxide (NO) and prostaglandin (PG) was hypothesized to be responsible for the jaundice-induced hypotension. Several studies in BDL rat found increased NO in serum and urine [114-116]. Li et al showed that NO in BDL rat was produced from the Kupffer cells in the liver [115]. Of interest, iNOS deficiency mice with BDL were less severe than wild type animals [117]. In parallel, the urinary PGs (thromboxane A2, PGE2, 6-keto PGF2 alpha) underwent along with serum angiotensin II in OJ patients and BDL rats [118, 119]. Additionally, the reduced intestinal PGE1 in BDL rats was associated with higher bacterial translocation [37, 120]. Nevertheless, the role of endotoxin reduction and vasoactive mediators control for attenuating hyperbilirubinemia-inuced renal dysfunction is still inconclusive.

Direct Nephrotoxicityof Biliary Products

Although several studies mentioned the association between the severity of hyperbilirubinemia and patient morbidity/mortality, including cardiac dysfunction when TB was above 15 mg/dL [27],higher mortality rate in preoperative patients with TB over 20 mg/dL [84], more AKI in cholangiocarcinoma patients with TB more than 26 mg/dL [25], the studies that showed the direct nephrotoxicity of the bile products were limited. Injection of 15 mL of bile to anesthetized dogs caused natriuresis with normal level of plasma bilirubin (0.6 mg/dL) [41]. The high dose of bile salts injection, despite resulting in supraphysiologic elevation of serum cholate and taurocholate, did not show nephrotoxic effect but induced more susceptibility to renal ischemia [121]. It seems that bile components, especially the usual concentrations detected in OJ patients, were toxic to heart, induced natriuresis, but did not possess the direct toxicity to renal glomeruli or tubular epithelial cell.

In summary, the major pathophysiology of hyperbilirubin-induced AKI is hemodynamic changes due to the negative inotropic/chronotropic effect, vasodilatation, and natriuresis-induced volume depletion. There is insufficient evidence of direct nephrotoxicity from bile components. However, the effects of endotoxinemia and several vasoactive mediators toward AKI in OJ are still debatable. In this regard, several endotoxin-lowering therapeutic strategies such as lactulose or oral bile salts administration seem to be disappointed.

Prevention and Treatment of Hyperbilirubinemia-Induced Renal Dysfunction

Restoration of the Circulating Extracellular Volume

Hyperbilirubinemic patients especially from OJ usually present with water and sodium depletion at the time of diagnosis. These patients are also susceptible to development of hypotension, have a poor vascular response to volume loss, and have impaired myocardial function, all of which result in poor renal perfusion. This may be worsened by translocation of plasma volume, increased insensible losses due to fever and sweating, and vasodilation due to severe sepsis in patients complicated with cholangitis. Hence, the maintenance of circulating extracellular volume, comprising rehydration by fluid replacement as well as anemia correction, is the mainstay of prophylaxis and treatment of AKI in hyperbilirubinemic patients. A previous study in patients with OJ showed that rehydration with 5% glucose solution 24 hours before operation could prevent postoperative AKI in all of 23 patients [122]. Although no randomized controlled trials (RCTs) have been conducted to find out the preferred type of fluid for rehydration and the proper monitoring methods in OJ patients, normal saline would be the best fluid-replacement solution since OJ patients usually loss isotonic or even slightly hypertonic fluid. The patients would be carefully monitored, even with central venous pressure, of diuresis to ensure the adequate fluid rehydration.

Low hematocrit is another important risk factor in this population, thus, normalization of the hematocrit is worthwhile. Williams et al. [11] demonstrated a reduction in mortality after preoperative blood transfusion. Moreover, careful monitoring and maintaining of fluid status and cardiac performance should be simultaneously carried out.

Force Diuresis

Force diuresis, especially by using osmotic diuretic agent such as manitol, has been proposed to preserve renal function based on the hypothesis that manitol does not only cause diuresis and natriuresis but also induces volume expansion, maintains renal blood flow at low perfusion pressure, and prevents endothelial cell swelling and tubular obstruction. Several studies demonstrated the beneficial effects of mannitol in both animal and human models. Firstly, mannitol administered in hyperbilirubinemic rats before renal ischemia could preserve renal function and reduce the mortality [123]

Similarly, hyperbilirubinemic dogs subjected to hemorrhage had a transient but marked decrease in glomerular filtration rate that could be avoided by pre-hemorrhage administration of mannitol [124]. However, the roles of mannitol in human studies remained inconclusive since then. In a nonrandomized study, Dawson demonstrated that the use of mannitol immediately before the operation and for the first 2-3 postoperative days could prevent the decline in renal function in OJ patients[82].

In contradiction, a prospective randomized study by Gubern et al.[76] revealed that preoperative mannitol administration did not prevent the decrement of renal function after surgery when compared with no mannitol group. Moreover, it might even be harmful to those patients with impaired preoperative renal function.

Another report also observed that the incidence of postoperative renal impairment was still high despite the mannitol use [75]. This might be the adverse effect of the osmotic diureticagent that induced water depletion on top of the reduction of extracellular water and intravascular volume in patients with hyperbilirubinemia.

Another agent introduced to force diuresis is furosemide. In a randomized trial, Parks et al.[122] compared a preoperative bolus of furosemide and furosemide plus dopamine for 48 hours after surgery in patients who already received preoperative hydration. There was no development of postoperative AKI in both groups with comparable postoperative renal function. This implies that a more careful maintenance of hydration and electrolyte balance must be more important than the specific effects of the medications given.

In conclusion, the role of forced diuresis either by mannitol or furosemide would be applied with caution and should be performed after restoring the effective extracellular volume.

Avoidance of Nephrotoxic Drugs

The hyperbilirubinemic patients are more susceptible to nephrotoxic drugs than normobilirubinemic patients. Both animals with BDL and human subjects with cirrhosis responded to nonsteroidal anti-inflammatory drugs with a marked decrease in both renal blood flow and glomerular filtration rate[5]. The incidence of aminoglycoside-induced AKI in patients with OJ was also higher than who did not receive the aminoglycoside. The incidence of impaired renal function in patients with OJ who received either gentamicin or tobramycin were 32% compared with 11% and 5.6% in patients with OJ who did not receive aminoglycosides and in non-OJ patients receiving aminoglycoside, respectively [125]. As such, the obvious risk factor for AKI in patients with OJ is the degree of hyperbilirubinemia.

Anti-Endotoxin Therapy

In OJ, endotoxemia caused by the lack of bile salts in the intestine is at least partially responsible for the development of AKI. The attempt to reduce endotoxin absorption has been postulated to prevent and improve the perioperative AKI. Several studies demonstrated that the preoperative oral administration of sodium deoxycholate to hyperbiliruminemic patients could abrogate systemic as well as portal endotoxemia and prevent AKI [108, 109]. The reintroduction of bile salts to the intestine by internal biliary drainage provided benefit in mortality reduction as will be discussed later in the section of lowering bilirubin levels by biliary drainage. However, supplement of bile salts, biliary drainage, and other anti-endotoxin therapies including non-absorbable antibiotics and lactulose have not gained popularity and their potential benefits seem rather low [110, 111, 126, 127]. An obvious relationship between anti-endotoxin therapy, reduced endotoxin plasma levels, and improved postoperative renal function in the clinical setting has not been conclusively proven and requires the further confirmation for the benefit by well-designed controlled studies [77, 112].

Lowering Bilirubin Levels

Biliary Drainage

In OJ patients that required surgical treatment, the risk of postoperative AKI directly related to the degree of hyperbilirubinemia [82, 83]. Therefore, it is hypothesized that lowering bilirubin levels might be beneficial in reduction of the incidence of AKI. Two methods of biliary drainage, external percutaneous drainage and internal drainage were reported to test this hypothesis. The benefits of preoperative percutaneous biliary drainage in OJ were inconsistent among the studies. Most of the reports could not demonstrate the benefit [128-131]. However, it appears that an internal biliary drainage would result in a greater clinical benefit than external drainage. One animal study in rat with BDL illustrated that the mortality after inducing peritonitis was decreased from 83% to 25% by internal biliary drainage [132]. This benefit was also shown in a previous human study which reported the improved postoperative pancreatectomy outcome with prior internal biliary drainage[133]. The rationale that might explain the superior outcome of the internal over external drainage would be the additive benefit of reintroduction of bile salts to the intestine. As stated earlier, bile salts could neutralize endotoxin in the intestine. The absence of bile salts might play an important role in the pathogenesis of AKI in OJ. Several physiologic functions were improved after internal biliary drainage including hepatic function[133-136], immunity[134], tolerability of the kidney to ischemia[110], fluid derangement [55], and intestinal barrier function[137]. An earlier human study demonstrated that after endoscopic internal biliary drainage in 18 OJ patients, the prior contracted extracellular volume was increased significantly along with the increase in ANP concentration and aldosterone level [55]. The volume of fluid gain was correlated with diuresis after biliary decompression and the increase of the fractional sodium excretion.

Although it is still inconclusive regarding the definite indications for preoperative internal drainage, however, the procedure is very appealing and has already been implemented as a routine management in some countries [136]. Indeed, patients with choledecholithiasis with cholangitis clearly received benefit from this intervention[80].

Extracorporeal Liver Support

Extracorporeal liver support systems (ECLS) have been innovated and applied to liver failure patients. Basically, ECLS could replace liver functions which can be divided into two main categories including detoxification and synthetic or metabolic function. The bilirubin should decrease after these kinds of treatment.

ECLS can be classified into two broad groups: non-cell based and cell-based systems (Table 2). Non-cell based systems do not incorporate tissue and provide only detoxification function by using membranes and adsorbents. Cell-based systems aim to provide the excretory, synthetic, and metabolic functions of the liver by using the living liver cells [138].

Table 2. Extracorporeal liver support systems

Non-cell based systems
    Hemoperfusion techniques
    Hemodiabsorption
    Plasma exchange techniques
    Plasmapheresis
    Plasma filtration absorption
    Selective Plasma Filtration Technology (SEPET)
    Albumin dialysis
    Molecular Adsorbent Recirculating System (MARS)
    Single Pass Albumin Dialysis (SPAD)
    Prometheus
Cell based systems (Bioartifical liver support systems)
    Human hepatocytes
    Porcine hepatocytes

Hemoperfusion was the first ECLS technique that was introduced to treat the hyperbilirubinemic liver failure patients in the 1960s. The blood was circulated over a sorbent material for the purpose of removing protein-bound toxins. Initial studies and clinical reports of hemoperfusion with charcoal demonstrated improved neurological status. Nevertheless, a subsequent RCT study could not exhibit survival benefit in fulminant hepatic failure patients [139]. Moreover, the technique was associated with severe complications, including thrombocytopenia and impaired coagulation parameters.

Hemodiabsorption is a technique combining hemodialysis (HD) and hemoperfusion. The blood is passed through a hemodialyzer containing a suspension of sorbents on the dialysate side that provides the capacity of adsorption. Previous controlled trials showed conflicting results [140, 141]. Although the largest prospective RCT study in 56 patients illustrated improvement in neurological status and blood pressure in patients treated with this method [142], there was no survival benefit. This device is no longer commercially available.

Plasmapheresis or plasma exchange has been used in liver failure since the 1970s. Plasma is separated from the cellular blood components and discarded while the equivalent volume of fresh plasma is replaced. Numerous small uncontrolled trials have been performed with various forms of plasmapheresis and could show improvement in hepatic encephalopathy and hemodynamic parameters but no difference in mortality was observed[143]. Furthermore, the usefulness of this approach is limited by the large amount of plasma needed, ineffective removal of intracellular protein- and tissue- bound toxins, and several complications, comprising pulmonary edema, infection, and hypocalcemia from citrate intoxication.

Other plasma therapy techniques, including high volume plasmapheresis [144-146], plasmapheresis with hemodiafiltration [147], and plasma filtration-absorption have been innovated and applied for liver failure patients. Plasma filtration-absorption(Figure 6A) was firstly used in the 1989 for the treatment of liver failure with intractable jaundice [148]. Since then, multiple uncontrolled observations have been published [149-152].Unfortunately, none of these above techniques have shown definite survival benefit despite improvements in neurological function and coagulation indices.

Selective plasma filtration therapy (SEPET) is a method for removing a specific plasma fraction containing substances within a specific molecular weight. During SEPET, the patient's blood is circulated through a hollow fiber plasma filter with a molecular weight cut-off at 100 kDa. A plasma fraction containing several accumulated toxins (small MW toxins and free cytokines) is discarded after passing the membrane but the majority of albumin and most clotting factors are retained. The fluid loss is replaced by a combination of electrolyte solution, human albumin solution, and fresh frozen plasma, and the system is in line with the standard HD machine [153]. Although no human clinical data have yet been reported, the FDA has approved the phase I trial of safety, tolerability, and efficacy of SEPET in liver failure.

Recently, various novel techniques based on the principle of albumin dialysis have been innovated and widely used. Basically, albumin binds to many toxins that accumulate in liver failure. Thus, albumin mixed in the dialysate is utilized to pull and carry albumin-bound toxin out of the blood compartment of the extracorporeal circuit. As a result, different albumin dialysis methods to remove albumin-bound toxins have been investigated. Most data are available for the "Molecular Adsorbent Recirculating System" (MARS), followed by "Prometheus", and "Single Pass Albumin Dialysis" (SPAD).

MARS is commercially available (Figure 6B). This extracorporeal technique recycles albumin containing dialysate using an anion exchange resin and active charcoal adsorption to extract and remove the albumin-bound toxins from the patient. Blood is circulated through a dialysis filter. Albumin-containing dialysate circulates in the extracorporeal circuit counter-current to the blood, allowing albumin-bound toxins in the blood to cross the membrane and bind to the albumin in the dialysate. The albumin-bound toxins are later removed by the anion exchange resin and the charcoal adsorber. The toxin-free albumin is recirculated to the extracorporeal circuit. Standard HD, either intermittent or continuous, is integrated to remove water-soluble toxins. Although, some small RCT studies have revealed the survival and clinical markers of disease severity benefit including hyperbilirubinemic level in acute liver failure patients, a meta-analysis based on four small RCTs and two selected nonrandomized trials in patients with acute-on-chronic liver failure concluded that MARS offered no significant survival benefit over the standard medical therapy [154]. Although MARS therapy has not demonstrated a definite survival benefit in patients with liver disease, it has been shown to improve hepatic encephalopathy and other metabolic parameters including hyperbilirubinemia. In acute liver failure condition, the procedure would be a potential benefit in sustaining patients until either liver transplantation can be performed or recovery occurs. Controlled clinical trials will be necessary before any firm recommendations for use in acute liver failure can be made. Regardless of this extreme liver condition, MARS could provide benefit in other liver conditions such as Wilson's disease, primary biliary cirrhosis, drug-induced liver injury, and cholestasis condition.

Prometheus is the system based on fractional plasma separation and adsorption (FPSA) and HD (Figure 6C). This technique uses a membrane with a cut-off of 250 kDa which is permeable to albumin. Toxin-laden albumin crosses the membrane and passes a neutral resin adsorber and an anion exchanger; toxins are cleared from the albumin and the toxin-free albumin is recirculated to the patient. The method is combined with additional HD to remove water-soluble and albumin-bound toxins [155]. The majority of studies regarding Prometheus are limited to patients with acute fulminant liver failure [156]. Ten patients with hepatorenal syndrome underwent two consecutive Prometheus treatments with significant improvements in renal and hepatic function markers, but the patient survival was not reported [157].

SPAD is a modification of MARS in which the albumin-containing dialysate undergoes a single pass through the dialysate compartment of the high-flux dialyzer counter-current to the blood, and is then discarded (Figure 6D) [158]. SPAD has been used to bridge patients toliver transplantation [159]. One study found that SPAD was as effective as MARS in removing albumin-bound toxins and more efficient to remove ammonia [160].

The inability of non-biological liver support techniques to replace the multiple functions of the liver leads to a rationale for the development bioartificial liver systems (BALs) in which living cells perform synthetic and regulatory functions [161]. Primary porcine hepatocytes, immortalized human cells, and cells derived from hepatic tumors are the most common cell sources used in BALs[162]. There are limitations and specific concerns associated with each cell type used in cell-based therapies. Human hepatocytes are difficult to grow in cultures and rapidly lose their liver-specific functions [163]. Cells from discarded human donor livers are in short supply and have low viability. Immortalized hepatoma cells might present a risk of tumorgenicity and the cells of animal origin are feared to spread xenozoonoses or to provoke immunological responses in patients. The cell type most utilized at present is of porcine origin because it is widely available and can be easily cryopreserved [162]. Cell-containing liver support systems have mainly been tested in single-center phase I and II trials to provide proof of principle that the systems improve clinical and biochemical parameters [162, 164, 165]. Up to now, no recommendations on their use can be made. While the safety of the cell-based devices remains unestablished, there are no uniform standards of efficacy. Consensus is needed in clinical trial design, choice of end points, use of controls, and indications for enrollment.

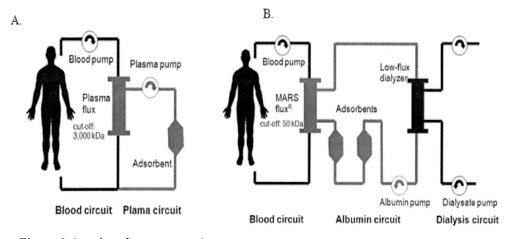

Figure 6. (continued on next page)

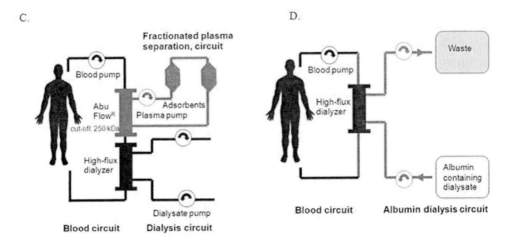

Figure 6. Schematic depiction of the different non-cell based extracorporeal liver support systems [A] Plasma filtration-absorption [B] MARS^R, molecular adsorbents recirculating system [C] Prometheus system^R or fractional plasma separation and adsorption [D] SPAD, Single Pass Albumin Dialysis.

In conclusion, prevention and treatment of AKI related to hyperbilirubinemia is firstly focused on fluid restoration with careful monitoring. Forcing diuresis could be applied after effective circulatory volume could be maintained. The potential nephrotoxic drugs must be rationally avoided. Hyperbilirubinemia should be reduced especially in patients who are going to have surgery.

Preoperative biliary drainage along with appropriate rehydration and metabolic support appears to be a potential adjunct to improve the condition of patients with OJ, particularly those with severe hyperbilirubinemia or sepsis. Finally, the extracorporeal liver support that have provided certain benefits in various metabolic and some clinical aspects in liver failure patients is a novel promising treatmentalthough the survival benefit and definite indication need further investigation.

References

[1] Menzies, JA. (1912). bservations on the Secretion and Composition of Human Bile. *Biochem J,* 6, 218-210.

[2] Hofmann, AF; Small, DM. (1967). Detergent properties of bile salts: correlation with physiological function. *Annu Rev Med,* 18, 376-333.

[3] Jacquemin, E; Hagenbuch, B; Stieger, B; Wolkoff, AW; Meier, PJ. (1994). Expression cloning of a rat liver Na(+)-independent organic anion transporter.*Proc Natl Acad Sci U S A,*91,.137-133

[4] Chen, CF; Kuo, CH. (1992). Toxic effect of bile acid ingestion in rats. *J Formos Med Assoc,*91,.746-743

[5] Green, J; Better, OS. (1995). hypotension and renal failure in obstructive jaundice-mechanistic and therapeutic aspects. *J Am Soc Nephrol,*5,.1871-1853

[6] Epstein, M (1994). Hepatorenal syndrome: emerging perspectives of pathophysiology and therapy. *J Am Soc Nephrol,*4,.1753-1735

[7] Armstrong, CP; Dixon, JM; Taylor, TV; Davies, GC. (1984). Surgical experience of deeply jaundiced patients with bile duct obstruction. *Br J Surg*,71,.238-234
[8] Dawson, JL. (1968). Acute post-operative renal failure in obstructive jaundice. *Ann R Coll Surg Engl*,42,.181-163
[9] Dixon, JM; Armstrong, CP; Duffy, SW; Davies, GC. (1983). Factors affecting morbidity and mortality after surgery for obstructive jaundice: a review of 373 patients. *Gut*,24,.852-845
[10] Walker, JG. (1962). Renal failure in jaundice. *Proc R Soc Med*,55,.570
[11] Williams, RD; Elliott, DW; Zollinger, RM. (1960). The effect of hypotension in obstructive jaundice. *Arch Surg*,81,.340-334
[12] Wittenstein, BH; Giacchino, JL; Pickleman, JR; Geis, WP; Rajagopalan, AE; Hadcock, WE; Freeark, RJ. (1981). Obstructive jaundice: the necessity for improved management. *Am Surg*,47,.120-116
[13] Zollinger, RM; Williams, RD. (1956). Surgical aspects of jaundice. *Surgery*,39,-1016 .1030
[14] Padillo, FJ; Cruz, A; Briceno, J; Martin-Malo, A; Pera-Madrazo, C; Sitges-Serra, A. (2005). Multivariate analysis of factors associated with renal dysfunction in patients with obstructive jaundice. *Br J Surg*,92,.1392-1388
[15] Padillo, FJ; Rodriguez, M; Gallardo, JM; Andicoberry, B; Naranjo, A; Martin-Malo, A; Mino, G; Sitges-Serra, A; Pera-Madrazo, C. (1999). Preoperative assessment of body fluid disturbances in patients with obstructive jaundice. *World J Surg*,23,687-681 ; discussion .687
[16] Magoula, I; Tsapas, G; Kountouras, J; Paletas, K. (1991). Cholangiocarcinoma and severe renal hypouricemia: a study of the renal mechanisms. *Am J Kidney Dis*,18,-514 .519
[17] Schlosstein, L; Kippen, I; Bluestone, R; Whitehouse, MW; Klinenberg, JR. (1974). Association between hypouricaemia and jaundice. *Ann Rheum Dis*,33,.312-308
[18] Bairaktari, E; Liamis, G; Tsolas, O; Elisaf, M. (2001). Partially reversible renal tubular damage in patients with obstructive jaundice. *Hepatology*,33,.1369-1365
[19] Schenker, S; McCandless, DW. (1964). Renal Disposition of Conjugated Bilirubin in Aglomerular Fish. *Nature*,202,.1344
[20] Odell, GB. (1959). Studies in kernicterus. I. The protein binding of bilirubin. *J Clin Invest*,38,.833-823
[21] Fulop, M; Katz, S; Lawrence, C. (1971). Extreme hyperbilirubinemia. *Arch Intern Med*,127,.258-254
[22] Klatskin, G. (1961). Bile pigment metabolism. *Annu Rev Med*,12,.250-211
[23] Fulop, M. (1967). Bilirubinemia and renal failure. *N Engl J Med*,276,.1209-1208
[24] Wait, RB; Kahng, KU. (1989). Renal failure complicating obstructive jaundice. *Am J Surg*,157,.263-256
[25] Sitprija, V; Kashemsant, U; Sriratanaban, A; Arthachinta, S; Poshyachinda, V. (1990). Renal function in obstructive jaundice in man: cholangiocarcinoma model. *Kidney Int*,38,.955-948
[26] Mairiang, P; Bhudhisawasdi, V; Borirakchanyavat, V; Sitprija, V. (1990). Acute renal failure in obstructive jaundice in cholangiocarcinoma. *Arch Intern Med*,150,.2360-2357

[27] Padillo, J; Puente, J; Gomez, M; Dios, F; Naranjo, A; Vallejo, JA; Mino, G; Pera, C; Sitges-Serra, A. (2001). Improved cardiac function in patients with obstructive jaundice after internal biliary drainage: hemodynamic and hormonal assessment. *Ann Surg*,234,.656-652

[28] Carlson, E; Zukoski, CF; Campbell, J; Chvapil, M. (1977). Morphologic, biophysical, and biochemical consequences of ligation of the common biliary duct in the dog. *Am J Pathol*,86,.320-301

[29] McCalden, TA; Eidelman, BH; Bloom, D. (1976). Abnormal cerebrovascular response to altered PaCO2 in baboons with obstructive jaundice. *Stroke*,7,.193-190

[30] Rous, P; Larimore, LD. (1920). The Biliary Factor in Liver Lesions. *J Exp Med*,32,-249 .272

[31] Wright, JE; Braithwaite, JL. (1964). The Effects of Ligation of the Common Bile Duct in the Rat. *J Anat*,98,.233-227

[32] Macgregor, CA. (1953). Nature of liver failure due to complete biliary obstruction. *AMA Arch Surg*,67,.901-878

[33] Masumoto, T; Masuoka, S. (1980). Kidney function in the severely jaundiced dog. *Am J Surg*,140,.430-426

[34] Ohlsson, EG; Rutherford, RB; Haalebos, MM; Wagner, HN, Jr.; Zuidema, GD. (1970). The effect of biliary obstruction on hepatosplanchnic blood flow in dogs. *J Surg Res*,10,.208-201

[35] Symeonidis, A; Trams, EG. (1957). Morphologic and functional changes in the livers of rats after ligation or excision of the common bile duct. *Am J Pathol*,33,.27-13

[36] Unikowsky, B; Wexler, MJ; Levy, M. (1983). Dogs with experimental cirrhosis of the liver but without intrahepatic hypertension do not retain sodium or form ascites. *J Clin Invest*,72,.1604-1594

[37] Alon, U; Berant, M; Mordechovitz, D; Hashmonai, M; Better, OS. (1982). Effect of isolated cholaemia on systemic haemodynamics and kidney function in conscious dogs. *Clin Sci (Lond)*,63,.64-59

[38] Hardison, WG; Hatoff, DE; Miyai, K; Weiner, RG. (1981). Nature of bile acid maximum secretory rate in the rat. *Am J Physiol*,241, G.343-337

[39] Mor, L; Blendis, LM; Mordehovich, D; Sideman, S; Brandes, JM; Better, OS. (1985). Serum bilirubin constituents in different experimental models of conjugated hyperbilirubinemia. *Biochem Med*,33,.267-256

[40] Byers, SO; Friedman, M; Biggs, MW; Gunning, B. (1953). Observations concerning the production and excretion of cholesterol in mammals. IX. The mechanism of the hypercholesteremic effect of cholic acid. *J Exp Med*,97,.524-511

[41] Topuzlu, C; Stahl, WM. (1966). Effect of bile infusion on the dog kidney. *N Engl J Med*,274,.763-760

[42] Adrogue, HJ; Madias, NE. (2000). Hyponatremia. *N Engl J Med*,342,.1589-1581

[43] Martinez-Rodenas, F; Oms, LM; Carulla, X; Segura, M; Sancho, JJ; Piera, C; Fernandez-Espina, MR; Sitges-Serra, A. (1989). Measurement of body water compartments after ligation of the common bile duct in the rabbit.*Br J Surg*,76,-461 .464

[44] Oms, L; Martinez-Rodenas, F; Valverde, J; Jimenez, W; Sitges-Serra, A. (1990). Reduced water and sodium intakes associated with high levels of natriuretic factor following common bile duct ligation in the rabbit. *Br J Surg*,77,.755-752

[45] Valverde, J; Martinez-Rodenas, F; Pereira, JA; Carulla, X; Jimenez, W; Gubern, JM; Sitges-Serra, A. (1992). Rapid increase in plasma levels of atrial natriuretic peptide after common bile duct ligation in the rabbit. *Ann Surg*,216,.559-554

[46] Sellinger, M; Haag, K; Burckhardt, G; Gerok, W; Knauf, H. (1990). Sulfated bile acids inhibit Na(+)-H+ antiport in human kidney brush-border membrane vesicles. *Am J Physiol*,258, F.991-986

[47] Better, OS; Guckian, V; Giebisch, G; Green, R. (1987). The effect of sodium taurocholate on proximal tubular reabsorption in the rat kidney. *Clin Sci (Lond)*,72,-139 .141

[48] Green, J; Beyar, R; Sideman, S; Mordechovitz, D; Better, OS. (1986). The "jaundiced heart": a possible explanation for postoperative shock in obstructive jaundice. *Surgery*,100,.20-14

[49] Gillett, DJ. (1971). The effect of obstructive jaundice on the blood volume in rats. *J Surg Res*,11,.449-447

[50] Sitges-Serra, A; Carulla, X; Piera, C; Martinez-Rodenas, F; Franch, G; Pereira, J; Gubern, JM. (1992). Body water compartments in patients with obstructive jaundice. *Br J Surg*,79,.556-553

[51] Antunes-Rodrigues, J; de Castro, M; Elias, LL; Valenca, MM; McCann, SM. (2004). Neuroendocrine control of body fluid metabolism. *Physiol Rev*,84 ,.208-169

[52] Baxter, GF. (2004). The natriuretic peptides. *Basic Res Cardiol*,99,.75-71

[53] Pereira, JA; Torregrosa, MA; Martinez-Rodenas, F; Claria, J; Pallares, L; Gubern, JM; Ruano-Gil, D; Sitges-Serra, A. (1994). Increased cardiac endocrine activity after common bile duct ligation in the rabbit. Atrial endocrine cells in obstructive jaundice. *Ann Surg*,219,.78-73

[54] Martinez-Rodenas, F; Pereira, JA; Jimenez, W; Gubern, JM; Sitges-Serra, A. (1998). Circulating bile is the main factor responsible for atrial natriuretic peptide release in experimental obstructive jaundice. *Br J Surg*,85,.484-480

[55] Gallardo, JM; Padillo, J; Martin-Malo, A; Mino, G; Pera, C; Sitges-Serra, A. (1998). Increased plasma levels of atrial natriuretic peptide and endocrine markers of volume depletion in patients with obstructive jaundice. *Br J Surg*,85,.31-28

[56] Raaschou, F; Trautner, K. (1953). Obstruction of the common bile duct in experimental renal hypertension in dogs. *Scand J Clin Lab Invest*,5,.235-223

[57] Bomzon, A; Rosenberg, M; Gali, D; Binah, O; Mordechovitz, D; Better, OS; Greig, PD; Blendis, LM. (1986). Systemic hypotension and decreased pressor response in dogs with chronic bile duct ligation. *Hepatology*,6,.600-595

[58] Finberg, JP; Syrop, HA; Better, OS. (1981). Blunted pressor response to angiotensin and sympathomimetic amines in bile-duct ligated dogs. *Clin Sci (Lond)*,61,.539-535

[59] Bomzon, A; Finberg, JP; Tovbin, D; Naidu, SG; Better, OS. (1984). Bile salts, hypotension and obstructive jaundice. *Clin Sci (Lond)*,67,.183-177

[60] Bomzon, A; Gali, D; Better, OS; Blendis, LM. (.(1985Reversible suppression of the vascular contractile response in rats with obstructive jaundice. *J Lab Clin Med*,105,-568 .572

[61] Jacob, G; Said, O; Finberg, J; Bomzon, A. (1993). Peripheral vascular neuroeffector mechanisms in experimental cholestasis. *Am J Physiol*,265, G.586-579

[62] Joubert, P. (.(1978An in vivo investigation of the negative chronotropic effect of cholic acid in the rat. *Clin Exp Pharmacol Physiol*,5,.8-1

[63] Joubert, P. (1978). Cholic acid and the heart: in vitro studies of the effect on heart rate and myocardial contractility in the rat. *Clin Exp Pharmacol Physiol*,5,.16-9

[64] Binah, O; Rubinstein, I; Bomzon, A; Better, OS. (1987). Effects of bile acids on ventricular muscle contraction and electrophysiological properties: studies in rat papillary muscle and isolated ventricular myocytes. *Naunyn Schmiedebergs Arch Pharmacol*,335,.165-160

[65] Better, OS; Aisenbrey, GA; Berl, T; Anderson, RJ; Handelman, WA; Linas, SL; Guggenheim, SJ; Schrier, RW. (1980). Role of antidiuretic hormone in impaired urinary dilution associated with chronic bile-duct ligation.*Clin Sci (Lond)*,58,.500-493

[66] Hishida, A; Honda, N; Sudo, M; Nagase, M. (1980). Mechanisms of altered renal perfusion in the early stage of obstructive jaundice. *Kidney Int*,17,.230-223

[67] Shasha, SM; Better, OS; Chaimovitz, C; Doman, J; Kishon, Y. (1976). Haemodynamic studies in dogs with chronic bile-duct ligation. *Clin Sci Mol Med*,50,.537-533

[68] Bairaktari, ET; Kakafika, AI; Pritsivelis, N; Hatzidimou, KG; Tsianos, EV; Seferiadis, KI; Elisaf, MS. (2003). Hypouricemia in individuals admitted to an inpatient hospital-based facility. *Am J Kidney Dis*,41,.1232-1225

[69] Liamis, G; Bairaktari, E; Elisaf, M. (2002). Hypokalemia due to Fanconi syndrome in a patient with obstructive jaundice. *Nephron*,92,.712-711

[70] Tinsatul, U; Sitprija, V. (1970). Hyponatremia and hypokalemia in obstructive jaundice. A case study. *J Med Assoc Thai*,53,.442-437

[71] Bastis-Maounis, B; Matsaniotis, N; Maounis, F. (1973). Serum alkaline phosphatase in infants with obstructive jaundice: relation to vitamin D supplementation. *J Pediatr*,82,.72-68

[72] Kimura, S; Seino, Y; Harada, T; Nose, O; Yamaoka, K; Shimizu, K; Tanaka, H; Yabuuchi, H; Fukui, Y; Kamata, S; et al. (1988). Vitamin D metabolism in biliary atresia: intestinal absorptions of -25hydroxyvitamin D3 and -1,25dihydroxyvitamin D.3*J Pediatr Gastroenterol Nutr*,7,.346-341

[73] Chiba, T; Okimura, Y; Inatome, T; Inoh, T; Watanabe, M; Fujita, T. (1987). Hypocalcemic crisis in alcoholic fatty liver: transient hypoparathyroidism due to magnesium deficiency. *Am J Gastroenterol*,82,.1087-1084

[74] Wimmer, LE. (1974). Neonatal hypomagnesemia: report of a case. *J Am Osteopath Assoc*,74,.243-241

[75] Allison, ME; Prentice, CR; Kennedy, AC; Blumgart, LH. (1979). Renal function and other factors in obstructive jaundice. *Br J Surg*,66,.397-392

[76] Gubern, JM; Sancho, JJ; Simo, J; Sitges-Serra, A. (1988). A randomized trial on the effect of mannitol on postoperative renal function in patients with obstructive jaundice. *Surgery*,103,.44-39

[77] Pain, JA; Cahill, CJ; Bailey, ME. (1985). Perioperative complications in obstructive jaundice: therapeutic considerations. *Br J Surg*,72,.945-942

[78] Thompson, JN; Edwards, WH; Winearls, CG; Blenkharn, JI; Benjamin, IS; Blumgart, LH. (1987). Renal impairment following biliary tract surgery. *Br J Surg*,74,.847-843

[79] Bismuth, H; Kuntziger, H; Corlette, MB. (1975). Cholangitis with acute renal failure: priorities in therapeutics. *Ann Surg*,181,.887-881

[80] Lai, EC; Mok, FP; Tan, ES; Lo, CM; Fan, ST; You, KT; Wong, J. (1992). Endoscopic biliary drainage for severe acute cholangitis. *N Engl J Med*,326,.1586-1582

[81] Dawson, JL. (1965). The Incidence of Postoperative Renal Failure in Obstructive Jaundice. *Br J Surg*,52,.665-663
[82] Dawson, JL. (1965). Post-Operative Renal Function in Obstructive Jaundice: Effect of a Mannitol Diuresis.*Br Med J*,1,.86-82
[83] Evans, HJ; Torrealba, V; Hudd, C; Knight, M. (1982). The effect of preoperative bile salt administration on postoperative renal function in patients with obstructive jaundice.*Br J Surg*,69,.708-706
[84] Amerio, A; Campese, VM; Coratelli, P; Dagostino, F; Micelli, M; Passavanti, G; Petrarulo, F. (1981). Prognosis in acute renal failure accompanied by jaundice. *Nephron*,27,.154-152
[85] Holmes, TW, Jr. (1953). The histologic lesion of cholemic nephrosis. *J Urol*,70,-677 .685
[86] Sant, SM; Purandare, NM. (1965). Cholemic Nephrosis--an Autopsy and Experimental Study. *J Postgrad Med*,11,.89-79
[87] Leelahavanichkul, A; Yasuda, H; Doi, K; Hu, X; Zhou, H; Yuen, PS; Star, RA. (2008). Methyl--2acetamidoacrylate, an ethyl pyruvate analog, decreases sepsis-induced acute kidney injury in mice. *Am J Physiol Renal Physiol*,295, F.1835-1825
[88] Tirapelli, LF; Barione, DF; Trazzi, BF; Tirapelli, DP; Novas, PC; Silva, CS; Martinez, M; Costa, RS; Tucci, S, Jr.; Suaid, HJ; Cologna, AJ; Martins, AC. (2009). Comparison of two models for evaluation histopathology of experimental renal ischemia. *Transplant Proc*,41,.4087-4083
[89] Kawaguchi, K; Koike, M. (1987). Glomerular alterations associated with obstructive jaundice.*Hum Pathol*,18,.1154-1149
[90] Uslu, A; Tasli, FA; Nart, A; Postaci, H; Aykas, A; Bati, H; Coskun, Y. (2010). Human kidney histopathology in acute obstructive jaundice: a prospective study. *Eur J Gastroenterol Hepatol*,22,.1465-1458
[91] Bailey, ME. (1976). Endotoxin, bile salts and renal function in obstructive jaundice. *Br J Surg*,63,.778-774
[92] Ingoldby, CJ; McPherson, GA; Blumgart, LH. (1984). Endotoxemia in human obstructive jaundice. Effect of polymyxin B. *Am J Surg*,147,.771-766
[93] Fox, ES; Broitman, SA; Thomas, P. (1990). Bacterial endotoxins and the liver. *Lab Invest*,63,.741-733
[94] Nolan, JP. (1975). The role of endotoxin in liver injury. *Gastroenterology*,69,-1346 .1356
[95] Nolan, JP. (1981) Endotoxin, reticuloendothelial function, and liver injury. *Hepatology*,1,.465-458
[96] Nolan, JP. (1985). Spontaneous endotoxinemia. *J Pediatr Gastroenterol Nutr*,4,.8-7
[97] Drivas, G; James, O; Wardle, N. (1976). Study of reticuloendothelial phagocytic capacity in patients with cholestasis. *Br Med J*,1,.1569-1568
[98] Fletcher, MS; Westwick, J; Kakkar, VV. (1982). Endotoxin, prostaglandins and renal fibrin deposition in obstructive jaundice. *Br J Surg*,69,.629-625
[99] Katz, S; Grosfeld, JL; Gross, K; Plager, DA; Ross, D; Rosenthal, RS; Hull, M; Weber, TR. (1984). Impaired bacterial clearance and trapping in obstructive jaundice. *Ann Surg*,199,.20-14
[100] McCuskey, RS; Urbaschek, R; Urbaschek, B. (1996). The microcirculation during endotoxemia. *Cardiovasc Res*,32,.763-752

[101] Ito, Y; Machen, NW; Urbaschek, R; McCuskey, RS. (2000). Biliary obstruction exacerbates the hepatic microvascular inflammatory response to endotoxin. *Shock*,14,.604-599

[102] Kigawa, G; Nakano, H; Kumada, K; Kitamura, N; Takeuchi, S; Hatakeyama, T; Yamaguchi, M; Nagasaki, H; Boudjema, K; Jaeck, D. (2000). Improvement of portal flow and hepatic microcirculatory tissue flow with N-acetylcysteine in dogs with obstructive jaundice produced by bile duct ligation. *Eur J Surg*,166,.84-77

[103] Matsumoto, Y; Niimoto, S; Katayama, K; Hirose, K; Yamaguchi, A; Torigoe, K. (2002). Effects of biliary drainage in obstructive jaundice on microcirculation, phagocytic activity, and ultrastructure of the liver in rats. *J Hepatobiliary Pancreat Surg*,9,.366-360

[104] Okaya, T; Nakagawa, K; Kimura, F; Shimizu, H; Yoshidome, H; Ohtsuka, M; Morita, Y; Miyazaki, M. (2008). Obstructive jaundice impedes hepatic microcirculation in mice. *Hepatogastroenterology*,55,.2150-2146

[105] Wardle, EN; Wright, NA. (1978). Endotoxin and acute renal failure associated with obstructive jaundice. *Br Med J*,4,.474-472

[106] Westwick, J; Fletcher, MS; Kakkar, VV. (1983). Inhibition of thromboxane formation prevents endotoxin-induced renal fibrin deposition in jaundiced rats. *Adv Prostaglandin Thromboxane Leukot Res*,12,.91-83

[107] Gatta, A; Milani, L; Merkel, C; Zuin, R; Amodio, P; Caregaro, L; Ruol, A. (1982). Lack of correlation between endotoxaemia and renal hypoperfusion in cirrhotics without overt renal failure. *Eur J Clin Invest*,12,.422-417

[108] Cahill, CJ. (1983). Prevention of postoperative renal failure in patients with obstructive jaundice--the role of bile salts. *Br J Surg*,70,.595-590

[109] Cahill, CJ; Pain, JA; Bailey, ME. (1987). Bile salts, endotoxin and renal function in obstructive jaundice. *Surg Gynecol Obstet*,165,.522-519

[110] Greve, JW; Maessen, JG; Tiebosch, T; Buurman, WA; Gouma, DJ. (1990). Prevention of postoperative complications in jaundiced rats. Internal biliary drainage versus oral lactulose. *Ann Surg*,212,.227-221

[111] Pain, JA; Cahill, CJ; Gilbert, JM; Johnson, CD; Trapnell, JE; Bailey, ME. (1991). Prevention of postoperative renal dysfunction in patients with obstructive jaundice: a multicentre study of bile salts and lactulose. *Br J Surg*,78,.469-467

[112] Thompson, JN; Cohen, J; Blenkharn, JI; McConnell, JS; Barr, J; Blumgart, LH. (1986). A randomized clinical trial of oral ursodeoxycholic acid in obstructive jaundice. *Br J Surg*,73,.636-634

[113] Van Bossuyt, H; Desmaretz, C; Gaeta, GB; Wisse, E. (1990). The role of bile acids in the development of endotoxemia during obstructive jaundice in the rat. *J Hepatol*,10,.279-274

[114] Inan, M; Sayek, I; Tel, BC; Sahin-Erdemli, I. (1997). Role of endotoxin and nitric oxide in the pathogenesis of renal failure in obstructive jaundice. *Br J Surg*,84,.947-943

[115] Li, W; Chan, AC; Lau, JY; Lee, DW; Ng, EK; Sung, JJ; Chung, SC. (2004). Superoxide and nitric oxide production by Kupffer cells in rats with obstructive jaundice: effect of internal and external drainage. *J Gastroenterol Hepatol*,19 ,.165-160

[116] Yuksel, BC; Tanriverdi, P; Ozel, H; Avsar, FM; Topaloglu, S; Iskit, AB. (2003). The effects of nitric oxide synthase blockers on mesenteric blood flow with bile duct ligation. *Hepatogastroenterology*,50 *Suppl* 2, ccxix-ccxxi.

[117] Hong, JY; E, FS; Nishikawa, T; Hiramoto, K; Inoue, M. 92009). Effect of Obstructive Jaundice and Nitric Oxide on the Profiles of Intestinal Bacterial Flora in Wild and iNOS Mice. *J Clin Biochem Nutr*,44,.167-160

[118] O'Neill, P; Wait, RB; Kahng, KU. (1990). Obstructive jaundice and renal failure in the rat: the role of renal prostaglandins and the renin-angiotensin system. *Surgery*,108,-356 .362

[119] Uemura, M; Tsujii, T; Fukui, H; Takaya, A; Tsukamoto, N; Nakayama, M; Matsumoto, M; Uemura, N; Fujimoto, M; Tamura, M. (1989). Urinary prostaglandins and renal function in obstructive jaundice. *Scand J Gastroenterol*,24,.715-705

[120] Gurleyik, E; Coskun, O; Ustundag, N; Ozturk, E. (2006). Prostaglandin E1 maintains structural integrity of intestinal mucosa and prevents bacterial translocation during experimental obstructive jaundice. *J Invest Surg*,19,.289-283

[121] Aoyagi, T; Lowenstein, LM. (1968). The effect of bile acids and renal ischemia on renal function. *J Lab Clin Med*,71,.692-686

[122] Parks, RW; Diamond, T; McCrory, DC; Johnston, GW; Rowlands, BJ. (1994). Prospective study of postoperative renal function in obstructive jaundice and the effect of perioperative dopamine. *Br J Surg*,81,.439-437

[123] Dawson, JL. (1964). Jaundice and Anoxic Renal Damage: Protective Effect of Mannitol. *Br Med J*,1,.811-810

[124] Dawson, JL. (1966). The effect of haemorrhage on renal function in obstructive jaundice.*Br J Surg*,53,.685-679

[125] Lucena, MI; Andrade, RJ; Cabello, MR; Hidalgo, R; Gonzalez-Correa, JA; Sanchez de la Cuesta, F. (1995). Aminoglycoside-associated nephrotoxicity in extrahepatic obstructive jaundice. *J Hepatol*,22,.196-189

[126] Ingoldby, CJ. (1980). The value of polymixin B in endotoxaemia due to experimental obstructive jaundice and mesenteric ischaemia. *Br J Surg*,67,.567-565

[127] Pain, JA; Bailey, ME. (1986). Experimental and clinical study of lactulose in obstructive jaundice. *Br J Surg*,73,.778-775

[128] McPherson, GA; Benjamin, IS; Hodgson, HJ; Bowley, NB; Allison, DJ; Blumgart, LH. (1984). Pre-operative percutaneous transhepatic biliary drainage: the results of a controlled trial. *Br J Surg*,71,.375-371

[129] Hatfield, AR; Tobias, R; Terblanche, J; Girdwood, AH; Fataar, S; Harries-Jones, R; Kernoff, L; Marks, IN. (1982). Preoperative external biliary drainage in obstructive jaundice. A prospective controlled clinical trial.*Lancet*,2,.899-896

[130] Pitt, HA; Gomes, AS; Lois, JF; Mann, LL; Deutsch, LS; Longmire, WP, Jr. (1985). Does preoperative percutaneous biliary drainage reduce operative risk or increase hospital cost? *Ann Surg*,201,.553-545

[131] Gundry, SR; Strodel, WE; Knol, JA; Eckhauser, FE; Thompson, NW. (1984). Efficacy of preoperative biliary tract decompression in patients with obstructive jaundice. *Arch Surg*,119,.708-703

[132] Gouma, DJ; Coelho, JC; Schlegel, JF; Li, YF; Moody, FG. (1987). The effect of preoperative internal and external biliary drainage on mortality of jaundiced rats. *Arch Surg*,122,.734-731

[133] Trede, M; Schwall, G. (1988) .The complications of pancreatectomy. *Ann Surg*,207,-39 .47

[134] Thompson, RL; Hoper, M; Diamond, T; Rowlands, BJ. (1990). Development and reversibility of T lymphocyte dysfunction in experimental obstructive jaundice. *Br J Surg*,77,.1232-1229

[135] Suzuki, H; Iyomasa, S; Nimura, Y; Yoshida, S. (1994). Internal biliary drainage, unlike external drainage, does not suppress the regeneration of cholestatic rat liver after partial hepatectomy. *Hepatology*,20,.1322-1318

[136] Kawarada, Y; Higashiguchi, T; Yokoi, H; Vaidya, P; Mizumoto, R. (1995). Preoperative biliary drainage in obstructive jaundice. *Hepatogastroenterology*,42,-300 .307

[137] Parks, RW; Clements, WD; Smye, MG; Pope, C; Rowlands, BJ; Diamond, T. (1996). Intestinal barrier dysfunction in clinical and experimental obstructive jaundice and its reversal by internal biliary drainage. *Br J Surg*,83,.1349-1345

[138] Cerda, J; Tolwani, A; Gibney, N; Tiranathanagul, K. (2011). Renal replacement therapy in special settings: extracorporeal support devices in liver failure. *Semin Dial*,24,-197 .202

[139] O'Grady, JG; Gimson, AE; O'Brien, CJ; Pucknell, A; Hughes, RD; Williams, R. (1988). .Controlled trials of charcoal hemoperfusion and prognostic factors in fulminant hepatic failure. *Gastroenterology*,94,.1192-1186

[140] Hughes, RD; Pucknell, A; Routley, D; Langley, PG; Wendon, JA; Williams, R. (1994). Evaluation of the BioLogic-DT sorbent-suspension dialyser in patients with fulminant hepatic failure. *Int J Artif Organs*,17,.662-657

[141] Ellis, AJ; Hughes, RD; Nicholl, D; Langley, PG; Wendon, JA; O'Grady, JG; Williams, R. (1999). Temporary extracorporeal liver support for severe acute alcoholic hepatitis using the BioLogic-DT. *Int J Artif Organs*,22,.34-27

[142] Wilkinson, AH; Ash, SR; Nissenson, AR. (1998). Hemodiabsorption in treatment of hepatic failure. *J Transpl Coord*,8,.50-43

[143] Lepore, MJ; Stutman, LJ; Bonanno, CA; Conklin, EF; Robilotti, JG, Jr.; McKenna, PJ. (1972). Plasmapheresis with plasma exchange in hepatic coma. II. Fulminant viral hepatitis as a systemic disease. *Arch Intern Med*,129,.907-900

[144] Kondrup, J; Almdal, T; Vilstrup, H; Tygstrup, N. (1992). High volume plasma exchange in fulminant hepatic failure. *Int J Artif Organs*,15,.676-669

[145] Larsen, FS; Hansen, BA; Jorgensen, LG; Secher, NH; Kirkegaard, P; Tygstrup, N. (1994). High-volume plasmapheresis and acute liver transplantation in fulminant hepatic failure. *Transplant Proc*,26,.1788

[146] Clemmesen, JO; Gerbes, AL; Gulberg, V; Hansen, BA; Larsen, FS; Skak, C; Tygstrup, N; Ott, P. (1999). Hepatic blood flow and splanchnic oxygen consumption in patients with liver failure. Effect of high-volume plasmapheresis. *Hepatology*,29,.355-347

[147] Yoshiba, M; Sekiyama, K; Iwamura, Y; Sugata, F. (1993). Development of reliable artificial liver support (ALS)--plasma exchange in combination with hemodiafiltration using high-performance membranes. *Dig Dis Sci*,38,.476-469

[148] Morimoto, T; Matsushima, M; Sowa, N; Ide, K; Sawanishi, K. (1989). Plasma adsorption using bilirubin-adsorbent materials as a treatment for patients with hepatic failure. *Artif Organs*,13,.452-447

[149] Senf, R; Klingel, R; Kurz, S; Tullius, S; Sauer, I; Frei, U; Schindler, R. (2004). Bilirubin-adsorption in 23 critically ill patients with liver failure. *Int J Artif Organs*,27,.722-717

[150] Geiger, H; Klepper, J; Lux, P; Heidland, A. (1992). Biochemical assessment and clinical evaluation of a bilirubin adsorbent column BR-350 in critically ill patients with intractable jaundice. *Int J Artif Organs*,15,.39-35

[151] Ott, R; Rupprecht, H; Born, G; Muller, V; Reck, T; Hohenberger, W; Kockerling, F. (1998). Plasma separation and bilirubin adsorption after complicated liver transplantation: a therapeutic approach to excessive hyperbilirubinemia. *Transplantation*,65,.437-434

[152] Tabei, K; Akai, Y; Takeda, S; Homma, S; Kusano, E; Asano, Y. (1991). Application of plasma perfusion in hepatic failure. *Biomater Artif Cells Immobilization Biotechnol*,19,.201-193

[153] Rozga, J; Umehara, Y; Trofimenko, A; Sadahiro, T; Demetriou, AA. (2006). A novel plasma filtration therapy for hepatic failure: preclinical studies. *Ther Apher Dial*,10,.144-138

[154] Khuroo, MS; Khuroo, MS; Farahat, KL. (2004). Molecular adsorbent recirculating system for acute and acute-on-chronic liver failure: a meta-analysis. *Liver Transpl*,10,.1106-1099

[155] Vienken, J; Christmann, H. (2006). How can liver toxins be removed? Filtration and adsorption with the Prometheus system. *Ther Apher Dial*,10,.131-125

[156] Grodzicki, M; Kotulski, M; Leonowicz, D; Zieniewicz, K; Krawczyk, M. (2009). Results of treatment of acute liver failure patients with use of the prometheus FPSA system. *Transplant Proc*,41,.3081-3079

[157] Rifai, K; Ernst, T; Kretschmer, U; Hafer, C; Haller, H; Manns, MP; Fliser, D. (2005). The Prometheus device for extracorporeal support of combined liver and renal failure. *Blood Purif*, 23,.302-298

[158] Boonsrirat, U; Tiranathanagul, K; Srisawat, N; Susantitaphong, P; Komolmit, P; Praditpornsilpa, K; Tungsanga, K; Eiam-Ong, S. (2009). Effective bilirubin reduction by single-pass albumin dialysis in liver failure. *Artif Organs*,33,.653-648

[159] Karvellas, CJ; Bagshaw, SM; McDermid, RC; Stollery, DE; Bain, VG; Gibney, RT. (2009). A case-control study of single-pass albumin dialysis for acetaminophen-induced acute liver failure. *Blood Purif*, 28,.158-151

[160] Sauer, IM; Goetz, M; Steffen, I; Walter, G; Kehr, DC; Schwartlander, R; Hwang, YJ; Pascher, A; Gerlach, JC; Neuhaus, P. (2004). In vitro comparison of the molecular adsorbent recirculation system (MARS) and single-pass albumin dialysis (SPAD). *Hepatology*,39,.1414-1408

[161] Hui, T; Rozga, J; Demetriou, AA. (2001). Bioartificial liver support. *J Hepatobiliary Pancreat Surg*,8,.15-1

[162] Sauer, IM; Kardassis, D; Zeillinger, K; Pascher, A; Gruenwald, A; Pless, G; Irgang, M; Kraemer, M; Puhl, G; Frank, J; Muller, AR; Steinmuller, T; Denner, J; Neuhaus, P; Gerlach, JC. (2003). Clinical extracorporeal hybrid liver support--phase I study with primary porcine liver cells. *Xenotransplantation*,10,.469-460

[163] Sauer, IM; Zeilinger, K; Obermayer, N; Pless, G; Grunwald, A; Pascher, A; Mieder, T; Roth, S; Goetz, M; Kardassis, D; Mas, A; Neuhaus, P; Gerlach, JC. (2002). Primary human liver cells as source for modular extracorporeal liver support--a preliminary report. *Int J Artif Organs*,25,.1005-1001

[164] Sussman, NL; Gislason, GT; Conlin, CA; Kelly, JH. (1994). The Hepatix extracorporeal liver assist device: initial clinical experience. *Artif Organs*,18,.396-390

[165] van de Kerkhove, MP; Di Florio, E; Scuderi, V; Mancini, A; Belli, A; Bracco, A; Dauri, M; Tisone, G; Di Nicuolo, G; Amoroso, P; Spadari, A; Lombardi, G; Hoekstra, R; Calise, F; Chamuleau, RA. (2002). Phase I clinical trial with the AMC-bioartificial liver. *Int J Artif Organs*,25,..959-950

In: Bilirubin: Chemistry, Regulation and Disorder
Editors: J. F. Novotny and F. Sedlacek

ISBN: 978-1-62100-911-5
© 2012 Nova Science Publishers, Inc.

Chapter V

Serum Bilirubin and the Genetic Epidemiology of Complex Disease

*Phillip E. Melton**
Texas Biomedical Research Institute, San Antonio TX

Abstract

There has been increased interest in the underlying genetic mechanisms influencing serum bilirubin due to its antioxidant properties that have been shown to be protective against the development of cardiovascular disease and certain types of neoplasms, including Hodgkin's lymphoma and endometrial cancer. Recent research has shown genetic variation between ethnic populations and these differences have important implications for pharmacogenetic development, due to the role of serum bilirubin in the removal of biochemical toxins through glucuronidation.

This chapter provides a detailed review of genes previously identified that control serum bilirubin and an overview of genetic variation and epidemiology studies that demonstrate an association of these loci with complex diseases and bilirubin. These genes include uridine diphosphate glucuronosyltransferase (*UGT1A1*), heme oxygenase-1 (*HMOX1*), biliverdin reductase (*BLVR-A* and *BLVR-B*), and the solute carrier organic anion transporters (*SLCO1B1* and *SLCO1B3*). Epidemiological studies have consistently demonstrated an inverse relationship between a promoter region polymorphism in *UGT1A1* and complex diseases in European populations.

However, this protective *UGT1A1* promoter association has not been demonstrated in other global populations including Africans, Asians, and American Indians, suggesting that either other variants in this gene or other genetic factors may play a role in serum bilirubin levels in these communities.

Therefore, understanding this genetic variation between extant human populations has important implications given the importance of bilirubin levels in protecting against

---

* Corresponding Author: Phillip E. Melton, Ph.D.Texas Biomedical Research Institute, Department of Genetics, P.O. Box 760549, San Antonio, TX 78245, Tele: 210-258-9788, Fax: 210-258-9444, E-mail: pmelton@txbiomedgenetics.org

certain complex diseases and may have profound significance for drug development and therapy. Genetic epidemiology is involved in understanding the underlying genetic components that contribute to human disease. Traditionally, this field focused on the identification of Mendelian or monogenic disorders though the investigation of families or populations isolates. More recently, these studies have been expanded to examine multiple genetic factors in common complex chronic diseases such as cardiovascular disease (CVD) and cancer.

This latter type of analysis has rapidly advanced due to increased sequencing technology and reduced costs allowing for large-scale studies of genetic association with common complex diseases and intermediate risk factors using large numbers of single nucleotide polymorphisms (SNPs) [1]. One of these intermediate phenotypes where there has been growing interest is in genes involved in serum bilirubin production, due to its role in glucuronidation and its antioxidant properties that have made it important potential protective risk factor for understanding the development of complex chronic diseases [2,3].

Serum bilirubin is the principal product of heme degradation and a powerful antioxidant that suppresses lipid oxidation and retards atherosclerosis formation. Historically, high levels of serum bilirubin have been viewed by clinicians as a marker of liver dysfunction [4]. While high levels of serum bilirubin are potentially toxic it is normally rendered harmless by tight binding to albumin and excretion through the liver [4]. Epidemiological research has demonstrated an inverse relationship between elevated serum bilirubin levels and the development of CVD [5-16] and some forms of cancer [17-24] due to its antioxidant properties [25]. This has led to increased interest in serum bilirubin in genetic epidemiology and pharmacogenetic studies [26-29]. This chapter reviews bilirubin genetics, its inherited disorders, genetic frequencies in different ethnic populations, and its protective components and risk factors in chronic diseases.

Bilirubin Biology

Serum bilirubin is the principal biological product resulting from the breakdown of old erythrocytes in the liver in mammals. Other forms of bilirubin are formed from the liver, additional hemo-proteins, and other internal organs. Serum bilirubin levels are regulated by four enzymes: (1) heme oxygenase-1 (HMOX1); (2) biliverdin reductase (BLVR), 3) solute carrier organic anion transporters (SLCO1B) and (3) uridine diphosphate-glucuronosyltransferase (UGT1A1) [30]. Figure 1 illustrates the biochemical process by which heme is broken down into bilirubin. Heme (ferroprotoporphyrin IX) is cleaved through selective oxidation of the α-methane bridge, catalyzed by HMOX1.

This biochemical reaction involves three molecules of O_2 and the reducing agent nicotinamide adenine dinucleotide phosphate (NADPH). This results in the formation of biliverdin and a single molecule of carbon dioxide (CO) while the iron (FE) molecule is released. In humans, biliverdin is catalyzed by BLVR into bilirubin. The resulting bilirubin is then transported in the blood with albumin and taken into the liver by SLCO1B1 within hepatocytes.

The solubility of bilirubin is increased by the addition of one or two molecules of glucuronic acid that is catalyzed by UGT1A1, which regulates the removal of circulating bilirubin into bile or urine [30]. This breakdown of hemoglobin accounts for 80% of serum bilirubin produced daily in the human body [31]. Additional hemo-proteins and unbound heme account for the remaining 20% of bilirubin production [4].

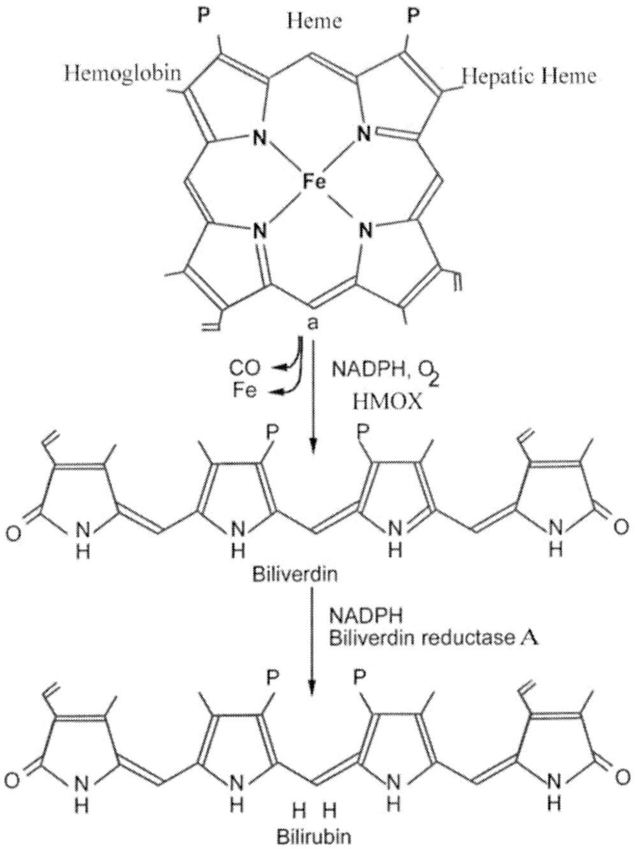

Figure 1. Production of bilirubin from Heme.

Research has demonstrated serum bilirubin to be an effective antioxidant both *in vitro* and *in vivo* [3]. Bilirubin antioxidant properties have shown the ability to suppress the oxidation of lipids and lipoproteins, especially LDL cholesterol, and are thought to directly be related to total serum antioxidant capacity in humans [32]. A number of recent epidemiological studies have demonstrated that bilirubin is more efficient at protecting lipids from oxidation than other water soluble antioxidants such as glutathione, which shield proteins from oxidation [33].

In addition, there is some support for the hypothesis that bilirubin has anti-inflammatory properties that may inhibit such biochemical products tumor-necrosis factor-α (TNF-α), vascular cell adhesion molecule 1 (VCAM-1) and intercellular adhesion molecule 1 (ICAM-1) [34]. Based on these antioxidant and anti-inflammatory properties, it is thought that elevated bilirubin levels could possibly prevent plaque concentration and subsequent atherosclerosis [2]. However, hyberbilirubenimea has also been implicated in adverse drug metabolism reactions [26,27].

The United States Food and Drug Administration (FDA) currently recommends genetic testing for individuals with Gilbert's syndrome (UGT1A1*28) prior to being prescribed irinotecan and nitonleb, both drugs for colorectal cancer [26,27]. Therefore, understanding variation in genes involved in serum bilirubin and different functional variants across different ethnic populations has important pharmacogenetic consequences [35].

Genes, Population Variation, and Bilirubin Production

In humans, serum bilirubin levels are highly heritable and a number of genes have been shown to be involved in its production. As previously mentioned, serum bilirubin results from the breakdown of heme-containing proteins. After breakdown, bilirubin conjugates are actively secreted into the canaliculi by the gene, membrane ATP-binding cassette family C (*ABBC2*) [4]. These along with genes involved in the life span of red blood cells, such as glucose-6-phosphate dehydrogenase (*G6PD*) on the X-chromosome, are important candidate genes for the control of serum bilirubin [36,37]. In addition to *ABCC2* and *G6PD*, serum bilirubin production also includes the previously mentioned genes; *HMOX1, BLVRA, SLCOB1,* and *UGT1A1*.

Heme-Oxygenase (Decycling) 1 (HMOX1)

In humans there are two genes involved in heme-degradation, *HMOX1* on chromosome 22 and *HMOX2* on chromosome 16. *HMOX2* is constitutive; whereas *HMOX1* may be induced by adverse environmental conditions contributing to oxidation within the body. Three previous studies have investigated association between *HMOX1* and bilirubin at the population level [38-40]. These studies have shown that carriers of a short microsatellite repeat (<25 GT) in the *HMOX1* promoter demonstrate higher bilirubin levels when compared with individuals with longer repeats in Asian populations. Endler et al. [38] did not find this polymorphism in Europeans, while Lin et al. [41] found this association only to be present in the Uyghurs out of three Asian populations that they investigated (Kazaks, Han, Uyghur). These results led the authors to suggest that *HMOX1* is not strongly associated with regulation of bilirubin levels in Asian populations and is highly susceptible to environmental factors [41].

Biliverdin reductase A (BLVRA)

Biliverdin is reduced to bilirubin through the actions of the gene biliverdin reductase A (*BLVRA*) on chromosome 12. *BLVRA* also binds to HMOX1, and regulates *HMOX1* activity. In addition, *BLVRA* regulates induction of *HMOX1* expression after oxidative stress. Only one study has investigated potential polymorphisms in this gene and their relationship to bilirubin levels. Lin et al. [39] investigated the common *BLVRA* coding region SNP rs699513 as being potentially related to elevated bilirubin levels in three Asian populations but did not identify any genetic association with this variant.

Solute Carrier Anion Transporter 1B (SLCO1B1)

SLCO1B1 is expressed at the basolateral membrane of hepatocytes and is responsible for transporting unconjugated bilirubin from the blood into the liver [42]. There have been three population-based studies that have investigated genetic association between *SLCO1B1* and bilirubin levels in Asian populations. One study of Taiwanese adults found that individuals with the minor allele of the coding region SNP rs2306283 (C>T) and the promoter region SNP rs4149056 (T>C) increased the risk of unconjugated bilirubin [43]. Another study investigated genetic association within Taiwanese neonates and individuals with the minor allele of rs2306283 were at a higher risk of developing severe hyperbilirubinemia than patients with just the minor allele for rs4149056 [44]. Lin et al. [41] investigated three *SLCO1B1* promoter SNPs (rs4149013, rs18328763, rs4149014) and two coding region SNPs (rs2306283, rs4149056) but did not find a significant association between any of these five SNPs and bilirubin in three different Asian populations.

Uridine Diphosphate-Glucurnosyltransferase (UGT1A1)

Genome-wide linkage and segregation studies of hyperbilirubinemia have identified a major linkage peak on chromosome 2q37 and *UGT1A1* is located within this region [45,46]. *UGT1A1* is responsible for the removal of circulating bilirubin through the bile or urine. Rare mutations within this gene are responsible for the congenital forms of hyperbilirubinemia and consist of Crigler-Najjar syndrome types 1 and 2 as well as the milder Gilbert's syndrome (see below for further description). A number of studies have investigated the promoter region repeat, UGT1A1*28, and serum bilirubin that is associated with Gilbert syndrome. Individuals homozygous for 7 TAA repeats (7/7) have higher levels of serum bilirubin when compared with heterozygous individuals (6/7) or individuals homozygous for 6 TAA repeats (6/6). A number of studies have also identified an additional coding region variant (rs4148323) in the first exon of *UGT1A1* and found that this SNP is associated with higher serum bilirubin levels in East Asian populations but is mostly monomorphic in Europeans [3]. The majority of these studies have examined individuals with adult or neonatal hyperbilirubinemia with fewer studies within the normal bilirubin range [46-48] or in healthy adults [39,49-51].

In European populations, the frequency of UGT1A1*28 is about 35% but varies dramatically among different ethnic populations with Africans demonstrating high frequencies and Asian the lowest (Table 1). African populations demonstrate the most variation of the *UGT1A1* locus with repeats between 5 and 8 being more common than in any other population [36].

Most individuals diagnosed with Gilbert's Syndrome are homozygous for UGT1A1*28. This repeat is known to reduce the transcription of the genes to 18 to 33% of normal, resulting in decreased enzymatic activity and higher levels of circulating bilirubin.

A conditional linkage scan on the *UGT1A1* repeat in the Framingham Heart Study demonstrated that this variant explained the linkage signal in Europeans [11].

An association study also found that when conditional on the *UGT1A1* repeat with 13 additional functional variants that none of these were statistically significant [52]. However, a conditional linkage study of *UGT1A1* and bilirubin in the Strong Heart Family Study of American Indians found that there was a probability of additional variants accounting for the observed linkage signal [46]. In addition, a variant in the T3297G phenobarbital-responsive enhancer region of *UGT1A1* has been found to reduce transcriptional activity and increase serum bilirubin levels [53] and a trinucleotide repeat (CAT) insertion has been found in 10% of Asian Indians that also elevated serum bilirubin levels [54].

G6PD and ABBC2

G6PD is the key enzyme in the pentose phosphate pathway and provides the NADPH essential for erythrocytes and other cells. G6PD deficiency reduces the life span of red blood cells and leads to increased bilirubin levels. G6PD deficiency is also known to be associated with hyperbilirubinemia in infants and elevated bilirubin levels have been observed in adults [55]. An association has been suggested between African populations with G6PD deficiency and high *UGT1A1* number of repeats as being related to malarial resistance [36]. However, the specific underlying genetic components demonstrating association between *G6PD* and bilirubin are unknown. One study has investigated the genetics of *ABBC2* and bilirubin levels but found no evidence of association [56].

Table 1. Global Frequencies of UGT1A1*28

Country*	UGT1A1*28 Frequency	Reference
Africa		
Morocco	0.35	36
Algeria	0.34	36
Senegal	0.34	36
Ghana	0.47	36
Nigeria	0.45	36
Cameroon	0.50	36
Ethiopia	0.43	36
Northern Sudan	0.36	36
Southern Sudan	0.48	36
Uganda	0.46	36
Tanzania	0.43	36
Malawi	0.33	36
Mozambique	0.37	36
Zimbabwe	0.35	36
South Africa	0.35	36
Asia		
Turkey	0.38	36
Yemen	0.32	36
Chinese Han	0.12	84
Chinese Dong	0.15	84
Chinese She	0.08	84
India	0.35	54
Malay	0.19	138
Thailand	0.16	139
Japan	0.11	140
Europe		
Britain	0.35	36
Italy	0.36	83
Scotland	0.34	141
Dutch	0.34	142
Spanish	0.35	143
North America		
European	0.33	47
African	0.43	144
Native American	0.35	46
South America		
European	0.32	145
African	0.41	145
Native American	0.33	145

*UGT1A1*28 frequencies are presented for country of origin, except in North and South America where they are divided by ethnicity.

Genome Wide Association Studies

Two genome-wide association studies (GWAS) have investigated serum bilirubin in European populations [37,57]. The first GWAS conducted a meta-analysis of three studies and investigated 2.5 million autosomal SNPs in 9464 individuals. This resulted in two SNPs rs6742078 ($p=5.0E^{-324}$) near UGT1A1*28 and rs4149056 ($p=6.7E^{-13}$) in *SLCO1B1* [57]. Another GWAS of 4300 Sardinians tested 500,000 SNPs including those on the X-chromosome and found three significant SNPs; rs887829 ($p=6.2.E^{-62}$) near the *UGT1A1* repeat, rs766420 ($p=9.4E^{-9}$) near *G6PD*, and rs2117032 ($p=4.7E^{-8}$) in *SLCO1B3* [37]. *SLCO1B3* is in the same gene family as *SLCO1B1* and the same region on chromosome 12. When this variant was investigated conditional on functional variants in *SLCO1B1* it remained highly significant, providing support that the *SLCO1B3* signal is not due to indirect effects of *SLCO1B1* variants [37].

Bilirubin and biliverdin are known substrates for both *SLCO1B1* and *SLCO1B3* and both have been shown to be expressed in the liver and they share 80% amino acid identity. The different involvement of these two transporter genes may be due to population-specific genetic features [3].

The majority of genetic studies of serum bilirubin identify *UGT1A1* as the primary contributor to variation for this important phenotype despite its significant heterogeneity. In addition, the effects of solute carrier transporters like *SLCO1B1* and *SLCO1B3* and genes involved in red blood life span, such as *G6PD* in serum bilirubin levels were confirmed. In both GWAS, test of epitasis between genes and the top SNPS were not significant suggesting that these three genes influence bilirubin independently [11]. The only other gene that appears to demonstrate association with bilirubin is *HMOX1* and the repeat in its promoter region at least in Asian and Europeans. However, based on the amount of genetic heterogeneity across populations it is probable that other unknown variants may also be involved with bilirubin production.

Genetic Epidemiology of Bilirubin

Bilirubin metabolism involves four distinct but interrelated stages: and deviations in any of these four processes may be the result of genetic effects and may result in increases or decreases in the amounts of circulating bilirubin in the blood stream. These fluctuations in bilirubin levels may have either severe, moderate, or lesser effects on a number of different chronic complex diseases. In addition, there a number monogenic disorders that have been shown to be inherited which are either associated with unconjugated or conjugated forms of bilirubin.

These include the unconjugated disorders; Crigler-Najjar Syndrome types 1 and 2, Gilbert's Syndrome, and the conjugated disorders; Roter Syndrome, Dubin Johnson Syndrome, and Alagille Syndrome. Of these disorders the most severe is Crigler-Najjar Syndrome type 1 with two others, Crigler-Najjar Syndrome Type 2 and Gilbert's Syndrome, being associated with mutations in *UGT1A1*. The underlying genetic components associated with conjugated hyperbilirubinemia are less well known but appear to be associated with the ATP-binding cassette (ABC) transporter pathway [4].

Genetic Disorders of Bilirubin Associated with Unconjugated Hyperbilirubinemia

Crigler-Najjar Syndrome Type 1

Crigler-Najjar Syndrome Type 1 is the most severe and debilitating of the common genetic disorders associated with overproduction of bilirubin [58]. In these cases hepatic liver *UGT1A1* activity is completely absent or almost minimal.

Until recently, absent medical treatment, this disorder was fatal during the first 18 months of life due to kernicterus. In a few exceptions, patients survived past puberty, but then died due to bilirubin encephalopathy in early adulthood. However, with modern medical treatment applying phototherapy and intermittent plasmapheresis during emergencies, survival until puberty is routine but the risk of bilirubin encephalopathy increases at this age [4]. Liver transplantation through orthopedic or auxiliary means is successful in treating this disorder.

Crigler-Najjar syndrome type 1 is clinically characterized by elevated bilirubin levels, which are usually between 20 to 25 mg/dL, but have known increase to as high as 50 mg/dL in cases of intermittent medical crises occurring in the patients [59]. All other laboratory tests appear normal. Evidence of hemolysis is also absent. Histology of the liver is normal except for the presence of "pigment plugs" in bile capillaries [4].

UGT1A1 is the major genetic isoform that contributes to bilirubin metabolism [60]. Genetic mutations within the coding region of this gene eliminate hepatic bilirubin glucuronidation, causing Crigler-Najjar syndrome type 1. Since 1952, when Crigler-Najjar syndrome type 1 was described [58], a number of genetic mutations, deletions, and insertion in or near *UGT1A1* coding regions have been shown to cause this genetic disorder [61-71] (Table 2). This genetic disorder is known to occur across all ethnic groups, but there is no single mutation that is common to any specific populations [4]. An exception to this may be among the Amish and Mennonite communities of Pennsylvania, where this disorder is common and all patients have the same mutations [72].

Crigler-Najjar Syndrome Type 2 (Arias Syndrome)

This variant of Crigler-Najjar syndrome is characterized by elevated serum bilirubin concentrations ranging from 7 to 20 mg/dL, and the prognosis due to this genetic disorder is considered much less severe. Treatment consists with administration of bilirubin-UGT inducing agents, such as phenobarbitol [73].

During fasting periods, serum bilirubin levels may be as high as 40 mg/dL or with associated periods of illness [74,75]. Cases of bilirubin encephalopathy are uncommon but are known to occur. Crigler-Najjar syndrome 2 is familial and inherited though an autosomal recessive pattern [4]. Similar to Crigler-Najjar type 1 this disease is caused only by mutations in *UGT1A1* coding region exons [76].

However, this differs from Crigler-Najjar type 1 syndrome type as all these mutations are only amino acid substitutions that result in higher levels of circulating serum bilirubin [77,78]. Some researchers argue that all point mutations within coding regions should be classified as Crigler-Najjar syndrome type 2 and any variations in the promoter region should be classified as Gilbert's syndrome [4].

Table 2. Selected UGT1A1 genetic variants implicated in different disorders associated with unconjugated hyperbilirubinemia [85]

Allele (UGT1A1*)	Nucleotides	Position	Gene Activity	Amino Acid
Crigler-Najjar Syndrome Type 1				
*2	877(T:A)/878-890 del	Exon 2	Absent	Frameshift/Del
*3	1124(C:T)	Exon 4	Inactive	S375
*4	1069(C:T)	Exon 3	Inactive	Q357X
*5	991(C:T)	Exon 2	Inactive	Q331X
*10	1021(C:T)	Exon 3	Absent	R341X
*11	923(G:A)	Exon 2	Absent	G308E
*13	508-510del	Exon 1	Inactive	F170del
*14	826(G:C)	Exon 1	Inactive	G276R
*15	529(T:C)	Exon 1	Inactive	C177R
*16	1070(A:G)	Exon 3	Absent	Q357R
*17	1143(C:G)	Exon 4	Absent	S381P
*18	1201(G:C)	Exon 4	Absent	A401P
*19	1005(G:A)	Exon 3	Absent	W335X
*20	1102(G:A)	Exon 4	Absent	A368T
*21	1223InsG	Exon4	Absent	Frameshift
*22	872(C:T)	Exon 2	Absent	A291V
*23	1282(A:G)	Exon 4	Absent	K426E
*24	1309(A:T)	Exon 5	Absent	K437X
*25	840(C:A)	Exon 1	Absent	C280X
Crigler-Najjar Syndrome Type 2				
*7	1456(T:G)	Exon 5	Reduced	Y486D
*8	625(C:T)	Exon 1	Reduced	R209W
*9	992(A:G)	Exon 2	Reduced	Q331R
*12	524(T:A)	Exon 1	Reduced	L175Q
*26	973delG	Exon 2	Absent	Frameshift
*37	A(TA)6TAA to A(TA)8TAA	Promoter	Reduced	N/A
Gilbert's syndrome				
*6	211(G:A)	Exon 1	Reduced	G71R
*27	686(C:A)	Exon 1	Reduced	P229Q
*28	A(TA)6TAA to A(TA)7TAA	Promoter	Reduced	N/A
*36	A(TA)6TAA to A(TA)5TAA	Promoter	Increased	N/A
*29	100(C:G)	Exon 4	Reduced	R367G
*60	-3279(T:G)	Promoter	None	N/A
*62	247(T:C)	Exon 1	None	F83L
*64	488-491dup	Exon 1	Reduced	Frameshift
*65	-1126(C:T)	Promoter	Reduced	N/A
*66	997-82(C:T)	Intron 2	None	N/A
*67	-85-83 ins(CAT)	Promoter	None	N/A
*68	-63(G:C)	Promoter	Reduced	N/A
*69	476(T:C)	Exon 1	Normal	I159T
*70	962(C:G)	Exon 2	Normal	A321G
*72	1075(G:A)	Exon 3	Normal	D359N
*73	1091(C:T)	Exon 4	Normal	P364L

N/A=not available.

Gilbert's Syndrome

The genetic disorder associated with hyperbilirubinemia that is the most prevalent across ethnic populations is Gilbert's syndrome [79]. This disorder is characterized by mild, chronic, and unconjugated hyperbilirubinemia. Levels of bilirubin fluctuate from normal to 3 mg/dL, and increase during fasting or intercurrent illnesses [80]. This disorder is caused by repeats within the promoter region of *UGT1A1* and has the dinucleotide repeat sequence A[TA]$_6$TAA (UGT1A1*28) [81]. TA repeats of seven or more have been found to be associated with Gilbert's syndrome. Homozygous UGT1A1*28 individuals for seven or greater repeats reduce the expression of *UGT1A1* and results in increased amounts of bilirubin within the blood. Approximately 9% of individuals of western European ancestry are homozygous for this promoter repeat variant but is at a much lower frequency in East Asian populations [35]. However, there is significant differences between ethnicities for UGT1A1*28 (Table 1).

In the vast majority of *UGT1A1* genetic studies, Western Europeans make up the majority of study subjects. While similar polymorphisms of *UGT1A1* exist across ethnic groups and result in the same physiological effect, the frequencies of these mutations are not the same across all populations (Table 1). African populations demonstrate the highest amount of diversity at the *UGT1A1* promoter locus with TA frequencies of 5 and 7 being common in several populations [36].

These five and seven *UGT1A1* promoter repeats are rare or absent in populations outside Africa. Even within populations there is differences in the frequency of UGT1A1*28 frequencies. In populations from Italy, UGT1A1*28 homozygous frequencies differed from 9 to 16% between groups [82,83]. In China, frequencies of UGT1A1*28 also differed across ethnic groups [84]. These findings indicate that people from the same country may have distinct genotypes and that Gilbert's Syndrome differs across similar geographic environments. Given that genetic testing for UGT1A1*28 is recommended by the US FDA [26,27] this has important implications for labeling drug warnings [35].

Recent advances in molecular genetics have now allowed for further characterization of these three *UGT1A1* genetic disorders associated with unconjugated hyperbilirubinemia. These studies would suggest that these three disorders are not distinct but rather a continuum of a single disorder ranging from the severe structural mutations of Crigler-Najjar Syndrome Type 1 to the mild effects of Gilbert syndrome [85]. Others have argued that these differences are semantic and that the correlating genetic disorders were classified prior to the molecular findings [4].

These latter authors propose that absence of *UGT1A1* activity due to structural defects from insertion/deletions in the gene should be classified as Crigler-Najjar Syndrome Type 1, when expression is lowered due to point mutations in coding region exons it should be classified as Crigler-Najjar Syndrome Type 2, and mutations within the promoter region should be classified as Gilbert's syndrome.

However, often times these mutations can occur both within the coding region and the promoter region together and these can lead to higher levels of hyperbilirubinemia and are associated with Gilbert's syndrome [85]. Given the current FDA recommendations regarding UGT1A1*28 drug testing, the different frequencies of this variant, and the numerous mutations within the gene suggest that this should be classified a single genetic disorder with different manifestations of severity.

Genetic Disorders of Bilirubin Associated with Conjugated Hyperbilirubinemia

Conjugated bilirubin accumulates in the plasma due to leakage from liver cells, such as in hepatocellular diseases, from disordered canalicular excretion, and biliary obstruction. In all of these cases both unconjugated and conjugated bilirubin accumulates in blood plasma. Molecular genetics has identified three disorders that can lead to the accumulation of conjugated in bilirubin. These include Dubin-Johnson syndrome, and Rotor syndrome. However, the molecular basis of Rotor syndrome remains unknown. In addition, genetic mutations that impact the biological formation and lead to hepatocellular excretory abnormalities in bile ducts can cause cholestasis. The genetic basis of one of these disorders, Alagille syndrome has been discovered [4].

Dubin-Johnson Syndrome

Dubin-Johnson Syndrome is a genetic disorder that is characterized by conjugated hyperbilirubinemia and black pigmentation of the liver without increasing levels of other liver enzymes. This disorder is rare except in Jewish populations from the Middle East, where the frequency is 1 in 1,300 [86]. In these Jewish populations, Dubin-Johnson syndrome is loosely linked with clotting factor VII deficiency.

Serum bilirubin levels are about normal and range between 2 and 5 mg/dL, but may increase to as high as 20 to 25 mg/dL during illness, oral contraceptive use, and pregnancy [87]. The gene responsible for Dubin-Johnson syndrome has been mapped to chromosome 10q23-q24 and is ATP-binding cassette sub-family C member 2 (*ABCC2*) [88-90].

Research suggests that a mutation within an intronic splice site resulting in abnormal expression of the ABCC2 transcript in a patient with Dubin-Johnson syndrome [90]. A number of other mutations within coding regions have also been shown to be associated (Table 3) with this genetic disorder.

Table 3. Mutations identified in the gene *ABCC2* and identified in patients with Dubin-Johnson Syndrome [4]

Nucleotides	Position	Amino Acid
1669-1815del	Exon13	Frameshift
Splice Site	Intron 15	Frameshift
2302(C:T)	Exon 18	R368W
272-2349 del	Exon 18	Frameshift
3196(C:T)	Exon 23	R1066X
4175-4180del	Exon 31	Frameshift

ABCC2= ATP-binding cassette sub-family C member 2.

Roter Syndrome

Rotor syndrome is an autosomal recessive disease of unknown genetic origin that is characterized by accumulation in conjugated bilirubin in the presence of normal liver function tests [91]. The only difference between Rotor syndrome and Dubin-Johnson syndrome is that the liver is not heavily pigmented. Researchers suggest that Dubin-Johnson is representative of a canilicular excretion disorder whereas Rotor syndrome is a disorder of hepatic storage and may be identical with familial hepatic storage disease [4].

Alagille Syndrome

Alagille syndrome is characterized by a low number or lack of small bile ducts, resulting in progressive intrahepatic cholestasis along with abnormalities in the eye, heart, and spinal column. The gene, jagged 1 (*JAG1*) on chromosome 20p12 encodes an unidentified ligand that bind to the notch receptor, which is critical for cell development in mammals. In some cases of Alagille syndrome this gene is deleted and in other cases, point mutations abolish expression of this gene leading to increased conjugated bilirubin in the plasma. This condition is autosomal dominant, so only one copy of the affected allele is necessary for it to be expressed [92].

Bilirubin and Chronic Disease

CVD and Bilirubin

Both retrospective and prospective epidemiological studies have demonstrated an inverse relationship between serum bilirubin and CVD. Recent studies have also shown that this inverse relationship with serum bilirubin extends to other heart related disease risk factors such as diabetes, metabolic syndrome, peripheral artery disease (PAD), carotid intima-media thickening, and stroke [2]. These findings are important as they suggest that serum bilirubin or some other factor or factors related to bilirubin aid in preventing CVD. A recent prospective study in the Framingham Offspring Study (FOS) evaluated the association between serum bilirubin and myocardial infarction (MI), coronary heart disease (CHD), and CVD mortality found that higher levels of serum bilirubin reduced risk of all three in males but not in females [47]. One meta-analysis of 11 studies found this inverse relationship in males between serum bilirubin levels and atherosclerosis but not in females. This study all also found for each 1.0μmol/L increase of bilirubin as associated with a 6.5% decrease in CVD risk [93]. However, other studies have not been found able to demonstrate an association between CVD and hyperbilirubinemia. This may be due to underlying genetic variation that differentially effects the expression of bilirubin. To date, few studies have investigated the relationship between genes associated with bilirubin production and CVD. These studies have primarily focused on *HMOX* 1with a more recent interest in *UGT1A1* in European populations [2]. Very few of these studies have investigated this relationship in other ethnic groups.

Heme-Oxgenase (HMOX1) and CVD

As previously mentioned the number of GT repeats in the promoter region of *HMOX1* modulates the gene transcription levels [94,95]. A number of studies have examined the relationship of individuals with varying GT repeats to risk for CAD, PAD, and diabetes [96-100]. Kaneda et al [96] found that patients with shorter GT (<25) repeats were less likely to have CAD than patients with longer GT (>29) repeats. Other researchers have also investigated this repeat in diabetes, speculating that type 2 diabetic individuals with longer repeats would have higher oxidative stress and increased susceptibility to CAD. This study found that serum thiobarbituric acid-reactive substances, a measure of lipid peroxidation, was significantly higher in subjects with GT (>32) repeats. Among type 2 diabetic subjects, frequencies these high GT repeat patients were significantly higher in those with CAD than in those without CAD. The adjusted odds ratio for CAD in these type 2 diabetic patients with >32 repeats was 4.7 (p=0.001) [97]. Two previous studies have investigated the relationship of long GT (>25) repeats with PAD [99,100]. Schillinger and colleagues [100] found that shorter GT (<25) repeats had a reduced risk for restenosis after femoropopliteal balloon dilation, potentially related to differences in C-reactive protein levels. A 21 month follow up to this study in 472 patients with PAD found that individuals with shorter GT (<25) repeats had a reduced risk of myocardial infarction [99]. Another variant (T:-413:A) of *HMOX1* is associated with increased hypertension in women [101]. Other *HMOX1* variants and a microsatellite and association with kidney transplant outcome demonstrated no relationship [102].

UGT1A1 and CVD

Given that *UGT1A1* has been shown to have a predominant effect on bilirubin levels and individuals with higher bilirubin concentrations have been found to have lower risk of CVD a number of studies have investigated this relationship. The Rotterdam study investigated the relationship between myocardial infarction and UGT1A1*28 homozygous individuals and detected no significant protective effect [5]. These authors did recognize that they may have missed any protective effect because of a lack of power to detect such an effect. The Etude Cas Temoins de l'Infarctus du Myocarde (ECTIM) study involving two European male populations demonstrated similar results and found no significant protective effect between myocardial infarction and UGT1A1*28 [8]. In the PAD study, Cardiovascular Disease in Patients with Intermittent Claudication (CAVASIC) there was a significant association in both cases and controls between UGT1A1*28 and bilirubin and individuals with PAD has lower bilirubin levels [13]. However, cases and controls showed no differences with respect to genotype frequencies. An additional case-control study looked at both the UGT1A1*28 and the T3279G phenobarbital-responsive enhancer polymorphism with CAD. These authors also found an association between *UGT1A1* variants but not with CAD [12]. The majority of these previous studies have focused on European populations. A more recent case-control study focused on a small Chinese sample and also found no association between UGT1A1*28 and CAD [103]. There a number of potential problems with these studies that may have prevented them from detecting association of UGT1A1*28 with CVD risk including not accounting for underlying liver disease, low penetrance of UGT1A*28 allele homozygosity, and not stratifying analysis by *UGT1A1* genotype according to bilirubin status [3].

An investigation of 1780 participants of the FOS found an association between CVD and *UGT1A1* genotypes [47]. Serum bilirubin and other CVD risk factors were measured at baseline from the 1st FOS examination and CVD and CAD events were followed up after 24 years. This was the 1st prospective study to demonstrate an association between CVD and CAD for UGT1A1*28 homozygote individuals. These UGT1A1*28 homozygote FOS participants were found to have significantly higher bilirubin levels and 1/3 the risk of CVD and CAD than either carriers who were heterozygous for UGT1A1*28 or those with 6 or less repeats. They detected a similar trend between these homozygote individuals and myocardial infarction but this association was not significant. In this study, the UGT1A1*28 allele was significantly associated with bilirubin and explained up to 18% of the variation in serum bilirubin. The inclusion of serum bilirubin and UGT1A1*28 in a Cox hazard regression models demonstrated only the bilirubin effect remained significant and the UGT1A1*28 was not considered significant. This result would suggest that serum bilirubin levels are more closely associated with CVD and CAD than just the UGT1A1*28 genotype [11,47]. Another recent prospective study of association between UGT1A1*28, serum bilirubin and first time myocardial infarction in three European Swedish cohorts detected no significant association between either the genotype or serum bilirubin [104]. This study concluded that lower bilirubin may have been due to the decreased production, increased degradation or increased elimination of bilirubin. When these researchers looked at other variables the strongest association was found to be with serum iron levels. Iron is one of the byproducts of heme degradation mediated by *HMOX1* [95] and this may be representative of interaction between this gene and *UGT1A1*.

Only one prospective study has investigated the relationship between CVD, serum bilirubin and UGT1A1*28 in a non-European population. Lin et al. [105] investigated CAD risk in 1320 CAD and 1060 controls in a Chinese Han population looking at eight common Asian SNPs in *UGT1A1*. They found a significant association for the SNP G364A (located near the *UGT1A1* TA repeat region) with CAD in males only (p=0.0014). In this study, an inverse relationship between serum bilirubin levels and CAD was demonstrated in males but not in females, which consistent with one other study [38].

In the majority of these previous studies concerning serum bilirubin, *UGT1A1*, and CVD, a clear association has been demonstrated between UGT1A1*28 and bilirubin levels. Serum bilirubin levels have also been found to be associated with CVD outcomes in most of these studies, however the only significant association of UGT1A1*28 and CVD was detected in FOS [47] and the large Han Chinese case/control study [105]. The major difference between FOS and the other studies is that it is a population-based cohort study, whereas the other studies are case/control studies with greater mean ages at recruitment. Therefore, a survival bias has been suggested by some to account for these differing results [3]. These authors have also suggested that genetic heterogeneity at UGT1A1*28 are clearly evident. If study participants have differing genetic profiles and only 1 genetic variant is tested, no association may be detected. Finally, lack of statistical power may not be adequate, in case/control studies larger sample sizes are needed to identify significant association genetic effects. This would suggest that only the FOS [47], the Han Chinese case/control [105], and the Swedish myocardial infarction meta-analysis [104] were adequately powered to detect association. Another underlying drawback to these studies is the have been predominately done in European populations and therefore differing genetic profiles in other ethnic populations may identify other variants that may also be associated with CVD in different groups.

Bilirubin and Cancer

Free oxygen radical scavenging and repair of DNA are central mechanistic processes inhibiting carcinogenesis. Hence, there is interest on whether bilirubin levels are related to increased or decreased cancer risk. A number of epidemiological studies have investigated this but there is limited evidence regarding the issue. Two prospective studies investigated the relationship between serum bilirubin levels and colorectal cancer [17,18]. The earliest of these studies investigated 118 cases and controls matched according to sex, age and date of blood draw among US residents and found an odds ratio (OR) of 0.83 between the highest and lowest quartiles but found that it was not significant [17]. The second prospective study investigated 110 cases of colorectal cancer over 20 years from National Health and Nutrition Examination Study (NHANES) I and did not find any difference between quartiles [18].

However, a study on NHANES II found that colorectal cancer risk was lowered by 0.257 per 1.0 m/dL increase in total bilirubin and also found a slightly weaker association (OR=0.80) for nongastrointestinal cancer [19]. A Belgian mortality study investigated the relationship between bilirubin and cancer and found an inverse relationship for males for all cancers but this association did not hold up for females [20].

One Australian epidemiological study did find an inverse relationship between breast cancer and serum bilirubin prior to treatment but did not account for underlying liver disease which may have biased results [21]. As with CVD epidemiological studies the differing results may be due to genetic variation within genes regulating bilirubin production. Also similar to CVD, only two genes (*HMOX1* and *UGT1A1*) have been investigated using candidate gene studies.

HMOX1 and Cancer Risk

Only the *HMOX1* promoter repeat GT variant has been examined for association between serum bilirubin and cancer including oral [22], esophageal [23], lung [106], gastric [107,108], melanoma [109], and breast [110] cancers. In most of these studies, the six *HMOX1* GT genotypes were classified into three categories; short, medium or large, based on the number of repeats. In general longer GT (>27) repeats demonstrated a moderate increase in risk for oral [22], esophageal [23], lung [106], and gastric in Taiwan [107] cancers. However, decreased risk for gastric cancer in Japan was found for individuals with long repeats [108] and for those with melanoma in Austria [109]. The largest study of *HMOX1* GT repeat investigated the risk of post-menopausal breast cancer development along with iron supplement use. These authors found an increased risk of breast cancer associated with irons supplements with medium and long repeats [110].

UGT1A1 and Cancer Risk

The majority of studies investigating a relationship between cancer and *UGT1A1* have focused on the repeat region in the promoter region. Two case control studies did not find any of the *UGT1A1* repeat genotypes to be associated with colorectal cancer in a Dutch population [111] and European and African-Americans residing in the US [112]. However, the *UGT1A1* G211A variant was associated with increased risk of colorectal cancer in small case/control study from China [113]. In another Dutch case/control study of individuals homozygous for 6 *UGT1A1* promoter repeats, cases were found to be at increased risk for head and neck cancer [114]. Although this association may be somewhat spurious it was more pronounced among heavy smokers and heavy alcohol users [115].

A number of studies have investigated the role *UGT1A1* plays in estrogen related cancers due to its involvement in the metabolic process of glucuronidation. In a case/control study of premenopausal African American individuals with 7 or 8 *UGT1A1* repeats demonstrated 1.8 increased risk of developing breast cancer [116]. In contrast to this, a Nigerian study found a decreased risk of premenopausal breast cancer with greater number of repeats [117]. Other studies have either found no association for this *UGT1A1* polymorphism [118] or only a modest increase of risk [119] for premenopausal or postmenopausal breast cancer among females of European ancestry. A single Chinese case/control study did not find an increased risk of breast cancer associated with *UGT1A1*28* among women under 40 but not for women over 40 [120].

There is still little epidemiological evidence on the relationship between bilirubin and cancer. Except for very few genetic epidemiological studies on eitherUGT1A1*28 or the *HMOX1* GT variant there are no other large scale genetic studies investigating this relationship and often the results of these studies are contradictory with some demonstrating increased risk while other suggest protective factors. The same criticisms regarding CVD and genes involved in bilirubin production can be made for cancer. In this case, all of these studies are case/control studies with no population-based cohort investigations. The small sample size in a number of these studies may indicate lack of statistical power to detect genetic associations. As with CVD studies, the *UGT1A1* studies do not stratify by genotype. Once again these studies predominately focus on individuals of European descent with some studies having been conducted with Africans, African-Americans, or Asians. However, *UGT1A1* is known to impact drug metabolism of cancer related pharmaceuticals so there are important reasons for better understanding the underlying genetic association of these genes with cancer and other chronic diseases.

Drug Metabolism and Serum Bilirubin Production Genes

A direct impact of serum bilirubin genetic epidemiology that has materialized recently is in regards to pharmacogenetics with drugs that exhibit both a narrow therapeutic spectrum along with commonly prescribed drugs. This interaction between serum bilirubin, glucuronidation, and pharmaceuticals has a high potential for unwanted side effects. Therefore, the significance of serum bilirubin and its associated genetic variants has considerable potential for patient morbidity, drug development, and economic impact [85].

UGT1A1 is involved in the metabolic process of glucuronidation of several commonly prescribed drugs, including gemfibrocil, ezetimibe, simvastatin, atorvastatin, cerivastatin, ethinylestradiol, buprenorphine, ibuprofen, and ketoprofen [121-126]. One of the best examples of this gene-drug interaction and disposition to side effects is the relationship demonstrated between *UGT1A1* and the anticancer drug irinotecan [26,127]. Currently, US FDA recommends testing for the presence of UGT1A1*28, to predict irinotecan and nilotineb toxicity [26,27].

Concerning the pharmacogenetics of genes involved in serum bilirubin production and drug side effects, there are two excellent examples with a potential to have a significant impact on genetic testing and influence drug administration [85]. These two examples include: 1) irinotecan and the SN-38 metabolite, a *UGT1A1* substrate; and 2) simvastatin, statin induced myopathy, and the serum bilirubin gene *SCLOB1*. These are both excellent examples of pharmacogenetic influences on serum bilirubin production and drug safety.

Irinotecan Toxicity

Irinotecan (camptohecin) is a standard treatment option for metastatic colorectal cancer as well as other solid tumors [128,129]. The prodrug irinotecan is transformed to 7-ethyl-10-hydroxycamptohecin (SN-38) through universally expressed carboxylesterases. SN-38 is typically inactivated through UGT1A proteins that create nontoxic SN-38 glucuronides, which are excreted through the urine and bile. Irinotecan has a limited therapeutic ranges and leads to side effects, such as myelosuppression and diarrhea [130]. This side effect is observed in 29-44% of patients, which often results in the need to either discontinue or alter dosage of the drug [128].

These irinotecan detoxification side effects have been linked to the genetic variant UGT1A1*28 [26]. However, regarding the predisposition for irinotecan toxicity by UGT1A1*28, inconsistent results have been reported. In a study of 66 patients, individuals with UGT1A1*28 were not found to predict all cases of irinotecan-associated neutropenia [26]. Another study demonstrated that UGT1A1*28 produced an effect on irinotecan-induced neutropenia but not on gastrointestinal toxicity [131]. However, another study observed the opposite effect with UGT1A1*28 patients had increased diarrhea but not neutropenia [132]. These studies strongly indicate that additional genetic factors may contribute to irinotecan toxicity, including neighboring UGT1A genes that impact SN-38 activity [85]. There is also evidence that the dose of irinotecan is relevant [133] as well as *UGT1A1* variants in differing ethnicities (i.e. UGT1A1*6) may appear to be a significant risk factor in particular for neutropenia [134].

A number of studies have also indicated that the neighboring gene, *UGT1A7*, also interacts exhibit significant interaction with SN-38 and variants within this gene may be also be involved in irinotecan toxicity [135,136]. These results may aid in explaining results obtained with just UGT1A1*28 alone and may demonstrate a combination of markers may improve irinotecan toxicity.

Simvastatin, Statin Induced Myopathy, and SCLOB1

HMG-CoA reductase inhibitors (i.e. statins) are among the most widely prescribed drugs globally [137]. These drugs are generally considered safe and highly effective but on occasion serious side effects do occur. The most serious of these adverse effects is myopathy, which can be life threatening in a small number of individuals. *UGT1A1* and other UGT-related proteins have been shown to interact with simvastatin, atorvastatin, and cerivastatin. Recently, a GWAS study identified a *SLCOB1* SNP associated with risk for statin-induced myopathy [28].

In this study, 6031 patients enrolled in a drug-trial who were treated with 80mg does of the commonly prescribed simvastatin were surveyed to identify individuals who had myopathy. They found a total of 85 patients who had myopathy were matched with 90 controls from the same group of patients treated with simvastatin. The most significant finding was the intronic *SCLO1B1* SNP, rs4363657, which was highly significant (p=10e^{-9}) and had an odd ratio of 16.9 for homozygous individuals.

Due to the small statistical power, this was then replicated in an independent study that included 10,269 patients who had been treated with 40mg of simvastatin. This study showed an odd ration of 2.6 per copy of the variant allele. While this SNP is located in a non-coding region it is known to be in linkage disequilibrium with the *SCLO1B1* coding region SNP, rs4149056, which has been previously linked to statin metabolism [29]. Presently, it is unknown whether or not how this will impact pharmacogenetics but it does demonstrate the need for better understanding of genes involved in serum bilirubin production is necessary for understanding drug toxicity.

Conclusion

This chapter has provided an overview of genes and the genetic epidemiology of serum bilirubin, an important risk factor for several complex chronic diseases. The underlying findings presented in this chapter would suggest two important future directions for investigation. One of the most important of these would be to investigate genes involved in serum bilirubin production in individuals of ethnicities other than those individuals of European origin.

The second would be to focus on a genome-wide approach in order to better understand the complex interaction of genes involved in bilirubin production. A number of studies have investigated specific variants and their association with complex diseases but to-date no study has investigated multiple variants to determine if they have an additive effect on complex disease outcome. This may lead to identification of novel genetic variants that may have important implications for understanding drug toxicity and result increasingly beneficial public health outcomes.

References

[1] Teare M editor. *Genetic epidemiology* New York, NY : Humana Press,; 2011.
[2] Schwertner HA, Vitek L. Gilbert syndrome, UGT1A1*28 allele, and cardiovascular disease risk: possible protective effects and therapeutic applications of bilirubin. *Atherosclerosis* 2008 May;198(1):1-11.
[3] Lin JP, Vitek L, Schwertner HA. Serum bilirubin and genes controlling bilirubin concentrations as biomarkers for cardiovascular disease. *Clin. Chem.* 2010 Oct;56(10):1535-1543.
[4] Chowdury N, Arias I, Wolfoff A, Chowdury J. Disorders of Bilirubin Metabolism. In: Arias I, editor. The Liver: *Biology and Pathobiology*, Fifth Edition; 2009. p. 292.
[5] Bosma PJ. Inherited disorders of bilirubin metabolism. *J. Hepatol.* 2003 Jan;38(1):107-117.
[6] Breimer L.H., Wannamethee G., Ebrahim S., Shaper A.G. Serum bilirubin and risk of ischemic heart disease in middle-aged British men. *Clin. Chem.* 1995;41:1504-1504-1508.
[7] Djoussé L., Levy D., Cupples L.A., Evans J.C., D'Agostino R.B., Ellison R.C. Total serum bilirubin and risk of cardiovascular disease in the Framingham offspring study. *Am. J. Cardiol.* 2000;87:1196-1196-1200.
[8] Gajdos V, Petit FM, Perret C, Mollet-Boudjemline A, Colin P, Capel L, et al. Further evidence that the UGT1A1*28 allele is not associated with coronary heart disease: The ECTIM Study. *Clin. Chem.* 2006 Dec;52(12):2313-2314.
[9] Hopkins P.N., Wu L.L., Hunt S.C., James B.C., Vincent G.M., Williams R.R. Higher serum bilirubin is associated with decreased risk for early familial coronary artery disease. *Arterioscler. Thromb. Vasc. Biol.* 1996;16:250-250-255.
[10] Hunt S.C., Kronenberg F., Eckfeldt J.H., Hopkins P.N., Myers R.H., Heiss G. Association of plasma bilirubin with coronary heart disease and segregation of bilirubin

as a major gene trait: the NHLBI family heart study. *Atherosclerosis* 2001;154:747-747-754.

[11] Lin JP, Schwaiger JP, Cupples LA, O'Donnell CJ, Zheng G, Schoenborn V, et al. Conditional linkage and genome-wide association studies identify UGT1A1 as a major gene for anti-atherogenic serum bilirubin levels-The Framingham Heart Study. *Atherosclerosis* 2009 Mar 19.

[12] Lingenhel A, Kollerits B, Schwaiger JP, Hunt SC, Gress R, Hopkins PN, et al. Serum bilirubin levels, UGT1A1 polymorphisms and risk for coronary artery disease. *Exp. Gerontol.* 2008 Dec;43(12):1102-1107.

[13] Rantner B, Kollerits B, Anderwald-Stadler M, Klein-Weigel P, Gruber I, Gehringer A, et al. Association between the UGT1A1 TA-repeat polymorphism and bilirubin concentration in patients with intermittent claudication: results from the CAVASIC study. *Clin. Chem.* 2008 May;54(5):851-857.

[14] Schwertner H.A. Association of smoking and low serum bilirubin antioxidant concentrations. *Atherosclerosis* 1998;136:383-383-387.

[15] Schwertner H.A., Fisher Jr. J.R. Comparison of various lipid, lipoprotein, and bilirubin combinations as risk factors for predicting coronary artery disease. . *Atherosclerosis* 2000;150:381-381-387.

[16] Schwertner H.A., Jackson W.G , Tolan G. Association of low serum concentration of bilirubin with increased risk of coronary artery disease. *Clin. Chem.* 1994;40:18-18-23.

[17] Ko WF, Helzlsouer KJ, Comstock GW. Serum albumin, bilirubin, and uric acid and the anatomic site-specific incidence of colon cancer. *J. Natl. Cancer Inst.* 1994 Dec 21;86(24):1874-1875.

[18] Ioannou GN, Liou IW, Weiss NS. Serum bilirubin and colorectal cancer risk: a population-based cohort study. *Aliment Pharmacol. Ther.* 2006 Jun 1;23(11):1637-1642.

[19] Zucker SD, Horn PS, Sherman KE. Serum bilirubin levels in the U.S. population: gender effect and inverse correlation with colorectal cancer. *Hepatology* 2004 Oct;40(4):827-835.

[20] Temme EH, Zhang J, Schouten EG, Kesteloot H. Serum bilirubin and 10-year mortality risk in a Belgian population. *Cancer Causes Control* 2001 Dec;12(10):887-894.

[21] Ching S, Ingram D, Hahnel R, Beilby J, Rossi E. Serum levels of micronutrients, antioxidants and total antioxidant status predict risk of breast cancer in a case control study. *J. Nutr.* 2002 Feb;132(2):303-306.

[22] Chang KW, Lee TC, Yeh WI, Chung MY, Liu CJ, Chi LY, et al. Polymorphism in heme oxygenase-1 (HO-1) promoter is related to the risk of oral squamous cell carcinoma occurring on male areca chewers. *Br. J. Cancer* 2004 Oct 18;91(8):1551-1555.

[23] Hu JL, Li ZY, Liu W, Zhang RG, Li GL, Wang T, et al. Polymorphism in heme oxygenase-1 (HO-1) promoter and alcohol are related to the risk of esophageal squamous cell carcinoma on Chinese males. *Neoplasma* 2010;57(1):86-92.

[24] Lo SS, Lin SC, Wu CW, Chen JH, Yeh WI, Chung MY, et al. Heme oxygenase-1 gene promoter polymorphism is associated with risk of gastric adenocarcinoma and lymphovascular tumor invasion. *Ann. Surg. Oncol.* 2007 Aug;14(8):2250-2256.

[25] Sedlak TW, Saleh M, Higginson DS, Paul BD, Juluri KR, Snyder SH. Bilirubin and glutathione have complementary antioxidant and cytoprotective roles. *Proc. Natl. Acad. Sci. USA* 2009 Mar 31;106(13):5171-5176.

[26] Innocenti F, Undevia SD, Iyer L, Chen PX, Das S, Kocherginsky M, et al. Genetic variants in the UDP-glucuronosyltransferase 1A1 gene predict the risk of severe neutropenia of irinotecan. *J. Clin. Oncol.* 2004 Apr 15;22(8):1382-1388.

[27] Hoskins JM, Goldberg RM, Qu P, Ibrahim JG, McLeod HL. UGT1A1*28 genotype and irinotecan-induced neutropenia: dose matters. *J. Natl. Cancer Inst.* 2007 Sep 5;99(17):1290-1295.

[28] SEARCH Collaborative Group, Link E, Parish S, Armitage J, Bowman L, Heath S, et al. SLCO1B1 variants and statin-induced myopathy--a genomewide study. *N. Engl. J. Med.* 2008 Aug 21;359(8):789-799.

[29] Konig J, Seithel A, Gradhand U, Fromm MF. Pharmacogenomics of human OATP transporters. *Naunyn. Schmiedebergs. Arch. Pharmacol.* 2006 Mar;372(6):432-443.

[30] Sedlak TW, Snyder SH. Bilirubin benefits: cellular protection by a biliverdin reductase antioxidant cycle. *Pediatrics* 2004 Jun;113(6):1776-1782.

[31] London IM. The conversion of hematin to bile pigment. *J. Biol. Chem.* 1950 May;184(1):373-376.

[32] Sedlak TW, Saleh M, Higginson DS, Paul BD, Juluri KR, Snyder SH. Bilirubin and glutathione have complementary antioxidant and cytoprotective roles. *Proc. Natl. Acad. Sci. USA* 2009 Mar 31;106(13):5171-5176.

[33] Yesilova Z, Serdar M, Ercin CN, Gunay A, Kilciler G, Hasimi A, et al. Decreased oxidation susceptibility of plasma low density lipoproteins in patients with Gilbert's syndrome. *J. Gastroenterol Hepatol* 2008 Oct;23(10):1556-1560.

[34] Mazzone GL, Rigato I, Ostrow JD, Bossi F, Bortoluzzi A, Sukowati CH, et al. Bilirubin inhibits the TNFalpha-related induction of three endothelial adhesion molecules. *Biochem. Biophys. Res. Commun.* 2009 Aug 21;386(2):338-344.

[35] Marques SC, Ikediobi ON. The clinical application of UGT1A1 pharmacogenetic testing: gene-environment interactions. *Hum. Genomics* 2010 Apr;4(4):238-249.

[36] Horsfall LJ, Zeitlyn D, Tarekegn A, Bekele E, Thomas MG, Bradman N, et al. Prevalence of clinically relevant UGT1A alleles and haplotypes in African populations. *Ann. Hum. Genet.* 2011 Mar;75(2):236-246.

[37] Sanna S, Busonero F, Maschio A, McArdle PF, Usala G, Dei M, et al. Common variants in the SLCO1B3 locus are associated with bilirubin levels and unconjugated hyperbilirubinemia. *Hum. Mol. Genet.* 2009 May 6.

[38] Endler G, Exner M, Schillinger M, Marculescu R, Sunder-Plassmann R, Raith M, et al. A microsatellite polymorphism in the heme oxygenase-1 gene promoter is associated with increased bilirubin and HDL levels but not with coronary artery disease. *Thromb Haemost* 2004 Jan;91(1):155-161.

[39] Lin R, Wang X, Wang Y, Zhang F, Wang Y, Fu W, et al. Association of polymorphisms in four bilirubin metabolism genes with serum bilirubin in three Asian populations. *Hum. Mutat* 2009 Apr;30(4):609-615.

[40] D'Silva S, Borse V, Colah RB, Ghosh K, Mukherjee MB. Association of (GT)n repeats promoter polymorphism of heme oxygenase-1 gene with serum bilirubin levels in healthy Indian adults. *Genet Test. Mol. Biomarkers* 2011 Apr;15(4):215-218.

[41] Lin R, Wang X, Wang Y, Zhang F, Wang Y, Fu W, et al. Association of polymorphisms in four bilirubin metabolism genes with serum bilirubin in three Asian populations. *Hum. Mutat* 2009 Apr;30(4):609-615.

[42] Cui Y, Konig J, Leier I, Buchholz U, Keppler D. Hepatic uptake of bilirubin and its conjugates by the human organic anion transporter SLC21A6. *J. Biol. Chem.* 2001 Mar 30;276(13):9626-9630.

[43] Huang CS. Molecular genetics of unconjugated hyperbilirubinemia in Taiwanese. *J. Biomed. Sci.* 2005;12(3):445-450.

[44] Huang MJ, Kua KE, Teng HC, Tang KS, Weng HW, Huang CS. Risk factors for severe hyperbilirubinemia in neonates. *Pediatr. Res.* 2004 Nov;56(5):682-689.

[45] Lin JP, Cupples LA, Wilson PW, Heard-Costa N, O'Donnell CJ. Evidence for a gene influencing serum bilirubin on chromosome 2q telomere: a genomewide scan in the Framingham study. *Am. J. Hum. Genet* 2003 Apr;72(4):1029-1034.

[46] Melton PE, Haack K, Goring HH, Laston S, Umans JG, Lee ET, et al. Genetic influences on serum bilirubin in American Indians: The Strong Heart Family Study. *Am. J. Hum. Biol.* 2011 Jan-Feb;23(1):118-125.

[47] Lin JP, O'Donnell CJ, Schwaiger JP, Cupples LA, Lingenhel A, Hunt SC, et al. Association between the UGT1A1*28 allele, bilirubin levels, and coronary heart disease in the Framingham Heart Study. *Circulation* 2006 Oct 3;114(14):1476-1481.

[48] Sai K, Saeki M, Saito Y, Ozawa S, Katori N, Jinno H, et al. UGT1A1 haplotypes associated with reduced glucuronidation and increased serum bilirubin in irinotecan-administered Japanese patients with cancer. *Clin. Pharmacol. Ther.* 2004 Jun;75(6):501-515.

[49] Huang CS, Luo GA, Huang ML, Yu SC, Yang SS. Variations of the bilirubin uridine-diphosphoglucuronosyl transferase 1A1 gene in healthy Taiwanese. *Pharmacogenetics* 2000 Aug;10(6):539-544.

[50] Ki CS, Lee KA, Lee SY, Kim HJ, Cho SS, Park JH, et al. Haplotype structure of the UDP-glucuronosyltransferase 1A1 (UGT1A1) gene and its relationship to serum total bilirubin concentration in a male Korean population. *Clin. Chem.* 2003 Dec;49(12):2078-2081.

[51] Zhang D, Zhang D, Cui D, Gambardella J, Ma L, Barros A, et al. Characterization of the UDP glucuronosyltransferase activity of human liver microsomes genotyped for the UGT1A1*28 polymorphism. *Drug Metab. Dispos.* 2007 Dec;35(12):2270-2280.

[52] Hong AL, Huo D, Kim HJ, Niu Q, Fackenthal DL, Cummings SA, et al. UDP-Glucuronosyltransferase 1A1 gene polymorphisms and total bilirubin levels in an ethnically diverse cohort of women. *Drug Metab. Dispos.* 2007 Aug;35(8):1254-1261.

[53] Sugatani J, Yamakawa K, Yoshinari K, Machida T, Takagi H, Mori M, et al. Identification of a defect in the UGT1A1 gene promoter and its association with hyperbilirubinemia. *Biochem. Biophys. Res. Commun.* 2002 Mar 29;292(2):492-497.

[54] Farheen S, Sengupta S, Santra A, Pal S, Dhali GK, Chakravorty M, et al. Gilbert's syndrome: High frequency of the (TA)7 TAA allele in India and its interaction with a novel CAT insertion in promoter of the gene for bilirubin UDP-glucuronosyltransferase 1 gene. *World J. Gastroenterol* 2006 Apr 14;12(14):2269-2275.

[55] Huang YY, Huang CS, Yang SS, Lin MS, Huang MJ, Huang CS. Effects of variant UDP-glucuronosyltransferase 1A1 gene, glucose-6-phosphate dehydrogenase

deficiency and thalassemia on cholelithiasis. *World J. Gastroenterol.* 2005 Sep 28;11(36):5710-5713.

[56] Ieiri I, Suzuki H, Kimura M, Takane H, Nishizato Y, Irie S, et al. Influence of common variants in the pharmacokinetic genes (OATP-C, UGT1A1, and MRP2) on serum bilirubin levels in healthy subjects. *Hepatol. Res.* 2004 Oct;30(2):91-95.

[57] Johnson AD, Kavousi M, Smith AV, Chen MH, Dehghan A, Asplund T, et al. Genome-wide association meta-analysis for total serum bilirubin levels. *Hum. Mol. Genet.* 2009 May 4.

[58] Crigler JF,Jr, Najjar VA. Congenital familial nonhemolytic jaundice with kernicterus. *Pediatrics* 1952 Aug;10(2):169-180.

[59] Arias IM, Gartner LM, Cohen M, Ben-Ezzer J, Levi AJ. Chronic nonhemolytic unconjugated hyperbilirubinemia with glucuronyl transferase deficiency: evidence for genetic heterogeneity. *Trans. Assoc. Am. Physicians.* 1968;81:66-75.

[60] Bosma PJ, Seppen J, Goldhoorn B, Bakker C, Oude Elferink RP, Chowdhury JR, et al. Bilirubin UDP-glucuronosyltransferase 1 is the only relevant bilirubin glucuronidating isoform in man. *J. Biol. Chem.* 1994 Jul 8;269(27):17960-17964.

[61] Bosma PJ, Chowdhury NR, Goldhoorn BG, Hofker MH, Oude Elferink RP, Jansen PL, et al. Sequence of exons and the flanking regions of human bilirubin-UDP-glucuronosyltransferase gene complex and identification of a genetic mutation in a patient with Crigler-Najjar syndrome, type I. *Hepatology* 1992 May;15(5):941-947.

[62] Ritter JK, Yeatman MT, Ferreira P, Owens IS. Identification of a genetic alteration in the code for bilirubin UDP-glucuronosyltransferase in the UGT1 gene complex of a Crigler-Najjar type I patient. *J. Clin. Invest.* 1992 Jul;90(1):150-155.

[63] Ciotti M, Obaray R, Martin MG, Owens IS. Genetic defects at the UGT1 locus associated with Crigler-Najjar type I disease, including a prenatal diagnosis. *Am. J. Med. Genet.* 1997 Jan 20;68(2):173-178.

[64] Rosatelli MC, Meloni A, Faa V, Saba L, Crisponi G, Clemente MG, et al. Molecular analysis of patients of Sardinian descent with Crigler-Najjar syndrome type I. *J. Med. Genet.* 1997 Feb;34(2):122-125.

[65] Ritter JK, Yeatman MT, Kaiser C, Gridelli B, Owens IS. A phenylalanine codon deletion at the UGT1 gene complex locus of a Crigler-Najjar type I patient generates a pH-sensitive bilirubin UDP-glucuronosyltransferase. *J. Biol. Chem.* 1993 Nov 5;268(31):23573-23579.

[66] Seppen J, Bosma PJ, Goldhoorn BG, Bakker CT, Chowdhury JR, Chowdhury NR, et al. Discrimination between Crigler-Najjar type I and II by expression of mutant bilirubin uridine diphosphate-glucuronosyltransferase. *J. Clin. Invest.* 1994 Dec;94(6):2385-2391.

[67] Labrune P, Myara A, Hadchouel M, Ronchi F, Bernard O, Trivin F, et al. Genetic heterogeneity of Crigler-Najjar syndrome type I: a study of 14 cases. *Hum. Genet.* 1994 Dec;94(6):693-697.

[68] Aono S, Yamada Y, Keino H, Sasaoka Y, Nakagawa T, Onishi S, et al. A new type of defect in the gene for bilirubin uridine 5'-diphosphate-glucuronosyltransferase in a patient with Crigler-Najjar syndrome type I. *Pediatr. Res.* 1994 Jun;35(6):629-632.

[69] Gantla S, Bakker CT, Deocharan B, Thummala NR, Zweiner J, Sinaasappel M, et al. Splice-site mutations: a novel genetic mechanism of Crigler-Najjar syndrome type 1. *Am. J. Hum. Genet.* 1998 Mar;62(3):585-592.

[70] Erps LT, Ritter JK, Hersh JH, Blossom D, Martin NC, Owens IS. Identification of two single base substitutions in the UGT1 gene locus which abolish bilirubin uridine diphosphate glucuronosyltransferase activity in vitro. *J. Clin. Invest.* 1994 Feb;93(2):564-570.

[71] Moghrabi N, Clarke DJ, Burchell B, Boxer M. Cosegregation of intragenic markers with a novel mutation that causes Crigler-Najjar syndrome type I: implication in carrier detection and prenatal diagnosis. *Am. J. Hum. Genet.* 1993 Sep;53(3):722-729.

[72] Deocharan B, Gantla S, Morton D, Chowdhury JR, Chowdury N. Interaction of a Crigler-Najjar syndrome type I mutation and a Gilbert type promoter defect results in two grades of hyperbilirubinemia in members of an Amish and a Mennonite kindred of Lancaster County, *Pennsylvannia. Gastroenterology* 1997;112:125A.

[73] Arias IM. Chronic unconjugated hyperbilirubinemia without overt signs of hemolysis in adolescents and adults. *J. Clin. Invest.* 1962 Dec;41:2233-2245.

[74] Gollan JL, Huang SN, Billing B, Sherlock S. Prolonged survival in three brothers with severe type 2 Crigler-Najjar syndrome. Ultrastructural and metabolic studies. *Gastroenterology 1*975 Jun;68(6):1543-1555.

[75] Gordon ER, Shaffer EA, Sass-Kortsak A. Bilirubin secretion and conjugation in the Crigler-Najjar syndrome type II. *Gastroenterology* 1976 May;70(5 PT.1):761-765.

[76] Jansen PL, Bosma PJ, Chowdhury JR. Molecular biology of bilirubin metabolism. *Prog. Liver Dis.* 1995;13:125-150.

[77] Seppen J, Bosma PJ, Goldhoorn BG, Bakker CT, Chowdhury JR, Chowdhury NR, et al. Discrimination between Crigler-Najjar type I and II by expression of mutant bilirubin uridine diphosphate-glucuronosyltransferase. *J. Clin. Invest.* 1994 Dec;94(6):2385-2391.

[78] Bosma PJ, Goldhoorn B, Oude Elferink RP, Sinaasappel M, Oostra BA, Jansen PL. A mutation in bilirubin uridine 5'-diphosphate-glucuronosyltransferase isoform 1 causing Crigler-Najjar syndrome type II. *Gastroenterology* 1993 Jul;105(1):216-220.

[79] Gilbert A, Lereboullet P. La cholamae simple familiale. *Semin. Med.* 1901;21:241.

[80] Berk PD, Bloomer JR, Howe RB, Berlin NI. Constitutional hepatic dysfunction (Gilbert's syndrome). A new definition based on kinetic studies with unconjugated radiobilirubin. *Am. J. Med.* 1970 Sep;49(3):296-305.

[81] Bosma PJ, Chowdhury JR, Bakker C, Gantla S, de Boer A, Oostra BA, et al. The genetic basis of the reduced expression of bilirubin UDP-glucuronosyltransferase 1 in Gilbert's syndrome. *N. Engl. J. Med.* 1995 Nov 2;333(18):1171-1175.

[82] Cecchin E, Innocenti F, D'Andrea M, Corona G, De Mattia E, Biason P, et al. Predictive role of the UGT1A1, UGT1A7, and UGT1A9 genetic variants and their haplotypes on the outcome of metastatic colorectal cancer patients treated with fluorouracil, leucovorin, and irinotecan. *J. Clin. Oncol.* 2009 May 20;27(15):2457-2465.

[83] Biondi ML, Turri O, Dilillo D, Stival G, Guagnellini E. Contribution of the TATA-box genotype (Gilbert syndrome) to serum bilirubin concentrations in the Italian population. *Clin. Chem.* 1999 Jun;45(6 Pt 1):897-898.

[84] Zhang A, Xing Q, Qin S, Du J, Wang L, Yu L, et al. Intra-ethnic differences in genetic variants of the UGT-glucuronosyltransferase 1A1gene in Chinese populations. *Pharmacogenomics J.* 2007;7(5):333-338.

[85] Strassburg CP. Gilbert-Meulengracht's syndrome and pharmacogenetics: is jaundice just the tip of the iceberg? *Drug. Metab. Rev.* 2010 Feb;42(1):168-181.

[86] Shani M, Seligsohn U, Gilon E, Sheba C, Adam A. Dubin-Johnson syndrome in Israel. I. Clinical, laboratory, and genetic aspects of 101 cases. *Q. J. Med.* 1970 Oct;39(156):549-567.

[87] Cohen L, Lewis C, Arias IM. Pregnancy, oral contraceptives, and chronic familial jaundice with predominantly conjugated hyperbilirubinemia (Dubin-Johnson syndrome). *Gastroenterology* 1972 Jun;62(6):1182-1190.

[88] Tsujii H, Konig J, Rost D, Stockel B, Leuschner U, Keppler D. Exon-intron organization of the human multidrug-resistance protein 2 (MRP2) gene mutated in Dubin-Johnson syndrome. *Gastroenterology* 1999 Sep;117(3):653-660.

[89] Wada M, Toh S, Taniguchi K, Nakamura T, Uchiumi T, Kohno K, et al. Mutations in the canilicular multispecific organic anion transporter (cMOAT) gene, a novel ABC transporter, in patients with hyperbilirubinemia II/Dubin-Johnson syndrome. *Hum. Mol. Genet.* 1998 Feb;7(2):203-207.

[90] Kajihara S, Hisatomi A, Mizuta T, Hara T, Ozaki I, Wada I, et al. A splice mutation in the human canalicular multispecific organic anion transporter gene causes Dubin-Johnson syndrome. *Biochem. Biophys. Res. Commun.* 1998 Dec 18;253(2):454-457.

[91] Rotor AB, Manahan L, Florentin A. Familial nonhemolytic jaundice with direct van dem Bergh reaction. *Acta. Med. Phil.* 1948;5:37.

[92] Oda T, Elkahloun AG, Pike BL, Okajima K, Krantz ID, Genin A, et al. Mutations in the human Jagged1 gene are responsible for Alagille syndrome. *Nat. Genet.* 1997 Jul;16(3):235-242.

[93] Novotny L, Vitek L. Inverse relationship between serum bilirubin and atherosclerosis in men: a meta-analysis of published studies. *Exp. Biol. Med.* (Maywood) 2003 May;228(5):568-571.

[94] Exner M, Schillinger M, Minar E, Mlekusch W, Schlerka G, Haumer M, et al. Heme oxygenase-1 gene promoter microsatellite polymorphism is associated with restenosis after percutaneous transluminal angioplasty. *J. Endovasc. Ther.* 2001 Oct;8(5):433-440.

[95] Idriss NK, Blann AD, Lip GY. Hemoxygenase-1 in cardiovascular disease. *J. Am. Coll. Cardiol.* 2008 Sep 16;52(12):971-978.

[96] Kaneda H, Ohno M, Taguchi J, Togo M, Hashimoto H, Ogasawara K, et al. Heme oxygenase-1 gene promoter polymorphism is associated with coronary artery disease in Japanese patients with coronary risk factors. *Arterioscler Thromb Vasc. Biol.* 2002 Oct 1;22(10):1680-1685.

[97] Chen YH, Lin SJ, Lin MW, Tsai HL, Kuo SS, Chen JW, et al. Microsatellite polymorphism in promoter of heme oxygenase-1 gene is associated with susceptibility to coronary artery disease in type 2 diabetic patients. *Hum. Genet.* 2002 Jul;111(1):1-8.

[98] Schillinger M, Minar E. Restenosis after percutaneous angioplasty: the role of vascular inflammation. *Vasc. Health Risk. Manag.* 2005;1(1):73-78.

[99] Dick P, Schillinger M, Minar E, Mlekusch W, Amighi J, Sabeti S, et al. Haem oxygenase-1 genotype and cardiovascular adverse events in patients with peripheral artery disease. *Eur. J. Clin. Invest.* 2005 Dec;35(12):731-737.

[100] Schillinger M, Exner M, Minar E, Mlekusch W, Mullner M, Mannhalter C, et al. Heme oxygenase-1 genotype and restenosis after balloon angioplasty: a novel vascular protective factor. *J. Am. Coll. Cardiol.* 2004 Mar 17;43(6):950-957.

[101] Ono K, Mannami T, Iwai N. Association of a promoter variant of the haeme oxygenase-1 gene with hypertension in women. *J. Hypertens.* 2003 Aug;21(8):1497-1503.

[102] Turpeinen H, Kyllonen LE, Parkkinen J, Laine J, Salmela KT, Partanen J. Heme oxygenase 1 gene polymorphisms and outcome of renal transplantation. *Int. J. Immunogenet.* 2007 Aug;34(4):253-257.

[103] Hsieh CJ, Chen MJ, Liao YL, Liao TN. Polymorphisms of the uridine-diphosphoglucuronosyltransferase 1A1 gene and coronary artery disease. *Cell. Mol. Biol. Lett.* 2008;13(1):1-10.

[104] Ekblom K, Marklund SL, Jansson JH, Osterman P, Hallmans G, Weinehall L, et al. Plasma bilirubin and UGT1A1*28 are not protective factors against first-time myocardial infarction in a prospective, nested case-referent setting. *Circ. Cardiovasc. Genet.* 2010 Aug;3(4):340-347.

[105] Lin R, Wang Y, Wang Y, Fu W, Zhang D, Zheng H, et al. Common variants of four bilirubin metabolism genes and their association with serum bilirubin and coronary artery disease in Chinese Han population. *Pharmacogenet Genomics* 2009 Apr;19(4):310-318.

[106] Kikuchi A, Yamaya M, Suzuki S, Yasuda H, Kubo H, Nakayama K, et al. Association of susceptibility to the development of lung adenocarcinoma with the heme oxygenase-1 gene promoter polymorphism. *Hum. Genet.* 2005 Apr;116(5):354-360.

[107] Lo SS, Lin SC, Wu CW, Chen JH, Yeh WI, Chung MY, et al. Heme oxygenase-1 gene promoter polymorphism is associated with risk of gastric adenocarcinoma and lymphovascular tumor invasion. *Ann. Surg. Oncol.* 2007 Aug;14(8):2250-2256.

[108] Sawa T, Mounawar M, Tatemichi M, Gilibert I, Katoh T, Ohshima H. Increased risk of gastric cancer in Japanese subjects is associated with microsatellite polymorphisms in the heme oxygenase-1 and the inducible nitric oxide synthase gene promoters. *Cancer Lett.* 2008 Sep 28;269(1):78-84.

[109] Okamoto I, Krogler J, Endler G, Kaufmann S, Mustafa S, Exner M, et al. A microsatellite polymorphism in the heme oxygenase-1 gene promoter is associated with risk for melanoma. *Int. J. Cancer* 2006 Sep 15;119(6):1312-1315.

[110] Hong CC, Ambrosone CB, Ahn J, Choi JY, McCullough ML, Stevens VL, et al. Genetic variability in iron-related oxidative stress pathways (Nrf2, NQO1, NOS3, and HO-1), iron intake, and risk of postmenopausal breast cancer. *Cancer Epidemiol. Biomarkers. Prev.* 2007 Sep;16(9):1784-1794.

[111] van der Logt EM, Bergevoet SM, Roelofs HM, van Hooijdonk Z, te Morsche RH, Wobbes T, et al. Genetic polymorphisms in UDP-glucuronosyltransferases and glutathione S-transferases and colorectal cancer risk. *Carcinogenesis* 2004 Dec;25(12):2407-2415.

[112] Girard H, Butler LM, Villeneuve L, Millikan RC, Sinha R, Sandler RS, et al. UGT1A1 and UGT1A9 functional variants, meat intake, and colon cancer, among Caucasians and African-Americans. *Mutat Res.* 2008 Sep 26;644(1-2):56-63.

[113] Tang KS, Chiu HF, Chen HH, Eng HL, Tsai CJ, Teng HC, et al. Link between colorectal cancer and polymorphisms in the uridine-diphosphoglucuronosyltransferase 1A7 and 1A1 genes. *World J. Gastroenterol.* 2005 Jun 7;11(21):3250-3254.

[114] Lacko M, Roelofs HM, Te Morsche RH, Voogd AC, Ophuis MB, Peters WH, et al. Genetic polymorphism in the conjugating enzyme UGT1A1 and the risk of head and neck cancer. *Int. J. Cancer* 2010 Dec 15;127(12):2815-2821.

[115] Ohnaka K, Kono S. Bilirubin, cardivascular disease and cancer: epidemiological perspectives. *Expert Review of Endocrinology and Metabolism* 2010;5(6):891.

[116] Guillemette C, Millikan RC, Newman B, Housman DE. Genetic polymorphisms in uridine diphospho-glucuronosyltransferase 1A1 and association with breast cancer among African Americans. *Cancer Res.* 2000 Feb 15;60(4):950-956.

[117] Huo D, Kim HJ, Adebamowo CA, Ogundiran TO, Akang EE, Campbell O, et al. Genetic polymorphisms in uridine diphospho-glucuronosyltransferase 1A1 and breast cancer risk in Africans. *Breast Cancer Res. Treat.* 2008 Jul;110(2):367-376.

[118] Guillemette C, De Vivo I, Hankinson SE, Haiman CA, Spiegelman D, Housman DE, et al. Association of genetic polymorphisms in UGT1A1 with breast cancer and plasma hormone levels. *Cancer Epidemiol. Biomarkers Prev.* 2001 Jun;10(6):711-714.

[119] MARIE-GENICA Consortium on Genetic Susceptibility for Menopausal Hormone Therapy Related Breast Cancer Risk. Genetic polymorphisms in phase I and phase II enzymes and breast cancer risk associated with menopausal hormone therapy in postmenopausal women. *Breast Cancer Res. Treat.* 2010 Jan;119(2):463-474.

[120] Adegoke OJ, Shu XO, Gao YT, Cai Q, Breyer J, Smith J, et al. Genetic polymorphisms in uridine diphospho-glucuronosyltransferase 1A1 (UGT1A1) and risk of breast cancer. *Breast Cancer Res. Treat.* 2004 Jun;85(3):239-245.

[121] Chouinard S, Tessier M, Vernouillet G, Gauthier S, Labrie F, Barbier O, et al. Inactivation of the pure antiestrogen fulvestrant and other synthetic estrogen molecules by UDP-glucuronosyltransferase 1A enzymes expressed in breast tissue. *Mol. Pharmacol.* 2006 Mar;69(3):908-920.

[122] Ebner T, Remmel RP, Burchell B. Human bilirubin UDP-glucuronosyltransferase catalyzes the glucuronidation of ethinylestradiol. *Mol. Pharmacol.* 1993 Apr;43(4):649-654.

[123] Ghosal A, Yuan Y, Hapangama N, Su AD, Alvarez N, Chowdhury SK, et al. Identification of human UDP-glucuronosyltransferase enzyme(s) responsible for the glucuronidation of 3-hydroxydesloratadine. *Biopharm. Drug Dispos.* 2004 Sep;25(6):243-252.

[124] Kuehl GE, Lampe JW, Potter JD, Bigler J. Glucuronidation of nonsteroidal anti-inflammatory drugs: identifying the enzymes responsible in human liver microsomes. *Drug Metab. Dispos.* 2005 Jul;33(7):1027-1035.

[125] Ogilvie BW, Zhang D, Li W, Rodrigues AD, Gipson AE, Holsapple J, et al. Glucuronidation converts gemfibrozil to a potent, metabolism-dependent inhibitor of CYP2C8: implications for drug-drug interactions. *Drug Metab. Dispos.* 2006 Jan;34(1):191-197.

[126] Prueksaritanont T, Subramanian R, Fang X, Ma B, Qiu Y, Lin JH, et al. Glucuronidation of statins in animals and humans: a novel mechanism of statin lactonization. *Drug Metab. Dispos.* 2002 May;30(5):505-512.

[127] Ando Y, Saka H, Asai G, Sugiura S, Shimokata K, Kamataki T. UGT1A1 genotypes and glucuronidation of SN-38, the active metabolite of irinotecan. *Ann. Oncol.* 1998 Aug;9(8):845-847.

[128] Douillard JY, Cunningham D, Roth AD, Navarro M, James RD, Karasek P, et al. Irinotecan combined with fluorouracil compared with fluorouracil alone as first-line treatment for metastatic colorectal cancer: a multicentre randomised trial. *Lancet* 2000 Mar 25;355(9209):1041-1047.

[129] Folprecht G, Kohne CH. The role of new agents in the treatment of colorectal cancer. *Oncology* 2004;66(1):1-17.
[130] Innocenti F, Vokes EE, Ratain MJ. Irinogenetics: what is the right star? *J. Clin .Oncol.* 2006 May 20;24(15):2221-2224.
[131] Rouits E, Boisdron-Celle M, Dumont A, Guerin O, Morel A, Gamelin E. Relevance of different UGT1A1 polymorphisms in irinotecan-induced toxicity: a molecular and clinical study of 75 patients. *Clin. Cancer Res.* 2004 Aug 1;10(15):5151-5159.
[132] Marcuello E, Altes A, Menoyo A, Del Rio E, Gomez-Pardo M, Baiget M. UGT1A1 gene variations and irinotecan treatment in patients with metastatic colorectal cancer. *Br. J. Cancer* 2004 Aug 16;91(4):678-682.
[133] Stewart CF, Panetta JC, O'Shaughnessy MA, Throm SL, Fraga CH, Owens T, et al. UGT1A1 promoter genotype correlates with SN-38 pharmacokinetics, but not severe toxicity in patients receiving low-dose irinotecan. *J. Clin. Oncol.* 2007 Jun 20;25(18):2594-2600.
[134] Onoue M, Terada T, Kobayashi M, Katsura T, Matsumoto S, Yanagihara K, et al. UGT1A1*6 polymorphism is most predictive of severe neutropenia induced by irinotecan in Japanese cancer patients. *Int J. Clin. Oncol.* 2009 Apr;14(2):136-142.
[135] Tallman MN, Miles KK, Kessler FK, Nielsen JN, Tian X, Ritter JK, et al. The contribution of intestinal UDP-glucuronosyltransferases in modulating 7-ethyl-10-hydroxy-camptothecin (SN-38)-induced gastrointestinal toxicity in rats. *J. Pharmacol. Exp. Ther.* 2007 Jan;320(1):29-37.
[136] Cecchin E, Innocenti F, D'Andrea M, Corona G, De Mattia E, Biason P, et al. Predictive role of the UGT1A1, UGT1A7, and UGT1A9 genetic variants and their haplotypes on the outcome of metastatic colorectal cancer patients treated with fluorouracil, leucovorin, and irinotecan. *J. Clin. Oncol.* 2009 May 20;27(15):2457-2465.
[137] Wang L. Pharmacogenomics: a systems approach. *Wiley Interdiscip. Rev. Syst. Biol. Med.* 2010;2(1):3.
[138] Jada SR, Lim R, Wong CI, Shu X, Lee SC, Zhou Q, et al. Role of UGT1A1*6, UGT1A1*28 and ABCG2 c.421C>A polymorphisms in irinotecan-induced neutropenia in Asian cancer patients. *Cancer Sci.* 2007 Sep;98(9):1461-1467.
[139] Boyd MA, Srasuebkul P, Ruxrungtham K, Mackenzie PI, Uchaipichat V, Stek M,Jr, et al. Relationship between hyperbilirubinaemia and UDP-glucuronosyltransferase 1A1 (UGT1A1) polymorphism in adult HIV-infected Thai patients treated with indinavir. *Pharmacogenet Genomics* 2006 May;16(5):321-329.
[140] Takeuchi K, Kobayashi Y, Tamaki S, Ishihara T, Maruo Y, Araki J, et al. Genetic polymorphisms of bilirubin uridine diphosphate-glucuronosyltransferase gene in Japanese patients with Crigler-Najjar syndrome or Gilbert's syndrome as well as in healthy Japanese subjects. *J. Gastroenterol. Hepatol.* 2004 Sep;19(9):1023-1028.
[141] Monaghan G, Ryan M, Seddon R, Hume R, Burchell B. Genetic variation in bilirubin UPD-glucuronosyltransferase gene promoter and Gilbert's syndrome. *Lancet* 1996 Mar 2;347(9001):578-581.
[142] Te Morsche RH, Zusterzeel PL, Raijmakers MT, Roes EM, Steegers EA, Peters WH. Polymorphism in the promoter region of the bilirubin UDP-glucuronosyltransferase (Gilbert's syndrome) in healthy Dutch subjects. *Hepatology* 2001 Mar;33(3):765.

[143] Fernandez Salazar JM, Remacha Sevilla A, del Rio Conde E, Baiget Bastus M. Distribution of the A(TA)7TAA genotype associated with Gilbert syndrome in the Spanish population. *Med. Clin. Barc.* 2000;115:540.

[144] Beutler E, Gelbart T, Demina A. Racial variability in the UDP-glucuronosyltransferase 1 (UGT1A1) promoter: a balanced polymorphism for regulation of bilirubin metabolism? *Proc. Natl. Acad. Sci. USA* 1998 Jul 7;95(14):8170-8174.

[145] Fertrin KY, Goncalves MS, Saad ST, Costa FF. Frequencies of UDP-glucuronosyltransferase 1 (UGT1A1) gene promoter polymorphisms among distinct ethnic groups from Brazil. *Am. J. Med. Genet.* 2002 Mar 1;108(2):117-119.

In: Bilirubin: Chemistry, Regulation and Disorder
Editors: J. F. Novotny and F. Sedlacek

ISBN: 978-1-62100-911-5
© 2012 Nova Science Publishers, Inc.

Chapter VI

Bilirubin Oxidase

Takeshi Sakurai and Kunishige Kataoka
Graduate School of Natural Science and Technology, Kanazawa University,
Kakuma, Kanazawa, Japan

Abstract

Bilirubin oxidase is classified into multicopper oxidase together with ceruloplasmin, laccase, ascorbate oxidase etc. due to having four copper ions, a type I copper, a type II copper, and a pair of type III coopers in the active site. The former copper site functions to oxidize the substrate, and the latter two types of copper sites function to reduce the final electron acceptor, O_2 to H_2O by forming a trinuclear center. Bilirubin oxidase has been found in bacteria such as *Myrothecium verrucaria* and *Bacillus subtilis*, and has been used in the assay of bilirubin in serum and the diagnosis of jaundice because it effectively catalyzes the oxidation of bilirubin to biliverdin and further purple pigment(s). Structure, reaction mechanisms, and properties of bilirubin oxidase are reviewed in this article, followed by applications of it to clinical tests, the cathodic catalyst for biofuel cells, and formation and degradation of pigments.

Introduction

Bilirubin is formed by the activity of biliverdin reductase on biliverdin, a green tetrapyrrolic bile pigment produced by the heme catabolism in the reticuloendothelial cells of the spleen. Since solubility of bilirubin in water is very low (unconjugated bilirubin), it is then bound to albumin (δ-bilirubin) and sent to liver, where bilirubin is mono- or di-conjugated with glucuronic acid by glucuronyltransferase, making it soluble in water (conjugated bilirubin). Conjugated bilirubin goes into the bile and thus is secreted into the small intestine. A part of conjugated bilirubin is present in the large intestine, and is converted into urobilinogen by colonic bacteria, finally oxidized to stercobilin via stercobilinogen. Some of urobilinogen is reabsorbed and is excreted into the urine after oxidation as urobin.

A small amount of bilirubin is usually excreted in the urine. However, the conjugated bilirubin leaks out of the hepatocytes and appears in the urine, when the function of the liver is impaired or when bilinary drainage is blocked. On the other hand, in disorders such as hemolytic anemia the amount of unconjugated bilirubin in the serum increases due to the breaking down of red blood cells. However, an increase in bilirubin in the urine is not seen since the solubility of the unconjugated bilirubin is very low in water, and an increase in urobilinogen is shown in urine. It becomes possible to distinguish disorders in metabolic systems from this difference between increases in bilirubin and urobilonogen in urine.

Bilirubin has an open-chain four pyrrole-like rings, being analogous to phycobilin which captures light energy by certain algae, phytochrome which senses light by plants, and bacteriophytochrome by anoxygenic bacteria. Since bilirubin absorbs light to isomerize some of the double bonds, it has also been utilized in the phototherapy of jaundiced newborns. It is also possible to revert bilirubin to biliverdin. Due to this reaction it is possible to utilize bilirubin functions as a cellular antioxidant [1,2]. Assays of bilirubin for clinical test have been performed according to non-enzymatic means of diagnosis. However, bilirubin oxidase (BOD) (bilirubin: oxygen oxidoreductase, EC 1.3.3.5) was first found by Murao and Tanaka [3] as the enzyme to catalyze the following reaction. BOD is a multicopper oxidase (MCO)

$$2\text{bilirubin} + O_2 \rightarrow 2\text{biliverdin} + 2H_2O$$

containing four copper ions in the active site together with laccase, ceruloplasmin, ascorbate oxidase etc [4-6]. In this review, the recently determined X-ray crystal structures of BOD [7,8] are shown, followed by its characterizations. The reaction mechanism is discussed including our recent unpublished data on the binding of substrates and four-electron reduction of dioxygen. Clinical applications of BOD to analyze the content of bilirubin in serum and urine will be mentioned briefly because it is discussed in more detail in the other chapters. Due to conversions of dioxygen to water without forming or releasing activated oxygen species, BOD has received wide attention, especially from electrochemists with the aim of utilizing it and related MCOs as an electrode catalyst for glucose biosensor and biofuel cell. BOD has also potential uses as a catalyst in the formation or degradation of pigments.

Occurrence

BOD activities have been discovered in filamentous fungus *Myrothecium* by screening a large number of microorganisms, and those isolated from *M. verrucaria* MT-1 in 1981 [3,9-11]. Later, BOD has been discovered from *Trachyderma tsunodae* K-2693 [12,13], *Penicillium japathinellus* [14], and *Pleurotus ostreatus* [15,16]. These BODs belong to multicopper oxidase in the cupredoxin super-family together with laccase [17]. Alkcaliphilic laccase from *M. verrucaria* 24G-4 [18] and an endospore laccase from *Bacillus subtilis* (CotA) have also been found to show high BOD activities. CotA is now commercially available as "BOD II" [19]. Although the possible CotA from *Bacillus pumilus* [20] is named BOD, it might not be appropriate to give the name BOD for laccases in spite of having high BOD activity to avoid confusion. Laccases with BOD activity show optimum activities at neutral to alkaline pHs. In addition, laccase should have a wide hydrophobic cavity to accommodate bulky substrates such as bilirubin and derivatives. Most of laccases show comparatively broad substrate specificities, but do not show high activities at neutral to alkaline pHs favorable for the oxidation of bilirubin. Therefore, it might be appropriate to give different names for BOD and laccase, because it is not necessarily easy to distinguish them from amino acid sequences and 3D structures [4-6]. The reason why *Myrothecium* has a MCO to show high activity for bilirubin has yet to be understood.

Heterologous Expression

Heterologous expression systems of *M. verrucaria* BOD (hereafter simply called BOD if not origin is specified) in *Saccharomyces cerevisiae* [21] and *Aspergillus oryzae* [22] have been established, and the latter expression system has been utilized to prepare various mutants in the early studies on BOD. Later, a higher level heterologous expression system to utilize *Pichia pastoris* as host was developed, and it therefore became easier to prepare BOD mutants [23].

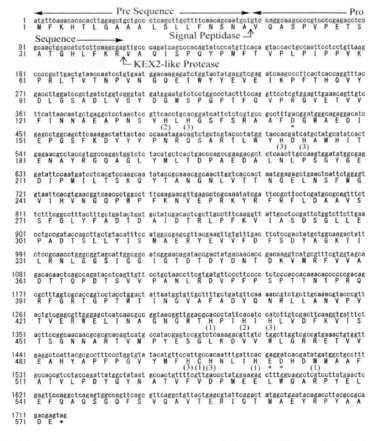

Figure 1. Nucleotide sequence of *Myrothecium verrucaria* bilirubin oxidase and the deduced amino acid sequence with Pre- and Pro-sequence. Amino acids involved in the binding to copper ions are Arabic-numbered according to their types. Asterisks are for the acidic amino acids involved in the reduction of O_2 to H_2O or for the stop codon.

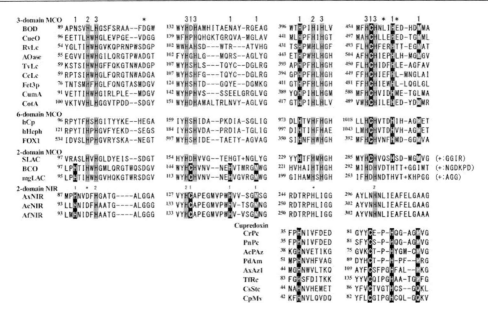

Figure 2. Alignment of amino acid sequences around the copper binding sites of MCO, nitrite reductase (NIR), and cupredoxines. 1, 2, 3, and asterisk represent T1, T2, T3 and acidic amino acid concerned in dioxygen reduction, respectively. Plus represents the three to six amino acids shown in the parenthesis. CueO, copper efflux oxidase from *E. coli*; RvLc, *Rhus vernicifera* laccase (Lc); AOase, ascorbate oxidase; TvLc, *Trametes versicolor* Lc; CcLc, *Coprinus cinereus* Lc; CumA, manganese oxidase from *Pseudomonas putida*; hCp, human ceruloplasmin; hHeph, human hephaestin; FOX1, *Chlamydomonas reinhardtii* ferroxidase; BCO, *Nitrosomonas europaea* blue copper oxidase; mgLAC, two-domain Lc from a metagenome; AxNIR, *Alcaligenes xylosoxidans* NIR; AcNIR, *Achromobacter cycloclastes* NIR; AfNIR, *Alcaligenes faecalis* NIR; CrPC, *Chlamydomonas reinhardtii* plastocyanin; PnPC, *Populus nigra* plastocyanin; AcPAz, *Achromobacter cycloclastes* pseudoazurin; PdAm, *Paracoccus denitrificans* amicyanin; AxAz1, *Alcaligenes xylosoxidans* azurin 1; TfRc, *Thiobacillus ferroxidans* rusticyanin; CsStc, *Cucumis sativus* stellacyanin; CpMv, *Cucurbita pepo* mavicyanin.

Structure and Spectral Properties

Since BOD is a secretory protein, the open reading frame encoding BOD contains pre-sequence (signal) and pro-sequence, each of which encode 19 amino acids [21]. The signal peptidase I and KEX2-like protease hydrolize off the pre-sequence and pro-sequence, respectively, before the preproprotein is matured with post-translational modification. Figure 1 shows whole nucleotide sequences of the BOD gene and the encoded amino acids of the most studied BOD from *M. verrucaria* MT-1. Arabic numbers, 1, 2, and 3 show the amino acids coordinated to type I (T1) copper, type II (T2) copper, and type III (T3) copper, respectively (*vide infra*).

It is considered that four copper ions are incorporated into an apoprotein in the Golgi apparatus, and the holoprotein to show BOD activity is completed, although any process on how BOD matures has not been traced. BOD shows high homologies with other MCOs in the amino acid sequence around the copper binding sites (Figure 2).

MCO can be classified into subgroups consisting of two-, three-, and six-domains. BOD belongs to the three-domain MCOs, the most widely distributed subgroup. The two-domain MCOs and copper-containing nitrite reductase having two domains form the hexagonal-like quaternary structures similar to the six-domain MCOs. Nevertheless, all these copper proteins evolved from a common ancestor called blue copper protein or cupredoxin, having a single copper center (T1 or blue copper) in a single domain protein molecule. T2 and T3 copper sites to bind and reduce O_2 have been constructed during molecular evolution by duplicating or triplicating cupredoxin domain and closely locating His residues in addition to the construction of the His-Cys-His triad to connect the T1 copper center and T3 copper centers, which is the unique sequence found in every MCO [24].

The crystal structure of BOD was determined in 2010 [7] and 2011 [8] independently by two groups. In the structure determined by Mizutani et al. for a twinned crystal [7], ca. 10% amino acid have been disordered, although β-structures to construct the flame of protein molecules and the structure of the active sites were the same with those shown by Cracknell et al. [8] (Figure 3).

T1 copper has a pseudo-trigonalbipyramidal geometry coordinated by Cys457, His398, and His462 in the equatorial plane and Met467 in an axial position. The main chain amide groups of Asn459 and Thr397 are located at the opposite axial position of Met467. The amide group in the side chain of Asn459 is in bonding distance to T1 copper, if the side chain is rotated. This has been realized by performing mutations at Met467 with the non-coordinating amino acids (Figure 4) [25]. T1 copper functions as the site to oxidize the substrate, and an imidazole edge of one of His resides coordinated to T1copper is exposed to the solvent in the cavity to bind substrate. A T2 copper and a pair of T3 coppers form a trinuclear copper center (TNC), where dioxygen, the final electron acceptor is bound and reduced to two water molecules. T2 copper is coordinated by His 94, His401, and a hydroxide ion in the resting form. One of T3 coppers is coordinated by His136, His403, and His456, and another by His96, His134, and His458, and both of the T3 coppers are bridged by a hydroxide ion.

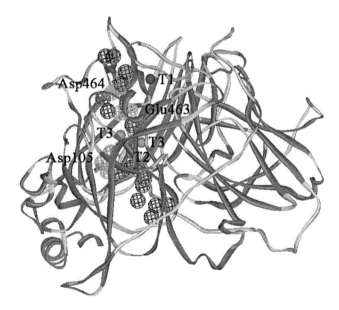

Figure 3. The crystal structure of *Myrothecium verrucaria* bilirubin oxidase (PDB code, 2XLL). T1, T2, and T3 coppers are shown as spheres. Hydrogen bond networks formed with acidic amino acids and water molecules lead form the exterior of the protein molecule to TNC, functioning as the proton relay and putative water excretion systems.

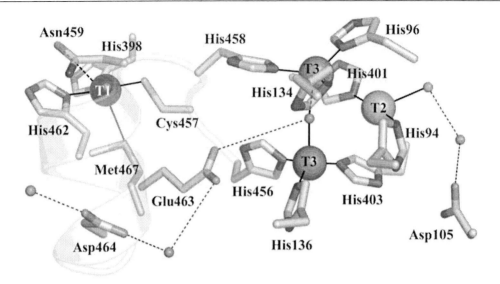

Figure 4. Structure of the active site in bilirubin oxidase (PDB code 3ABG). Asn459 is located at the opposite axial site of the axial ligand, Met467 to T1 copper (donated as T1). The proton relay pathway leads form the exterior of the protein molecule to the bridged OH⁻ between T3 coppers (denoted as T3) via Glu463 and Asp464 (A water molecule is not seen between Glu463 and the bridged OH⁻ in the crystal structure). Asp105, which is indirectly bound to T2 copper (denoted as T2) via a coordinated OH⁻ and a water molecule, is located in another channel constructed between domains 1 and 3.

Distances between each two copper center in TNC are in the range 4 to 5 Å, suggesting that copper centers are reduced to considerable extents in the crystal structures. Reduction of the copper centers easily take place due to irradiation of the MCO crystals to X-rays, and accordingly, it is not known whether the copper centers have been reduced in the crystals as isolated or were reduced during the collections of diffraction data by hydrated electrons.

Figure 3 shows the presence of the hydrogen bond networks leading from the outside of the protein molecule to TNC, in which the conserved acidic amino acids, Asp105 and Glu463, are located to concern in the binding of O_2 and proton relay, respectively. Figure 4 is the structure of the active site including these essential acidic amino acids required to realize the catalytic cyclings of the BOD reaction.

The Cu-S(Cys) bond in the T1 copper center is the origin of the blue color of BOD, and gives peculiar absorption, circular dichroism (CD), electron paramagnetic resonance (EPR), and resonance Raman spectra (Figure 5) [4-6]. Intensity of the blue color of BOD reversibly decreases with increasing pH due to an equilibrium of radial localization on T1 copper ions and Cys sulfur, Cu(II)-S⁻(Cys) ⇔ Cu(I)-S•(Cys) [26]. The change in the absorption intensity of BOD is not derived from autoreduction and autooxidation of T1 copper center depending on pH. Of the three copper ions to form TNC, only T2 copper gives the EPR signal with a normal magnitude of hyperfine splitting.

In contrast, T3 coppers are EPR-silent because of the antiferromagnetic interaction between them through the bridged hydroxide ion. This TNC structure gives a shoulder in the absorption spectrum at about 330 nm, which is characteristic to MCOs, although the band given by BOD is not very clear among MCOs. The authentic BOD and the recombinant BOD show the slightly different absorption and EPR spectra presumably due to a difference in the resting state or to a different content in the forms in the resting and isolated states [27].

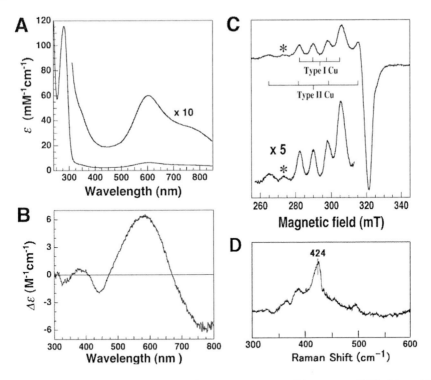

Figure 5. Absorption (A), CD (B), EPR (C), and resonance Raman (D) spectra of bilirubin oxidase. The asterisk in the EPR spectrum indicates the signal due to the magnetically uncoupled T3 copper.

Properties of Bilirubin Oxidase and Mutants

BOD has two *N*-linked glycosylation sites in addition to the *O*-linked glycosylation sites, but BOD is not necessarily a very stable enzyme among MCOs. The half inactivation time, $t_{1/2}$ of the authentic BOD, is 15 min in 0.1 M phosphate buffer (pH 6.0) at 60 °C, although that of the recombinant BOD expressed in *P. pastoris* is increased to 90 minutes due to the difference in the attached carbohydrates and content [23].

BOD shows high activities to oxidize bilirubin and ditaurobilirubin (corresponds to conjugated bilirubin), in addition to laccase substrates such as *p*-phenylenediamine, *N,N*-dimethy-*p*-phenylenediamine, *p*-aminodiphenylamine, and 2,2'-azino-bis(3-ethylbenzothiazoline-6-sulfonic acid) (ABTS), but oxidizing activities to phenolic compounds such as 2-methoxyphnol and *o*- or *p*-aminophenol are very low or not present. Discrimination of bilirubin and ditaurobilirubin with BOD is especially important for diagnosis (*vide infra*), and this becomes kinetically possible with BOD and mutants by properly selecting the measuring conditions. The amount of indirect bilirubin is calculated by subtracting the amount of direct bilirubin from the amount of total bilirubin. However, CotA (BOD II) shows markedly different activities to direct bilirubin and indirect bilirubin at different pH's: quantitative differentiation of them are possible only by performing assays of them at pH's 4 and 7 [19], and CotA seems to be superior to BOD for diagnosis use. BOD showed the optimum pH 7.3 for bilirubin, and optimum pH's for *p*-phenylenediamine, *o*-amino phenol, and ABTS correspond well to the optimum pH for the O_2 reduction activity of BOD with substrates [28].

For practical uses of BOD [29,30], catalytic activities have been studied in low water systems such as a water-in-oil microemulsion system [31], suspended lyophilized powder, and chloroform organic phase. Lyophilizations of BOD with dextran, polyvinylalcohol, poly(acrylic acid), and α,β-poly(N-hydroxyethyl)-L-aspartamide have been performed [32]. The lyophilized BOD with dextran exhibited higher stability with increasing water content. On the other hand, modification of BOD with a fluorescein derivative has also been performed [33].

The oxidase activity of BOD could be tuned by the substitutions of the axial ligand to T1 copper, Met467, which resulted in the change in the redox potential of T1 copper, the entry of electron in BOD. The redox potential of the Met467Gln mutant shifted to negative potential by 130 mV due to the binding of the amide O atom in the side chain in the place of the thioether in Met. This mutation resulted in the drastic reduction in the oxidation activity of bilirubin, although the oxidation activity of ferrocyanide was appreciably increased [34]. In the cases of other MCOs such as CueO, mutations of the axial ligand to non-coordinating amino acids resulted in positive shifts in the redox potential of T1 copper [35] in analogy with the T1 copper center in fungal laccases with high positive redox potential. Met467 in BOD has also been changed to the non-coordinating Gly and Arg and to the coordinating His, but less copper incorporations in the mutants made it difficult to assess the effect of these mutations at the axial ligand to T1 copper. The binding of an oxygen atom took place when Met467 was substituted with non-coordinating amino acids such as Phe and Leu [25]. This occurred as a result of the compensatory binding of Asn459 at the axial site opposite to Met467, as evidenced from the preparation of the double mutant Ans459Ala/Met467Phe. This compensatory binding of an adjacent amino acid has never been reported in other MCOs because no potential ligand is located in the corresponding position. Therefore, the unique redox properties of BOD that the reductive and oxidation titration curves do not overlap might be derived from this molecular architecture found only in BOD [36].

Mutations at the ligand groups to T2 and T3 coppers in BOD have been performed before other MCOs. The mutation at His94 coordinated to T2 copper by the non-coordinating Val resulted in the incomplete formation of the TNC structure, giving the EPR signals originated in the magnetically uncoupled T3 coppers [37]. The double mutation at His456 and His458 in the sequence specific to MCOs, His-Cys-His connects T1 copper and TNC with the non-coordinating Val residues resulted in the production of the completely vacant TNC. Thus, the spectra derived only from the T1 copper, which are usually overlapped with other bands or signals due to T2 and T3 coppers, were obtained. In contrast, TNC was fully occupied with copper ions in the mutations of His456 and His458 with the coordinating Lys and Asp, exhibiting the enzymatic activities, although they are considerably lowered [38]. These results have been derived from the fact that the affinity of TNC to O_2 was not completely lost by changing the imidazole ligand to a more electron-donating group. However, the redox potential of T3 coppers must have shifted to negative potentials, and the intramolecular electron transfer from T1 copper to T3 coppers would thermodynamically become more unfavorable. This accounts for the reductions in the enzymatic activities.

High specificities of BOD to bilirubin and derivatives owe to the wide substrate-binding pocket differing from many other MCOs. When Trp396 located on the protein surface near the substrate-binding pocket was substituted by Thr, the N-linked glycosylation sequence,

Asn-Gly-Thr was newly constructed, and the molecular mass of the mutant with this extra glycosylation site increased from 66 KDa to 79 KDa. The attached carbohydrates interfered the access of bilirubin and conjugated bilirubin to their binding cavity, inducing reductions in the oxidizing activities for them, especially the bulkier latter, although a change in polarity near T1 copper is not negligible. Since the activity for the latter was practically zero, discrimination of conjugated and unconjugated bilirubins has become possible in the diagnosis use of the mutant [39].

Redox potentials of T1 copper in *M. verrucaria* and *T. tsunodae* BODs have been reported to be in the range 490-670 mV and 615-710 mV by the mediated or direct spectroelectrochemical studies [36,40-42]. Difference in the redox potential came from a difference in the electrodes utilized, indicating that the BOD molecule might receive a deformation to a certain degree. Otherwise, polarity around T1 copper centers in BOD might easily change when the protein molecule was associated with the electrode surfaces. Many of the electrochemical studies on BOD have been performed with the aim of applying it as sensors or biofuel cells as summarized in the later sections.

Reaction Mechanism

The catalytic reaction of BOD is initiated with the binding of a substrate to the cavity on the protein surface, followed by the electron transfer from the substrate to T1 copper. The oxidation of substrate by BOD has already been completed at this step. However, BOD has TNC to utilize O_2 as the final electron acceptor. Thus, the later stage in the catalytic cycle of BOD involves the steps to convert an O_2 molecule to $2H_2O$. Since activated oxygen species are never formed or released from BOD, the turnover process of BOD has received wide attention.

A scheme of the reaction mechanism of BOD is shown in Figure 6. When TNC is fully reduced, O_2 is bound to it, and the intermediate I is formed. The intermediate I, also called the peroxide intermediate, has a very short lifetime, and accordingly, has never been detected during turnovers under physiological conditions.

However, when the vacant T1 copper center was formed by performing the mutation at the Cys residue coordinated to T1 copper, it became possible to trap and characterize the intermediate I with a variety of spectroscopies [43] similar to the cases of the plant laccase [44], CueO [45], and the ferroxidase Fet3p [46].

The succeeding intermediate II, also called the native intermediate, could be trapped at the final stage in a single turnover of the plant laccase [47] or BOD and CueO, of which a Glu mutant adjacent to TNC was mutated with Gln [45,48], and SLAC (small laccase) [49]. Since the Glu residue located in the hydrogen bond network leading from the exterior of the protein molecule to TNC functions to relay H^+ to O_2 bound to TNC (Figure 3), the mutation with Gln resulted in shutting down the function, and the decay of the intermediate II was prominently slowed down.

It has been revealed that the intermediate II has the O-centered TNC structure, in which all three copper(II) ions are magnetically coupled and yield an unusually broad EPR signal ($g < 2$) detectable only at cryogenic temperatures at less than 40 K. During the catalytic cycle of BOD, the O atoms receive protons from a water molecule with the assistance of Glu463.

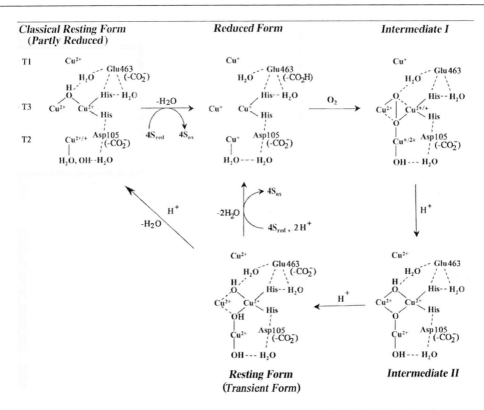

Figure 6. Reaction mechanism of bilirubin oxidase. Dioxygen is bund to the fully reduced TNC, and gives the intermediate I, which is promptly transformed to the intermediate II detectable at the single turnover conditions. The intermediate II is converted into a transient form by accepting a proton with an assistance of Glu463. Under turnover conditions, copper centers in TNC receive the stepwise reduction and release water molecules by accepting further two protons with the assistance of Glu463. Thus, proton and electron transfer processes are closely coupled, reaching the fully reduced form. When substrates are consumed up, BOD falls into the classical resting form, in which copper centers might be partly reduced because of the high redox potentials of the copper centers in BOD.

It has been demonstrated that Glu463 cycles protonation and deprotonation in accordance with changes in the redox state [50]. Another acidic amino acid, Asp464, located in the same hydrogen bond network, will also cycle protonation and deprotonation to relay protons from the exterior of the protein molecule to TNC. Finally, two H_2O molecules are released from T2 copper and one of the T3 copper centers upon stepwise reduction of TNC. An acidic amino acid Asp105 is located in another hydrogen bond network, the putative water efflux pathway, formed at the interface of domain 1 and domain 3. Asp 105 is indispensable for TNC to produce a high affinity to O_2, but other possible functions have yet to be revealed.

Clinical Applications

Diagnosis of bilirubin has been performed with the HPLC-method, diazo-method, chemical oxidation method using vanadate, and enzymatic method to utilize BOD. Literatures to analyze direct bilirubin (including conjugated bilirubin and δ-bilirubin), indirect bilirubin, and total bilirubin with BOD are summarized in Table 1.

Table 1. Bilirubin assay using bilirubin oxidase

Form of bilirubin	References
Conjugated bilirubin (direct bilirubin)	[87-96]
δ-Bilirubin	[97]
Unconjugated bilirubin (indirect bilirubin)	[98]
Protein(albumin)-unbound bilirubin	[99,100]
Total bilirubin, each fraction of bilirubin , or not specified	[101-109]

The diazo-method established by Malloy and Evelyn in 1937 [51] has been used all over the world, but the sensitivity of the chemical method is not necessarily very high. Furthermore, the diazo-method is susceptible to hemolysis, and its assay solution becomes turbid occasionally. These defects intrinsic to the diazo-method have been conquered by the BOD-method developed in the 1980s, and 70% of bilirubin diagnoses are now carried out in Japan with this method. Total bilirubin and direct bilirubin are independently analyzed from decreases in the absorption at 450 nm at pH 7.2 and pH 3.7, respectively, in the BOD-method.

In addition to the assay of bilirubin, BOD can be applicable to the assay of γ-glutamyltransferase for diagnosis of hepatic diseases [52], analysis of β-lactam antibiotics [53], and glucose in serum [54]. BOD is also applicable in eliminating the effect of bilirubin in the diagnoses of creatinine in serum and urine [55,56], the peroxidase/H_2O_2 detecting system [57,58], and the assays of fructosamine in serum [59], cholesterol in bile [60], and adenosine, inosine, and hypoxanthine/xanthine in tissue and plasma [61]. Further, BOD is used to eliminate hyperbilirubinemia [62], neonatal jaundice [63-65], obstructive jaundice [66], and the excretion of degradation products [67]. At the aim of constructing detoxification systems, BOD has been immobilized to agarose beads with a covalent attaching [66,68] or conjugated with polyethylene glycol [69].

Electrochemistry and Allications to Biosensor and Biofuel Cell

The redox potential of BOD has been determined by mediated or direct spectroelectrochemistry; here "mediated" and "direct" refer to the electric communications between electrode and proteins that are indirectly performed through a mediator to shuttle electrons and directly performed without a mediator, respectively [36,40,42,70]. Application of electrochemistry for the determination of the redox potential of MCOs has been limited because of the difficulty in realizing the electrochemistry of proteins with high molecular mass, and the electron transfer between MCOs and electrode is highly directional [71,72]. However, successful electrochemistries of oxidoreductases are increasing due to developments of electrodes, mediators, and promoters to modify the electrode surface, which enables electrical communication between a variety electrode systems and proteins. Thus, several MCOs have been utilized as bioelectrocatalysts for biofuel cells, and therefore, it is possible to utilize O_2 as the final electron acceptor at the cathode without forming or releasing activated oxygen species such as superoxide, peroxide, hydroxyl radical, which give fatal damages to electrochemical cells [41,73]. Among MCOs, BOD has been most widely utilized as a bioelectrocatalyst, presumably because BOD is commercially available, exhibits good electrochemical responses, and has comparatively high redox potential over a wide range of pH. In the construction of biofuel cells, BOD has been frequently coupled with the glucose sensor, which detects glucose concentrations from decreases in O_2 concentration or the formation of H_2O_2 in the conversion of β-D-glucose into D-glucono-1,5-lactone catalyzed by glucose oxidase (GOD) [74,75] (Figure 7). In the sensing of glucose, electrons are extracted from β-D-glucose by GOD at the anode according to the following reactions:

$$\beta\text{-D-glucose} + O_2 \rightarrow \text{D-glucono-1,5-lactone} + H_2O_2$$
$$H_2O_2 \rightarrow 2H^+ + O_2 + 2e^-$$

or

$$\beta\text{-D-glucose} \rightarrow \text{D-glucono-1,5-lactone} + 2e^-.$$

GOD has been frequently utilized as an anodic bioelectrocatalyst, but others such as glucose dehydrogenase utilized for pyroroquinoline quinone as the prosthetic group have also been available [76]. On the other hand, BOD functions as a cathodic catalyst, since BOD accepts electrons which flowed in from an anode through a leading wire. Four-electrons are transferred to O_2 by BOD to convert it into $2H_2O$ according to the following reaction:

$$4H^+ + 4e^- + O_2 \rightarrow 2H_2O$$

The overall reaction is the same with the combustion of hydrogen gas, but due to

$$2H_2 + O_2 \rightarrow 2H_2O$$

effective supplies of e^- and H^+ to O_2, the conversion of it to H_2O is realized by BOD under mild conditions. Electrochemistries of BOD have been determined by using bare electrodes such as carbon aerogel, pyrolytic graphite, and gold (Au(111)) electrodes or gold electrodes modified with appropriate promoters (Table 2). In addition, immobilizations of BOD as wired-electrodes have also been effective in achieving electric communication between BOD and electrodes. The immobilized BOD in a sol-gel processed silica film has been analyzed by scanning electrochemical microscopy [77].

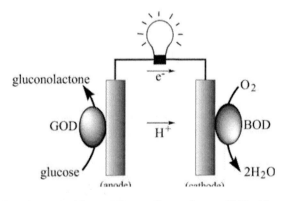

Figure 7. Biofuel cell using glucose oxidase as the anodic catalyst and bilirubin oxidase as the cathodic catalyst.

In addition to the uses of BOD for glucose sensors and biofuel cells, the electrochemical system used BOD as cathodic enzyme which has also been applied to detect DNA [78,79] and Hg^{2+} [80], and to monitor photosynthesis of living plants [81].

Table 2. Electrochemistry using bilirubin oxidase

Type of electrochemistry or electrode	References
Direct electrochemistry of BOD	
Carbon aerogel	110
Carbon film with thorn-like surface nanostructure	111
Au(111), Pt(111)	112
Cross lined onto a carbon nanotube-modified glassy carbon	
bis(sulfosuccinimidyl)suberate	113
Graphene nanosheet on Au	114
Mediator-assisted electrochemistry or immobilization of BOD on electrode surface	
6-Amino-2-naphthoic acid modified pyrolytic graphite (PGE)	112
$[Fe(CN)_6]^{3-/4-}$ Immobilized on the glassy carbon electrode	
with poly-L-Lys	115
1,1'- Dimethylferricinium-modified electrode	116
Cellulose-multiwall carbon nanotube on glassy carbon	117
Os(2,2'-bipyridine)$_2$Cl-polyvinylimidazole-modified	
teflon-shrouded glassy carbon	118
Single-walled carbon nanotube	119
N-[g-Maleimidobutyryloxy]sulfosuccinimide ester-coupled	
to the multi-walled carbon nanotube on Au	120
Buckypaper-based multi-walled carbon nanotube	121
Silica sol-gel/carbon nanotube encapsulated	122
DNA- or RNA-based self assembled multilayer consisting	
of poly(diallyldimethylammonium chloride)-wrapped	
carbon nanotube (PDDA-CNT) on indium tin oxide(ITO)	123
Electrostatic adduct of BOD and wired cationic copolymer with	
Os complex or without Os complex	124-133
$[W(CN)_6]^{4-}$, $[Os(CN)_6]^{4-}$, $[Mo(CN)_6]^{4-}$ at Pt mesh	40
C_3-SO_3H, Cn(n=2, 5, 7)-COOH modified Au	134
Multilayer architecture comprising of cytochrome c and BOD	135-137

Application as Biocatalyst

BOD has been applied to the degradations of organic dyes such as anthraquinone dye remazaol brilliant blue R, while BOD does not show activity to degrade lignin and related compounds [82,83]. In contrast, attempts have been made to use BOD as a catalyst to form dyes for hair-color together with other MCOs [84]. Further, it has been reported that BOD can be used for probing chromophore topography of phycocyanin, peridinin-chlorophyll a-protein, and phytochrome [85]. Electroactive polyaniline film, in which BOD, used to polymerize aniline, was still active and could be synthesized [86].

Conclusion

Research on BOD has two phases: basic research and application research, and the latter can be further divided into medical and electrochemical applications. Basic research on BOD has not been widely performed in spite of its discovery in 1982, although it has reached a new stage due to the determination of the crystal structure of the whole protein molecule, although the resolution was not very high and the detailed structure of the active site was still unclear. Nevertheless, mutation studies on BOD will be performed more adequately than before, and substrate specificity and reaction mechanisms will be studied in more detail in the near future. Most of the application studies on BOD for diagnosis were started before 1990, indicating the limitations of using BOD for the detection of direct bilirubin and indirect bilirubin, although the cost of the BOD-method is another defect to restrict a wider use of it for the medical applications. Successful modifications of BOD and the discovery of novel enzymes to exhibit BOD activity are also needed in the field. In contrast to clinical applications, electrochemical applications of BOD for sensors and biofuel cells have been expanding since the 1990s, and recently nano-structure materials have begun to be used in order to attain better electrochemical responses. Therefore, there is no doubt that the needs for BOD continue to expand in the field of electrochemistry in contrast to the biomedical fields. Further potential needs for BOD and related enzymes are as a catalyst to form dyes, for example for haircolor, which now use the non-enzymatic method to use hydrogen peroxide, but much safer methods to use BOD and other MCOs must be developed.

References

[1] Neuzil, J. and Stocker, R. (1994). Free and albumin-bound bilirubin are efficient co-antioxidants for α-tocopherol, inhibiting plasma and low density lipoprotein lipid peroxidation. *J. Biol. Chem.* **269**, 16712-16719.

[2] Gopinathan, V., Miller, N., J., Milner, A. D., and Rice-Evans, C. A. (1994) Bilirubin and ascorbate antioxidant activity in neonatal plasma. *FEBS Lett.* **349**, 197-200.

[3] Murao, M. and Tanaka N. (1981). A new enzyme "bilirubin oxidase" produced by *Myrothecium verrucaria* MT-1. *Agri. Biol. Chem.* **45**, 2383-2384.

[4] Sakurai, T. and Kataoka, K. (2007). Basic and applied features of multicopper oxidases, CueO, bilirubin oxidase, and laccase. *Chem. Rec. 7*, 220-229.

[5] Sakurai, T. and Kataoka, K. (2007). Structure and function of type I copper in multicopper oxidases. *Cell. Mol. Life Sci. 64*, 2642-2656.

[6] Sakurai, T. and Kataoka, K. (2011). Multicopper proteins, In K. Karlin and S. Itoh (Eds.), *Copper-oxygen chemistry* (pp. 131-168), John Wiley and Sons.

[7] Mizutani, K., Toyoda, M., Sagara, K., Takahashi, N., Sato. A., Kamitaka, Y., Tsujimura, S., Nakanishi, Y., Sugiura, T., Yamaguchi, S., Kano, K., and Mikami, B. (2010). X-ray analysis of bilirubin oxidase from *Myrothecium verrucaria* at 2.3 Å resolution using a twinned crystal. *Acta Cryst. F66*, 765-770.

[8] Cracknell, J. A., McNamara, T. P., Lowe, E. D., and Blanford, C. F. (2011) Bilirubin oxidase from *Myrothecium verrucaria*: X-ray determination of the complete crystal structure and a rational surface modification for enhanced electrocatalytic O_2 reduction. *Dalton Trans. 40*, 6668-6675.

[9] Murao, S. and Tanaka, N. (1982). Isolation and Identification of a microorganism producing bilirubin oxidase. *Agri. Biol. Chem. 46*, 2031-2034.

[10] Tanaka, N. and Murao, S. (1982). Purification and some properties of bilirubin oxidase of *Myrothecium verrucaria MT-1*. *46*, 2499-2503.

[11] Tanaka, N. and Murao, S. (1983). Difference between various copper-containing enzymes (*Polyporus* laccase, mushroom tyrosinase and cucumber ascorbate oxidase. *Agri. Biol. Chem. 47*, 1627-1628.

[12] Hiromi, K., Yamaguchi, Y., Sugiura, Y., Iwamoto, H., and Hirose, J. (1992). Bilirubin oxidase from *Trachyderma tsunodae* K-2593, a multi-copper enzyme. *Biosci. Biotechol. Biochem. 56*, 1349-1350.

[13] Hirose, J., Inoue, K., Sakuragi, H., Kikkawa, M., Minakami, M., Morikawa, T., Iwamoto, J., and Hiromi, K. (1998). Anion bindings to bilirubin oxidase from *Trachyderma tsunodae* K-2593 *Inorg. Chim. Acta 273*, 204-212.

[14] Seki, Y., Takeguchi, M., and Okura, I. (1996). Purification and properties of bilirubin oxidase from *Penicillium janthinellum*, *J. Biotechnol. 46*, 145-151.

[15] Masuda-Nishimura, I. Ichikawa, K., Hatamoto, O., Abe, K., and Koyama, Y. (1999). cDNA cloning of bilirubin oxidase from *Pleurotus ostreatus* strain Shinshu and its expression in *Aspergillus sojae*: an efficient screening of transformants, using the laccase activity of bilirubin oxidase. *J. Gen. Appl. Microbiol. 45*, 93-97.

[16] Pakhadnia, Y. G., Malinouski, N. I., and Lapko, A. G. (2009). Purification and characteristics of an enzyme with both bilirubin oxidase and laccase activities from mycelium of the basidiomycete *Pleurotus ostreatus*. *Biochemistry(Moscow), 74*, 1027-1034.

[17] Claus, H. (2004) Laccases: structure, reactions, distribution. *Micron 35*, 93-96.

[18] Sulistyaningdyah, W. T., Ogawa, J., Tanaka, H., Maeda, C., and Shimizu, S. (2004) Characterization of alkaliphilic laccase activity in the culture supernatant of *Myrothecium verrucaria* 24G-4 in comparison with bilirubin oxidase. *FEMS Microbiol. Lett., 230*, 209-214.

[19] Sakasegawa, S., Ishikawa, H., Imamura, S., Sakuraba, H., Goda, S., and Ohshima, T. (2006). Bilirubin oxidase activity of *Bacillus subtilis* CotA. *Appl. Environ. Microbiol., 72*, 972-975.

[20] Reiss, R., Ihssen, J., and Thöny-Meyer, L. (2011). *Bacillus pumilus* laccase: a heat stable enzyme with a wide substrate spectrum. *BMC Biotechnol., Jan 25; 11: 9*.

[21] Koikeda, S., Ando, K., Kaji, H., Inoue, T., Murao, S., Takeuchi, K., and Samejima, T. (1993). Molecular cloning of the gene for bilirubin oxidase from *Myrothecium verrucaria* and its expression in yeast. *J. Biol. Chem. 268*, 18801-18809.

[22] Yamaguchi, S., Takeuchi, K., Mase, T., and Matsuura, A. (1997). Effect of expression of mono- and diacylglycerol lipase gene from *Penicillium camembertii* U-150 in *Aspergillus oryzae* under the control of its own promoter. *Biosci. Biotechnol. Biochem. 61*, 800-805.

[23] Kataoka, K., Tanaka, K., Sakai, Y. and Sakurai, T. (2005). High-level expression of *Myrothecium verrucaria* bilirubin oxidase in *Pichia pastoris*, and its facile purification and characterization. *Prot. Expr. Pur. 41*, 77-83.

[24] Nakamura, K. and Go, N. (2005). Function and molecular evolution of multicopper blue proteins. *Cell. Mol. Life Sci. 62*, 2050-2066.

[25] Kataoka, K., Tsukamoto, K., Kitgawa, R., Ito, T., and Sakurai, T. (2008). Compensatory binding of an asparagine residue to the coordination-unsaturated type I Cu center in bilirubin oxidase mutants. *Biochem. Biophys. Res. Commun. 371*, 416-419.

[26] Zoppellaro, G., Sakurai, N., Kataoka, K., and Sakurai, T. (2004). The reversible change in the redox state of type I Cu in *Myrothecium verrucaria* bilirubin oxidase depending on pH. *Biosci. Biotechnol. Biochem. 68*, 1998-2000.

[27] Sakurai, T., Zhan, L., Fujita, T., Kataoka, K., Shimizu A., Samejima, T., and Yamaguchi, S. (2003). Authentic and recombinant bilirubin oxidase are in different resting forms. *Biosci. Biotechnol. Biochem. 67*, 1157-1159.

[28] Otsuka, K., Sugihara, T., Tsujino, Y., Osaki, T., and Tamiya, E. (2007). Electrochemical consideration on the optimum pH of bilirubin oxidase. *Anal. Biochem. 370*, 98-106.

[29] Aston, M. J. and Freedman, R. B. (2002). The water-dependence of the catalytic activity of bilirubin oxidase suspensions in low-water systems. *Biotechnol. Bioeng. 77*, 651-657.

[30] Skrika-Alexopoulos, W., Muir, J., and Freedman, R. B. (1993). Stability of bilirubin oxidase in organic solvent media: a comparative study in two low-water systems. *Biotechnol. Bioeng. 41*, 894-899.

[31] Oldfield, C. and Freedman, R. B. (1989). Kinetics of bilirubin oxidation catalyzed by bilirubin oxidase in a water-in-oil microemulsion system. *Eur. J. Biochem. 183*, 347-355.

[32] Yoshioka, S., Aso, Y., Kojima, S., and Tanimoto, T. (2000). Effect of polymer excipients on the enzyme activity of lyophilized bilirubin oxidase and β-galactosidase formulations. *Chem. Pharm. Bull. 48*, 283-285.

[33] Andreu, Y., Ostra, M., Ubide, C., Galbán, J., de Marcos, S., and Castillo, J. R. (2002). Study of a fluorometric-enzymatic method for bilirubin based on chemically modified bilirubin-oxidase and multivariate calibration. *Talanta 57*, 343-353.

[34] Shimizu, A., Sasaki, T., Kwon, J. H., Odaka, A., Satoh, T., Sakurai, N., Sakurai, T., Yamaguchi, S., and Samejima, T. (1999). Site-directed mutagenesis of a possible type I copper ligand of bilirubin oxidase; a Met467Gln mutants shows stellacyanin-like properties. *J. Biochem. 125*, 662-668.

[35] Kurose, S., Kataoka, K., Shinohara, N., Miura, Y., Tsutsumi, M., Tsujimura, S., Kano, K., and Sakurai, T. (2009). Modification of spectroscopic properties and catalytic activity of *Escherichia coli* CueO by mutants of methionine 510, the axial ligand to the type I Cu. *Bull. Chem. Soc. Jpn. 82*, 504-508.

[36] Christenson, A., Shleev, S., Mano, N., Heller, A., and Gorton, L. (2006). Redox potentials of the blue copper sites of bilirubin oxidases. *Biochim. Biophys. Acta 1757*, 1634-1641.

[37] Shimizu, A., Kwon, J.-H., Sasaki, T., Satoh, T., Sakurai, N., Sakurai, T., Yamaguchi, S., and Samejima, T. (1999). *Myrothecium verrucaria* bilirubin oxidase and its mutants for potential copper ligands. *Biochemistry 38*, 3034-3042.

[38] Shimizu, A., Samejima, T., Hirota, S., Yamaguchi, S., Sakurai, N., and Sakurai, T. (2003). Type III Cu mutants of *Myrothecium verrucaria* bilirubin oxidase. *J. Biochem. 133*, 767-772.

[39] Sakai, Y., Ito, T., Kataoka, K., and Sakurai, T. unpublished data.

[40] Tsujimura, S., Kuriyama, A., Fujieda, N., Kano, K., and Ikeda, T. (2005). Mediated spectrochemical titration of proteins for redox potential measurements by a separatorless one-compartment bulk electrolysis method. *Anal. Biochem. 337*, 325-331.

[41] Shleev, S., Tkac, A., Christenson, A., Ruzgas, T., Yaropolov. A. I., Whittaker, J. W., and Gorton, L. (2005). Direct electron transfer between copper-containing proteins and electrodes. *Biosens. Bioelectron. 20*, 2517-2554.

[42] Ivnitski, D., Artyushkova, K., and Atanassov, P. (2008). Surface characterization and direct electrochemistry of redox copper centers of bilirubin oxidase from fungi *Myrothecium verrucaria*. *Bioelectrochemistry 74*, 101-110.

[43] Kataoka, K., Kitagawa, R., Inoue, M., Naruse, D., Sakurai, T., and Huang, H-W. (2005). Point mutations at the type I copper ligand, Cys457 and Met467, and at the putative proton donor, Asp105, in *Myrothecium verrucaria* bilirubin oxidase and reactions with dioxygen. *Biochemistry, 44*, 7004-7012.

[44] Zoppellaro, G., Sakurai, T., and Huang, H.-W. (2001). A novel mixed valance form of *Rhus vernicifera* laccase and its reaction with dioxygen to give a peroxide intermediate bound to the trinuclear center. *J. Biochem. 129*, 949-953.

[45] Kataoka, K., Sugiyama, R., Hirota, S., Inoue, M., Minagawa, Y., Seo, D., and Sakurai, T. (2008). Four-electron reduction of dioxygen by a multicopper oxidase, CueO, and roles of Asp112 and Glu506 located adjacent to the trinuclear copper center. *J. Biol. Chem. 284*, 14405-14413.

[46] Augustine, A., Quintanar, L., Stoj, C. S., Kosman, D. J., and Solomon, E. I. (2007). Spectroscopic and kinetic studies of perturbed trinuclear copper clusters: the role of protons in reductive cleavage of the O-O bond in the multicopper oxidase Fet3p. *J. Am. Chem. Soc. 129*, 13118-13126.

[47] Huang, H.-W., Zopplellaro, G., and Sakurai, T. (1999). Spectroscopic and kinetic studies on the oxygen-centered radical formed during the four-electron reduction process of dioxygen by *Rhus vernifcifera* laccase. *J. Biol. Chem. 274*, 32718-32724.

[48] Kajikawa, T., Sugiyama, R., Kataoka, K., and Sakurai, T. unpublished data.

[49] Tepper, A. W., Milikisyants, S., Sottini, S., Vijgenboom, E. J., Groenen, E. J., and Canters, G. W. (2009). Identification of a radical intermediate in the enzymatic reduction of oxygen by a small laccase. *J. Am. Chem. Soc. 131*, 11680-11682.

[50] Iwaki, M., Kataoka, K., Kajino, T., Sugiyama, R., Morishita, H., and Sakurai, T. (2010). ATR-FTIR study of the protonation states of the Glu residue in the multicopper oxidases, CueO and bilirubin oxidase. *FEBS Lett.* 574, 4027-4031.

[51] Malloy, H. T. and Evelyn, K. A. (1937) The determination of bilirubin with the photometirc calorimeter. *J. Biol. Chem.* 119, 481-490.

[52] Satomura, S., Miki, Y., Hamanaka, T., and Sakata, Y. (1985). Kinetic assay of γ-glutamyltransferase with use of bilirubin oxidase as a coupled enzyme. *Clin. Chem.* 31, 1380-1393.

[53] Matsuura, A., Nagayama, T., and Kitagawa, T. (1988). Analytical studies on beta-lactam antibiotics. II. Displacement effect of beta-lactam antibiotics on bilirubin bound to human serum albumin. *Chemotherapy* 34, 345-353.

[54] Male, K. B., and Luong, J. H. (1996). An improved enzymatic assay for glucose determination in blood serum using 1,1'-dimethylferricinium dye. *Appl. Biochem. Biotechnol.* 61, 267-276.

[55] Artiss, J. D., McEnroe, R. J., and Zak, B. (1984). Bilirubin interference in a peroxide-coupled procedure for creatinine eliminated by bilirubin oxidase. *Clin. Chem.* 30, 1389-1392.

[56] Matsushita, M., Maehara, E., and Aoki, Y. (1986). Elimination of bilirubin interferences on creatinine determination (kinetic jaffé method) using bilirubin oxidase. *Rinsho Byori (Jpn. J. Clin. Pathol.).* 34, 694-698.

[57] Maguire, G. A. (1985). Elimination of the "chromogen oxidase" activity of bilirubin oxidase added to obviate bilirubin interference in hydrogen peroxide/peroxidase detecting systems. *Clin. Chem.* 31, 2007-2008.

[58] Maguire, G. A. and Bannister, M. W. (1987). The "Chromogen oxidase" activity of bilirubin oxidase. *Clin. Chem.* 33, 1304.

[59] Ohisa, Y., Kawamura, T., Abe, Y., and Ishimori, A. (1992). Effect of bilirubin on serum fructosamine value, and the elimination of bilirubin by bilirubin oxidase. *Rinsho Byori (Jpn. J. Clin. Pathol.).* 40, 194-198.

[60] Luhman, C. M., Galloway, S. T., and Beits, D. C. (1990). Simple enzymatic assay for determining cholesterol concentrations in bile. *Clin. Chem.* 36, 331-333.

[61] Jabs, C. M., Neglen, P., Eklof, B., and Thomas, E. J. (1990). Adenosine, inosine, and hypoxanthine/xanthine measured in tissue and plasma by a luminescence method. *Clin. Chem.* 36, 81-87.

[62] Eckmann, C. M., de Laaf, R. T. M., van Keulen, J. M., van Mourik, J. A., and de Laat, B. (2007.) Bilirubin oxidase as a solution for the interference of hyperbulirubinemia with ADAMTS-13 activity measurement by FRETS-VWF73 assay. *J. Thromb. Haemost.* 5, 1330-1331.

[63] Lavin, A. Sung, C., Klibanov, A. M., and Langer, T. (1985). Enzymatic removal of bilirubin from blood: a potential treatment for neonatal jaundice. *Science* 230, 543-545.

[64] Dennery, P. A. (2002). Pharmacological interventions for the treatment of neonatal jaundice. *Semin. Nenatol.* 7, 111-119.

[65] Mullon, C. J., Tosone, C., and Langer, R. (1989). Simulation of bilirubin detoxification in the newborn using an extracorporeal bilirubin oxidase reactor. *Pediatr. Res.* 26, 452-457.

[66] Kamisako, T., Miyawaki, S., Gabazza, E. C., Ishihara, T., Kamei, A., Kawamura, N., and Adachi, Y. (1998). Polyethylene glycol-modified bilirubin oxidase improves

hepatic energy charge and urinary prostaglandin levels in rats with obstructive jaundice. *J. Hepatol. 29*, 424-429.

[67] Kimura, M., Mitsumura, Y., Konno, T., Miyauchi, Y., and Maeda, H. (1990). Enzymatic removal of bilirubin by bilirubin oxidase in vitro and excretion of degradation products in vivo. *Proc. Soc. Exp. Biol. Med. 195*, 64-69.

[68] Sung, C., Lavin, A., Klibanov, A. M., and Langer, R. (1986). An immobilized enzyme reactor for the detoxification of bilirubin. *Biotechol. Bioeng.* 28, 1351-1319.

[69] Kimura, M., Matsumura, Y., Miyauchi, Y., and Maeda, H. (1988). A new tactic for the treatment of jaundice: an injectable polymer-conjugated bilirubin oxidase. *Proc. Soc. Exp. Biol. Med. 188*, 364-369.

[70] Ramirez, P., Mano, N., Andreu, R., Ruzgas, T., Heller, A., Gorton, L., and Shleev, S. (2008). Direct electron transfer from graphite and functionalized gold electrodes to T1 and T2/T3 copper centers of bilirubin oxidase. *Biochim. Biophys. Acta 1777*, 1364-1369.

[71] Sakurai, T. (1996) Cyclic voltammetry of cucumber ascorbate oxidase. *Chem. Lett.* 481-482.

[72] Kamitaka, Y., Tsujimura,S., Kataoka, K., Sakurai, T., Ikeda, T., and Kano, K. (2007). Effects of the axial mutation of the type I copper site in bulirubin oxidase on direct electron transfer-type bioelectrocatalytic reduction of dioxygen. *J. Electro. Anal. Chem. 601*, 119-124.

[73] Ikeda, T. (2004). A novel electrochemical approach to the characterization of oxidoreductase reactions. *Chem. Rec. 4*, 192-203.

[74] Mano, N. and Heller, A. (2005). Detection of glucose at 2fM concentration. *Anal. Chem. 77*, 729-732.

[75] Zhu, Z., Momeu, C., Zakhartsev, M., and Schwaneberg, U. (2006). Making glucose oxidase for biofuel cell applications by directed protein evolution. *Biosens. Bioelec. 21*, 2046-2051.

[76] Yuhashi, N., Tomiyama, M., Okuda, J., Igarashi, S., Ikebukuro, K., and Sode, K. (2005). Development of a novel glucose enzyme fuel cell system employing protein engineered PQQ glucose dehydrogenase. *Biosens. Bioelectron. 20*, 2145-2150.

[77] Nogala, W., Szot, K., Burchardt, M., Roelfs, F., Rogalski, J., Opallo, M., and Wittstock, G. (2010). Feedback mode SECM study of laccase and bilirubin oxidase immobilised in a sol-gel processed silicate film. *Analyst 135*, 2051-2058.

[78] Kim, H.-H., Zhang, Y., and Heller, A. (2004). Bilirubin oxidase label for an enzyme-linked affinity assay with O_2 as substrate in neutral pH NaCl solution. *Anal. Chem. 76*, 2411-2414.

[79] Zhang, Y., Pothukuchy, A., Shin, W., Kim, Y., and Heller, A. (2004). Detection of ~10^3 copies of DNA by an electrochemical enzyme-amplified sandwich assay with ambient O_2 as the substrate. *Anal. Chem. 76*, 4093-4097.

[80] Wen, D., Deng, L., Guo, S., and Dong, S. (2011). Self-powered sensor for trace Hg^{2+} detection. *Anal. Chem. 83*, 3968-3972.

[81] Flexer, V., and Mano, N. (2010). From dynamic measurements of photosynthesis in a living plant to sunlight transformation into electricity. *Anal. Chem. 82*, 1444-1449.

[82] Zhang, X., Liu, Y., Yan, K., and Wu, H. (2007). Decolorization of anthraquinone-type dye by bilirubin oxidase-producing nonlignilolytic fungus *Myrothecium sp.* IMER1. *J. Biosci. Bioeng. 104*, 104-110.

[83] Liu, Y., Huang, J., and Zhang, X. (2009). Decolorization and biodegradation of remazol brilliant blue R by bilirubin oxidase. *J. Biosci. Bioeng. 108*, 496-500.

[84] Kataoka, K., Tanaka, K., and Sakurai, T. (2010). Japan patent 4437652.

[85] Singh, B. R., Chol, J., Kwon, T.-L., and Song, P.-S. (1989). Use of bilirubin oxidase for probing chromophore topography in tetrapyrrole proteins. *J. Biochem. Biophys. Meth. 18*, 135-147.

[86] Aizawa, M., Wang, L. L., Shinohara, H., and Ikariyama, Y. (1990). Enzymatic synthesis of polyaniline film using a copper containing oxidoreduxtase: bilirubin oxidase. *J. Biotech. 14*, 301-309.

[87] Doumas, B. T., Perry, B., Jendrzejczak, B., and Davis, L. (1987). Measurement of direct bilirubin by use of bilirubin oxidase. *Clin. Chem. 33*, 1349-1353.

[88] Ihara, H., Aoki, Y., Aoki, T., and Yoshida, M. (1990). Light has a greater effect on direct bilirubin measured by the bilirubin oxidase method than by the diazo method. *Clin. Chem. 36*, 895-897.

[89] Nakayama, K. (1994). The main cause responsible for the disparity between enzymatic and diazo methods in measuring direct bilirubin from a viewpoint of bilirubin subfractionation. *Rinsho Byori (Jpn. J. Clin. Pathol.). 42*, 534-538.

[90] Palilis, L. P., Calokerinos, A. C., and Grekas, N. (1997). Chemiluminescence arising from the oxidation of bilirubin in aqueous media. *Biomed. Chromatogr. 11*, 71-72.

[91] Kurosaka, K., Senba, S., Tsubota, H., and Kondo, H. (1998). A new enzymatic assay for selectively measuring conjugated bilirubin concentration in serum with use of bilirubin oxidase. *Clin. Chim. Acta 269*, 125-136.

[92] Kimura, S., Iyama, S., Yamaguchi, Y., Hayashi, S., and Yanagihara, T. (1999). Enzymatic assay for conjugated bilirubin (Bc) in serum using bilirubin oxidase (BOD). *J. Clin. Lab. Anal. 13*, 219-223.

[93] Doumas, B. T., Yein, F., Perry, B., Jendrzejczak, B., and Kessner, A. (1999). Determination of the sum of bilirubin sugar conjugates in plasma by bulirubin oxidase. *Clin. Chem. 45*, 1255-1260.

[94] Morimoto, Y., Ishihara, T., Takayama, M., Kaito, M., and Adachi, Y. (2000). Novel assay for measuring serum conjugated bilirubin and its clinical relevance. *J. Clin. Lab. Anal. 14*, 27-31.

[95] Andreu, Y., Galbán, J., de Marcos, S., and Castillo, J. R. (2000). Determination of direct-bilirubin by a fluorimetric-enzymatic method based on bilirubin oxidase. *Fresenius J. Anal. Chem. 368*, 516-521.

[96] Kojima, R., Sasagawa, Y., Okazaki, Y., and Nagano, K. Japan Patent JP2007-202568.

[97] Umaji, M., Okumiya, T., Park, K., and Sasaki, M. (1989). A colorimetric method for determination of δ-bilirubin in serum using bilirubin oxidase (*Trachyderma tsunodae*). *Rinsho Byori (Jpn. J. Clin. Pathol.). 37*, 905-910.

[98] Cardenas-Vazquez, R., Yokosuka, S., and Billing, B. H. (1986). Enzymatic oxidation of unconjugated bilirubin by rat liver. *Biochem. J. 236*, 625-633.

[99] Yuno, T., Yamamoto, Y., and Nakamura, H. (1986). Determination of unbound bilirubin by the new enzyme "bilirubin oxidase". *Rinsho Byori (Jpn. J. Clin. Pathol.). 34*, 297-302.

[100] Goldfinch, M. E. and Maguire, G. A. (1988). Investigation of the use of bilirubin oxidase to measure the apparent unbound bilirubin concentration in human plasma. *Ann. Clin. Biochem. Jan;25 (Pt1)*, 73-77.

[101] Kosaka, A., Yamamoto, C., Morishita, Y., and Nakane, K. (1987). Enzymatic determination of bilirubin fractions in serum. *Clin. Biochem. 20*, 451-458.

[102] Otsuji, S., Mizuno, K., Ito, S., Kawahara, S., and Kai, M. (1988). A new enzymatic approach for estimating total and direct bilirubin. *Clin. Biochem. 21*, 33-38.

[103] Heinemann, G. and Vogt, W. (1988). The determination of bilirubin with a new enzymatic method (Dri-STAT bilirubin) using the Hitachi 704 elective analyzer. *J. Clin. Chem. Clin. Bochem. 26*, 391-397.

[104] Doumas, B. T. and Wu, T.-W. (1991). The measurement of bilirubin fractions in serum. *Crit. Rev. Clin. Lab. Sci. 28*, 415-445.

[105] Perry, B., Doumas, B. T., Buffone, G., Gllck, M., Ou, C.-N., and Ryder, K. (1986). Measurement of total bilirubin by use of bilirubin oxidase. *Clin. Chem. 32*, 329-332.

[106] Prezelj, M. (1988). Enzymatic determination of total bilirubin in serum with the BA-1000. *Clin. Chem. 34*, 176-177.

[107] Li, G., Liu, J., Yin, M., Guo, J., and Xie, S. (1996). Application of microbial enzymes in diagnostic analysis. *Ann. N. Y. Acad. Sci. 799*, 476-481.

[108] Zhou, X.-M., Liu, J.-W., Zou, X., and Chen, J.-J. (1999). Monitoring catalytic reaction of bilirubin oxidase and determination of bilirubin and bilirubin oxidase activity by capillary electrophoresis. *Electrophoresis 20*, 1916-1920.

[109] Ihara, N., Nakamura, H., Aoki, Y., Aoki, T., and Yshida, M. (1992). In vitro effects of light on serum bilirubin subfractions measured by high-performance liquid chromatography: comparison with four routine methods. *Clin. Chem. 38*, 2124-2129.

[110] Tsujmura, S., Kamitaka, Y., and Kano, K. (2007). Diffusion-controlled oxygen reduction on multi-copper oxidase-adsorbed carbon aerogel electrodes without mediator. *Fuel Cells 7*, 463-469.

[111] Ueda, A., Kato, R., Kamata, T., Inokuchi, H., Umemura, S., Hirono, S., and Niwa, O. (2011). Efficient direct electron transfer with enzyme on a nanostructured carbon film fabricated with a maskless top-down UV/ozone process. *J. Am. Chem. Soc. 133*, 4840-4846.

[112] dos Santos, L., Climent, V., Blanford, C. F., and Armstrong, F. A. (2010). Mechanistic studies of the blue Cu enzyme, bilirubin oxidase, as a highly efficient electrocatalyst for the oxygen reduction reaction. *Phys. Chem. Chem. Phys. 12*, 13962-13974.

[113] Yehezkeli, O., Tel-Vered, R., Raichlin, S., and Willner, I. (2011). Nano-engineered flavin-dependent glucose dehydrogenase/gold nanoparticle-modified electrodes for glucose sensing and biofuel cell applications. *ACS Nano 5*, 2385-2391.

[114] Liu, C., Alwarappan, S., Chen, Z., Kong, X., and Li., C.-Z. (2010). Membraneless enzymatic biofuel cells based on graphene nanosheets. *Biosens. Bioelectron. 25*, 1829-1843.

[115] Nakagawa, T., Tsujimura, S., Kano, K., and Ikeda T. (2003). Bilirubin oxidase and $[Fe(CN)_6]^{3-/4-}$ modified electrode allowing diffusion-controlled reduction of O_2 to water at pH 7.0. *Chem. Lett. 32*, 54-55.

[116] Luong, J. H. T., Masson, C., Brown, R. S., Male, K. B., and Nguyen, A.-L. (1994). Monitoring the activity of glucose oxidase during the cultivation of *Aspergillus niger* using novel amperometric sensor with 1,1'-dimethylferricinium as a mediator. *Biosens. Bioelectron. 9*, 577-584.

[117] Wu, X., Zhao, F., Varcoe, J. R., Thumser, A. E., Avignone-Rossa, C., and Slade, R. C. (2009). A one-compartment fructose/air biological fuel cell based on direct electron transfer. *Biosens. Bioelectron. 25*, 326-331.
[118] Jenkins, P. A., Boland, S., Kavanagh, P., and Leech, D. (2009). Evaluation of performance and stability of biocatalytic redox films constructed with different oxygenases and osmium-based redox polymers. *Bioelectrochemistry 76*, 162-168.
[119] Yan, Y.-M., Yehezkeli, O., and Willner, I. (2007). Integrated, electrically contacted NAD(P)$^+$-dependent-carbon nanotube electrodes for biosensors and biofuel cell applications. *Chem. Eur. 13*, 10168-10175.
[120] Tanne, C., Göbel, G., and Lisdat, F. (2010). Development of a (PQQ)-GDH-anode based on MWCNT-modified gold and its application in a glucose/O_2-biofuel cell. *Biosens. Bioelectron. 26*, 530-535.
[121] Hussein, L., Urban, G., and Krüger, M. (2011). Fabrication and characterization of buckylpaper-based nanostructured electrodes as a novel material for biofuel cell applications. *Phys. Chem. Chem. Phys. 13*, 5831-5839.
[122] Lim, J., Cirigliano, N., Wang, J., and Dunn, B. (2007). Direct electron transfer in nanostructured sol-gel electrodes containing bilirubin oxidase. *Phys. Chem. Chem. Phys. 9*, 1809-1814.
[123] Zhou, M., Du, Y., Chen, C., Li, B., Wen, D., Dong, S., and Wang, E. (2010). Aptamer-controlled biofuel cells in logic systems and used as self-powered and intelligent logic aptasensors. *J. Am. Chem. Soc. 132*, 2172-2174.
[124] Mano, N., Kim, H.-H., Zhang, Y., and Heller, A. (2002). An oxygen cathode operating in a physiological solution. *J. Am. Chem. Soc. 124*, 6480-6486.
[125] Mano, N., Mao, F., and Heller, A. (2002). A miniature biofuel cell operating in a physiological buffer. *J. Am. Chem. Soc. 124*, 12962-12963.
[126] Mano, N., Kim, H.-H., Heller, A. (2002). On the relationship between the characteristics of bilirubin oxidases and O_2 cathodes based on their "wiring". *J. Phys. Chem. B 106*, 8842-8848.
[127] Mano, N., Mao, F., and Heller, A. (2003). Characterizations of a miniature compartment-less glucose-O_2 biofuel cell and its operation in a living plant. *J. Am. Chem. Soc. 125*, 6588-6594.
[128] Mano, N., Fernandez, J. L., Kim, Y., Shin, W., Bard, A. J., and Heller, A. (2003). Oxygen is electroreduced to water on a "wired" enzyme electrode at a lesser overpotential than on platinum. *J. Am. Chem. Soc. 125*, 15290-15291.
[129] Kang, C., Shin, H., Zhang, Y., and Heller, A. (2004). Deactivation of bilirubin oxidase by a product of the reaction of urate and O_2. *Bioelectrochemistry 65*, 83-88.
[130] Kang, C., Shin, H., and Heller, A. (2006). On the stability of the "wired" bilirubin oxidase oxygen cathode in serum. *Bioelectrochemistry 68*, 22-26.
[131] Heller, A. (2006). Potentially implantable miniature batteries. *Anal. Bioanal. Chem. 385*, 469-473.
[132] Rowinski, P., Kang, C., Shin, H., and Heller, A. (2007). Mechanical and chemical protection of a wired enzyme oxygen cathode by a cubic phase lyotropic liquid crystal. *Anal. Chem. 79*, 1173-1180.
[133] Suraniti, E., Studer, V., Sojic, N., and Mano, N. (2011). Fast and easy enzyme immobilization by photoinitiated polymerization for efficient bioelectrochemical devices. *Anal. Chem. 83*, 2824-2828.

[134] Tominaga, M., Ohtani, M., and Taniguchi, I. (2008). Gold single-crystal electrode surface modified with self-assembled monolayers for electron tunneling with bilirubin oxidase. *Phys. Chem. Chem. Phys. 10*, 6928-6934.

[135] Dronov, R., Kurth, D. K., Möhwald, H., Scheller, F. W., and Lisdat, F. (2008). Communication in a protein stack; electron transfer between cytochrome *c* and bilirubin oxidase within a polyelectrolyte multilayer. *Angew, Chem. Int. Ed. 47*, 3000-3003.

[136] Lisdat, F., Dronov, R., Möhwald, H., Scheller, F. W., and Kurth, D. G. (2009). Self-assembly of electro-active protein architectures on electrodes for the construction of biomimetic signal chains. *Chem. Commun.* 274-283.

[137] Wegerich, F., Turano, P., Allegrozzi, M., Möhwald, H., and Lisdat, F. (2011). Electroactive multilayer assemblies of bilirubin oxidase and human cytochrome C mutants: insight in formation and kinetic behavior. *Langmuir 27*, 4202-4211.

In: Bilirubin: Chemistry, Regulation and Disorder
Editors: J. F. Novotny and F. Sedlacek
ISBN: 978-1-62100-911-5
© 2012 Nova Science Publishers, Inc.

Chapter VII

Brainstem Auditory Impairment in Infants with Hyperbilirubinemia

Ze Dong Jiang[*]
Department of Paediatrics, University of Oxford, John Radcliffe Hospital,
Oxford, United Kingdom

Abstract

The neonatal auditory system is sensitive to high level of serum bilirubin, and can be damaged in neonates who suffer hyperbilirubinemia. Irrespective of etiology, elevation in the levels of unconjugated bilirubin places infants at risk for the developing bilirubin encephalopathy, including brainstem auditory impairment. Early detection of bilirubin neurotoxicity to the brain is crucial for timing treatment to reduce the risk of occurring kernicterus. Examination of functional integrity of the brainstem auditory pathway can provide important information regarding the damage of hyperbilirubinemia to the neonatal auditory brainstem and the neonatal brain in general. The brainstem auditory evoked response has been used as an important tool to study and assess bilirubin neurotoxicity to the brain, specifically the auditory system. A considerable body of research has described changes in the response in infants with hyperbilirubinemia. Most authors found some abnormalities, although others did not. Recent studies further show that neonatal hyperbilirubinemia affects brainstem auditory evoked response. The abnormalities in the response reflect functional impairment of the auditory brainstem, including impaired neural conduction and depressed electrophysiology. The degrees of these abnormalities are related to the severity of neonatal hyperbilirubinemia, though may not precisely indicate the severity.

More recently, a relative new method - the maximum length sequence has been introduced to record and analyze brainstem auditory evoked response to further our understanding of functional integrity of the auditory brainstem. With this technique, recent studies demonstrated that hyperbilirubinemia does affect the functional integrity of

---

[*] Corresponding author: Telephone: ++44 1865 221364; Telefax: ++44 1865 221366. E-mail address: jiangzedong-oxshang@hotmail.com

the auditory brainstem, confirming the adverse effect of hyperbilirubinemia on the auditory brainstem. In addition, these studies revealed that neonatal hyperbilirubinemia not only damages the functional integrity of the more peripheral or caudal regions of the auditory brainstem, but also affects the more central or rostral regions. Compared with conventional response, the maximum length sequence brainstem auditory evoked response is more sensitive to bilirubin neurotoxicity to the more central regions. Therefore, this relatively new technique improves early detection of bilirubin encephalopathy, particularly for the impairment at the more central regions of the brainstem that may not be clearly shown by conventional response. An increase in BAER wave latencies and interpeak intervals and the reduction in wave amplitudes are useful indicators of bilirubin neurotoxicity to the neonatal auditory brainstem.

Introduction

In newborn infants hyperbilirubinemia is a most common condition requiring evaluation and treatment. As a well-known neurotoxin, bilirubin inhibits mitochondrial enzymes, affects DNA synthesis and ion exchange, disturbs neuroexcitatory signals and impairs neural function (Dennery et al., 2001; Hansen et al., 1988a,b; Shapiro, 2003; Shi et al., 2006). The neonatal brain, including the auditory system, is sensitive to high level of serum bilirubin, and can be damaged in hyperbilirubinemia (Watchko, 2006; Wennberg et al., 2006). Neonatal hyperbilirubinemia can be the results of various physiologic or pathologic etiologies. Irrespective of the etiology, elevation in the levels of unconjugated bilirubin places infants at risk for the developing bilirubin encephalopathy, including brainstem auditory impairment, or kernicterus. The brainstem auditory pathway has been found to be sensitive to bilirubin neurotoxicity. Examination of functional integrity of this pathway can provide important information regarding damage of hyperbilirubinemia to the neonatal brain and early detection of the damage, which is crucial for timing treatment to reduce the risk of occurring kernicterus.

As a non-invasive and objective test, the brainstem auditory evoked response (BAER) reflects functional integrity of the brainstem auditory pathway. The response is also known as brainstem auditory evoked potential (BAEPs) or auditory brainstem response (ABR). Previous studies have shown that BAER is highly sensitive to abnormalities of neuronal function such as conduction delay, desynchronization, and loss of cells, which occur in a number of pathologic conditions involving metabolic disorders, demyelination, and brain injury (Chiappa, 1990; Jiang, 2012; Wilkinson and Jiang, 2006). The response is electrical potential(s) that are evoked by acoustic stimulation and recorded from the scalp, either noninvasively using surface electrodes or invasively using needle electrodes. The evoked responses arise from neural generators in the brainstem auditory pathway, including auditory nerve, and brainstem auditory fiber tracts and nuclei. Over the last three decades, BAER testing has been the most reliable, objective method to study both peripheral and central (specifically brainstem) auditory function in infants (Jiang, 2012; Wilkinson and Jiang, 2006). The test has been widely used in both humans and animals to assess functional integrity of the brainstem auditory pathway. The study of BAER forms a common link between information available in animal models and humans.

BAER testing has been widely used as an important tool to study and assess bilirubin neurotoxicity to the brain, specifically the auditory system. A considerable body of previous reports has described changes in BAER in infants with hyperbilirubinemia (Ahlfors and Parker 2008; Ahlfors et al., 2009; Amin et al., 2001; Boo et al., 1994; Funato et al., 1994; Gupta and Mann 1998; Hung 1989, Jiang et al., 2007a, 2009e; Okumura et al., 2009; Smith et al., 2004; Soares et al., 1989; Streletz et al., 1986; Wennberg et al., 2006). Most found some abnormalities, while others did not. More recently, a relative new method - the maximum length sequence (MLS) has been introduced to study functional status of the human auditory brainstem. With this technique, recent studies demonstrate that hyperbilirubinemia does affect the functional integrity of the auditory brainstem, including impaired neural conduction and depressed electrophysiological activity of neurons (Jiang, 2008, 2010, 2012, Wilkinson et al., 2007). The MLS technique has also been shown to improve early detection of bilirubin encephalopathy in newborn infants. The main purpose of this chapter is to review recent findings in BAER, obtained with either conventional averaging techniques or the maximum length sequence technique in infants with hyperbilirubinemia to shed some lights on the influence of bilirubin neurotoxicity on the neonatal brain.

Bilirubin Pathophysiology and Encephalopathy

Bilirubin is formed from hemoglobin, about 75% from hemolysis and the remaining 25% from ineffective erythropoiesis. Unconjugated bilirubin is a breakdown product of the porphyrin ring of red blood cell haemoglobin and is lipid soluble, water insoluble, and neurotoxic. Taken up by liver cells, unconjugated bilirubin is conjugated in the liver by uridine diphosphate–glucuronosyltransferase (UDPGT) to a water-soluble, nontoxic glucuronide, known as conjugated bilirubin. After being excreted in bile, conjugated bilirubin is eliminated in stool. It is also broken down in the gut by bacteria to unconjugated bilirubin, which is then reabsorbed back into the bloodstream, forming the so-called enterohepatic circulation.

Because of its indirect reaction in the diazo assay used to measure bilirubin, unconjugated bilirubin is also called indirect bilirubin. Total serum bilirubin (TSB) is a combination of conjugated and unconjugated bilirubin. In human neonates, TSB is almost completely composed of unconjugated bilirubin and bound to protein in the blood, mainly albumin. Unconjugated bilirubin is a natural antioxidant at low levels, but neurotoxic at high levels. It is nonpolar and insoluble in water, and is bound to serum albumin. Only little unconjugated bilirubin exists in the form of unbound or "free" unconjugated bilirubin (B_f). B_f freely enters brain, interstitial fluid, and cerebrospinal fluid, resulting in neurotoxicity. Normally, albumin-bound bilirubin does not enter the brain. When the blood binding capacity is exceeded, or when other substances, such as sulfonamides, compete for binding sites, unconjugated bilirubin enters brain tissue as unbound bilirubin or B_f. In neonates, particularly those born prematurely, the relative immaturity of UDPGT and an increase in hemoglobin load result in physiologic jaundice.

The brain is unique by having a blood–brain barrier that slows the equilibrium between plasma and brain. Normally, the blood–brain barrier prevents albumin-bound bilirubin from entering into the brain. When the barrier is disrupted, albumin-bound bilirubin moves rapidly into the extracellular space of brain (Levin et al., 1982). This results in diffuse yellow staining but without specific pattern of kernicterus. In contrast, B_f moves easily through the blood–

brain barrier. In pathological condition, if the barrier is disrupted, sufficiently high B_f bilirubin will produce immediate global neurotoxicity (Ives et al., 1989; Levine et al., 1982; Wennberg and Hance 1986; Wennberg et al., 1991). It is the unbound unconjugated bilirubin or B_f and not the albumin-bound bilirubin in the brain leads to neurotoxicity. B_f is found to predict central nervous system dysfunction better than TSB (Amin et al., 2001; Funato et al., 1994). Conjugated bilirubin, though not neurotoxic, also binds to albumin, and competes with unconjugated bilirubin for albumin binding sites. Thus, in infants with high conjugated bilirubin resulting from hepatocellular disease or other causes, the TSB is a better indicator of risk of kernicterus than unconjugated bilirubin.

As a result of B_f entering the brain, bilirubin neurotoxicity occurs, producing selective damage of the central nervous system. Clinical symptoms of classical, chronic bilirubin encephalopathy, are known as kernicterus. The classical sequelae of excessive neonatal hyperbilirubinemia comprise a tetrad (athetoid cerebral palsy, deafness or hearing loss, impairment of upward gaze, and enamel dysplasia of the primary teeth). These correspond to pathologic lesions in the globus pallidus and subthalamic nucleus, auditory, and oculomotor brainstem nuclei. In addition, bilirubin also damages the cerebellum and the hippocampus.

Acute bilirubin toxicity in neonates results in acute bilirubin encephalopathy. Clinically, the encephalopathy features a progression of symptomatology, beginning with lethargy and decreased feeding and then progressing to variable or fluctuating hypotonia and hypertonia, high pitched cry, retrocollis and opisthotonus, impairment of upward gaze (setting sun sign), fever, seizures, and even death. In acute, severe neonatal encephalopathy, seizures may also occur and resolve over time. In acute bilirubin encephalopathy, BAER is often abnormal or even absent (Amin et al., 2001; Funato et al., 1994; Jiang et al., 2007a,2009e).

In chronic bilirubin encephalopathy, the clinical features range from deafness and severe athetoid cerebral palsy, seizures or even death from kernicterus, to mild mental retardation and subtle cognitive disturbances (Shapiro, 2010). Classical kernicterus, the most extreme brain injury due to hyperbilirubinemia, results in extrapyramidal abnormalities, spasticity, gaze abnormalities especially impairment of upward gaze, and enamel hypoplasia of deciduous (baby) teeth, hearing loss or deafness, and auditory neuropathy. However, intellect is usually unaffected.

In addition to typical bilirubin encephalopathy, there is a subtle degree of bilirubin encephalopathy. With subtle bilirubin encephalopathy, brain injury or encephalopathy and cognitive disturbances are less severe, with mild neurologic abnormalities (Johnson and Boggs, 1974; Rubin et al., 1979; Naeye, 1978). The encephalopathy also causes isolated neural hearing loss and auditory neuropathy, which may be discovered years later (Bergman et al., 1985; Rance et al., 1999; Salamy et al., 1989; Simmons and Beauchaine, 2000). It is difficult to determine the extent to which subtle bilirubin encephalopathy contributes to the prevalence of learning disabilities, central auditory processing disorders, and hearing loss. Even moderate hyperbilirubinemia may be associated with a significant increase in minor neurologic dysfunction (Soorani-Lunsing et al., 2001).

There may be subtypes of clinical kernicterus, including auditory predominant and motor predominant kernicterus. The pattern of involvement may relate to various factors, e.g. gestation, severity and duration of hyperbilirubinemia, and associated clinical conditions. Because the auditory system tends to develop earlier than extrapyramidal motor control pathways, early exposure to bilirubin toxicity during development may preferentially affect the auditory system.

Bilirubin Neurotoxicity and the Auditory System in Infants with Hyperbilirubinemia

Like the basal ganglia, the auditory pathways have been recognized as being particularly vulnerable to high level of bilirubin. For example, by using whole-cell voltage-clamp recordings, Shi et al (2006) observed the effect of bilirubin on inhibitory postsynaptic currents (IPSC) in postnatal 13–15-day-old neurons dissociated from lateral superior olive nuclei (LSO) — a brainstem auditory nucleus that is known to be highly vulnerable to bilirubin. In LSO neurons, bilirubin facilitates inhibitory synaptic transmission. Inhibitory synaptic transmission is increased in response to acute bilirubin, which may have implications for neurotoxicity and the impairment of auditory transduction seen in hyperbilirubinemia.

Hyperbilirubinemia is known to be an important risk factor for neonatal auditory impairment (Shapiro, 2003; Joint Committee on Infant Hearing, 2000; Newton, 2001; Kountakis et al., 2002; Wilkinson and Jiang, 2006). In infants who have had hyperbilirubinemia during the neonatal period, auditory function is found to be abnormal by many investigators (Dubilin, 1986; Rhee et al., 1999; Sano et al., 2005; Shapiro et al., 2001). Audiometric examination show that there is a predominantly high-frequency bilateral and symmetric hearing loss with recruitment and abnormal loudness growth functions, and that the hearing loss correlates with pathologic lesions in the cochlear nuclei (Dubilin, 1986). There is decreased binaural fusion, auditory aphasia and imperception, word deafness. Some patients appear to be "deaf". However, in objective tests, their hearing thresholds are normal. It appears that the site of a lesion in auditory impairment caused by hyperbilirubinemia in such patients is retrocochlear, with the cochlea unaffected (Sano et al., 2005). In some others, lesions causing auditory impairment after hyperbilirubinemia may include the organ of Corti, especially at the outer hair cells and the cochlear nerve (Rhee et al., 1999).

Postmortem examination of infants with kernicterus revealed central auditory pathology or lesions (Dublin, 1976; Ahdab-Barmada and Moossy, 1984). These lesions often involve brainstem auditory structures, including the dorsal and ventral cochlear nuclei, superior olivary complex, nuclei of the lateral lemniscus, and inferior colliculi, that are the origins of BAER, but no significant abnormalities are found in the eighth nerve or inner ear structures. There is an association between moderate-to-severe hearing loss and central auditory dysfunction and elevated bilirubin levels in the infant in the absence of kernicterus (Bergman et al., 1985). The central auditory pathology is related to the severity and duration of hyperbilirubinemia (Bergman et al., 1985).

The ototoxicity of hyperbilirubinemia on the neonatal brain has been studied mainly using BAER. In infants with neonatal hyperbilirubinemia, the BAER typically shows an increase in interpeak intervals and a decrease in wave amplitudes (Chisin et al., 1979; Jiang et al., 2007a,2009e; Kaga et al., 1979; Kotagal et al., 1981; Lenhardt et al., 1984; Nakamura et al., 1985; Nwaesei et al., 1984; Perlman et al., 1983). In contrast to the much more common sensorineural hearing loss, the damage of hyperbilirubinemia to the auditory system is mainly at the level of the auditory nerve and brainstem. This auditory dysfunction may lead to auditory neuropathy, or auditory dys-synchrony. Functionally, the neuropathy is defined as absent or abnormal BAER with evidence of normal inner ear function, assessed by either otoacoustic emissions or cochlear microphonic responses. More than half of the patients with auditory neuropathy are associated with significant hyperbilirubinemia during the neonatal period. In children and adults who have chronic kernicteric bilirubin encephalopathy, BAER was found to be abnormal, suggesting functional abnormality in the auditory nerve and the auditory brainstem. This was evident even in patients with normal cochlear microphonic recording (Chisin et al., 1979; Kaga et al., 1979). In any case, BAER can be affected by hyperbilirubinemia, reflecting functional impairment of the auditory pathway up to the brainstem level.

Impaired Brainstem Auditory Conduction in Infants with Hyperbilirubinemia

In infants with hyperbilirubinemia, the majority of investigators reported that the typical BAER abnormality is an increase in wave latencies and I-V interpeak interval (Ahlfors and Parker 2008; Ahlfors et al., 2009; Amin et al., 2001; Boo et al., 1994; Funato et al., 1994; Gupta and Mann 1998; Hung 1989, Jiang et al., 2007a; Okumura et al., 2009; Smith et al., 2004; Wennberg et al., 2006). Some investigators proposed that BAER abnormality in infants with hyperbilirubinemia can be a marker of bilirubin ototoxicity or neurotoxicity. On the other hand, a few others found no significant BAER abnormalities in infants with hyperbilirubinemia. For example, Soares et al (1989) reported that in their 72 preterm and term infants with hyperbilirubinemia, waves I, III, and V in BAER were always present, and that BAER thresholds were normal in all subjects. Only eight infants manifested abnormal wave latencies. There was no significant difference in I-V interpeak interval between infants with a higher level of bilirubin and those with a lower level of bilirubin. Streletz et al (1986) found an abnormal increase in BAER wave latencies in infants with hyperbilirubinemia, but no abnormality in the I-V interval.

We examined BAER in term neonates who had hyperbilirubinemia due to various perinatal conditions and required phototherapy or exchange transfusion during the first 2 weeks of life (Jiang et al., 2007a). BAER threshold was significantly higher than in control subjects. The latencies of BAER waves I, III, and V were significantly increased, compared with those in age-matched normal control subjects. The I-V interval in neonates with hyperbilirubinemia was also significantly increased.

Of our 90 term neonates with hyperbilirubinemia, 10% had an elevated BAER threshold (greater than 20 dB normal hearing level). A significant or abnormal increase in wave I latency (greater than 2.5 standard deviations of the mean value in normal control subjects) was seen in 9% of the neonates with hyperbilirubinemia. Some neonates had both an elevation in BAER threshold and an increase in wave I latency. These conditions led to a total of 16% of our neonates exhibiting BAER abnormalities (elevation in BAER threshold or an increase in wave I latency) that suggest peripheral auditory impairment. Twenty percent of the 90 neonates with hyperbilirubinemia had a significant increase in wave V latency, with about half associated with an elevation of BAER threshold (Jiang et al., 2007a). This abnormality appears to be somewhat similar to the prevalence (22%) of auditory impairment previously reported by other investigators (Boo et al., 1994).

We found that the increase in BAER latency in the neonates with hyperbilirubinemia was more significant for the later waves than for the earlier waves, suggesting a possible abnormality in central auditory function (Jiang et al., 2007a). This was further confirmed by the finding of a significant increase in the I-V interval in these neonates. An abnormal increase in I-V interval was seen in 18% of the neonates. Apparently, in addition to peripheral auditory impairment, there is also central auditory impairment after neonatal hyperbilirubinemia, which accounts for about 18% of neonates with hyperbilirubinemia. Some of these neonates had more than one BAER abnormality. In total, 28% of the neonates had BAER abnormalities, suggesting auditory impairment peripherally and/or centrally.

Recovery of BAER Abnormalities after Treatment

The abnormalities in BAER in infants with hyperbilirubinemia usually promptly return to normal following exchange transfusions (Hung, 1989; Nwaesei et al., 1984; Perlman et al., 1983; Wennberg et al., 1982; Yilmaz et al., 2001). Hung studied BAER in 75 jaundiced patients. Retrospective ABR study in most of the 10 known kernicteric patients (group I) showed elevation of the hearing threshold, increase in wave I latency and prolongation of I-V interpeak interval. Six jaundiced infants (group II) with ABR testing before and after blood exchange transfusion showed decrease in wave latencies and increase in wave amplitude after blood exchange transfusion. There were additional 20 infants with blood exchange transfusion (group III) and 39 with phototherapy (group IV) who received BAER testing after the therapeutic procedures. Prospective follow-up in groups II, III and IV showed normalization of the BAER in all except one patient in the following months. These findings demonstrate the nature of bilirubin neurotoxicity and BAER testing is sensitive in reflecting the effect of hyperbilirubinemia, and provides a valuable guide for the early recognition and close follow-up of bilirubin neurotoxicity.

In our study, some of the infants with hyperbilirubinemia were tested repeatedly with BAER several days and weeks after the initial BAER recording. The results indicated that after phototherapy or exchange transfusion, the BAER abnormalities in neonates with hyperbilirubinemia recovered quickly, suggesting that the auditory impairment is largely transient. This is similar to findings by other investigators (Wong et al., 2006). Infants or children who had hyperbilirubinemia were often associated with adverse neurodevelopmental outcomes after prompt treatment (Newman et al., 2006). Obviously, the toxic effect of hyperbilirubinemia on auditory system could be transient, provided that prompt treatment is initiated. Similar findings are also seen in visual evoked potentials. For instance, Chen and Wong (2006) reported that after prompt treatment the abnormalities in visual evoked potentials quickly recovered, suggesting that the toxic effect of hyperbilirubinemia to the visual system is transient.

Infants or children who have severe hyperbilirubinemia may not be associated with adverse neurodevelopmental outcomes after the prompt treatment with phototherapy or exchange transfusion (Funato et al., 1996). A few authors reported prognostic value of BAER testing in infants with neonatal hyperbilirubinemia, (Hung, 1989; Yilmaz et al., 2001). Yilmaz et al (2001) examined the prognostic value of BAER the neurologic outcome in 22 infants with neonatal indirect hyperbilirubinemia. At 12 months of age, 2 of these patients showed BAER abnormalities that are consistent with auditory neuropathy, but had no neurologic finding except a lack of speech. Two other patients had neurologic sequelae, one showing severe dyskinetic cerebral palsy, the other mild hypotonia and motor retardation, but their BAER was normal. The authors concluded that BAER might not provide reliable information for the neurologic prognosis. Neurologic disturbances resulting from bilirubin neurotoxicity can be seen in patients with a normal BAER, but patients with an abnormal BAER may not have any neurologic dysfunction apart from speech retardation. Further studied are warranted for the prognostic value of BAER in infants with hyperbilirubinemia.

Suppressed Brainstem Auditory Electrophysiology in Infants with Hyperbilirubinemia

In the BAER, in addition to wave latency and interpeak interval, the amplitudes or magnitude of BAER wave components may also be affected by bilirubin neurotoxicity. The amplitudes reflect electrophysiology of brainstem auditory neurons following acoustic stimulation. Earlier animal experiments carried out by Hansen and colleagues revealed that bilirubin decreased phosphorylation of synapsin I, a synaptic vesicle-associated neuronal phosphoprotein, in intact synaptosomes from rat cerebral cortex (Hansen et al., 1988a). In rat hippocampal slices, bilirubin also reduced the amplitude of extracellularly recorded synaptic field potentials (Hansen et al., 1988b). In their further experiments, the authors noticed that in bilirubin-infused piglets there was a significant reduction in the amplitudes of BAER waves II-V, while there were no changes in wave latencies (Hansen et al., 1992). By administering biliverdin, the immediate precursor of bilirubin, in 15- to 17-day-old Gunn rat pups, Rice and Shapiro (2006) created an improved model of bilirubin-induced neurological dysfunction. They found that a single biliverdin injection produced a significant decrease in the amplitude of BAER wave III.

In human hyperbilirubinemia, the typical abnormality is an increase in BAER wave latencies and interpeak intervals, particularly wave V latency and I–V interval—the major BAER variables that mainly reflect brainstem auditory function. In addition, some investigators found that the amplitudes of BAER wave components were reduced in neonate with hyperbilirubinemia (Funato et al., 1994,1996; Jiang et al., 2009e), although others did not. Amplitude reduction also occurs in other modalities of evoked potentials in neonates with hyperbilirubinemia. For example, Chen et al (1995) reported a reduction in the amplitudes of visual evoked potentials during the first week of life in neonates with hyperbilirubinemia.

The author and his colleagues found that all BAER wave amplitudes in neonates with hyperbilirubinemia tended to be smaller than those in the controls (Jiang et al., 2009e). The later waves, i.e. those with longer latency, were reduced more than the earlier waves, i.e. those with shorter latency. The amplitude of wave I was reduced slightly and did not differ significantly from that in the controls. The amplitudes of waves III and V were both reduced significantly. The V/I and V/III amplitude ratios in the neonates with hyperbilirubinemia were decreased significantly, compared with the controls. These findings are generally in agreement with those found by Funato, et al (1994,1996), and suggest that the neonatal auditory system is damage by hyperbilirubinemia, reflecting an acute toxic effect of hyperbilirubinemia on the auditory system.

In brain evoked potentials, the recorded wave amplitude is known to be proportional to the number of synchronously active elements contributing (Chiappa, 1990, Jiang and Wilkinson, 2012; Jiang et al., 2008). A reduction in the recorded amplitude is the result of a decreased average contribution from a given population of neurons. The reduction in BAER wave amplitude in our neonates with hyperbilirubinemia reflects neuronal damage in the auditory brainstem. It appears that bilirubin toxicity reduces the synchrony with which the auditory neurons are activated and the average membrane potential of the auditory neurons (so the driving potential for current flow is reduced). As a result, neuronal function in the auditory brainstem is suppressed in hyperbilirubinemia.

The amplitude reduction in BAER wave components was more significant for the later waves than for the earlier waves (Jiang et al., 2009e). Both the V/I and V/III amplitude ratios were also decreased. These findings indicate that there is an acute toxic effect of hyperbilirubinemia on the auditory system, resulting in auditory, but mainly central auditory, impairment. This is in agreement with our conclusion in the study of BAER wave latencies and interpeak intervals (Jiang et al., 2007a). These findings in infants with hyperbilirubinemia are essentially comparable with those reported in animal models (Hansen et al., 1988a,b; Shapiro, 2006). It appears that the reduction in wave III and V amplitudes and the decrease in the V/I and V/III amplitude ratio are useful indicators of bilirubin toxicity in infants with hyperbilirubinemia. The degrees of these abnormalities are associated with the severity of neonatal hyperbilirubinemia.

Recovery of BAER Amplitude Abnormalities after Treatment

In our neonates who had hyperbilirubinemia, several days and/or weeks after the first recording of BAER, reported here, and treatment with phototherapy and/or exchange transfusion, some were re-tested with BAER for any recovery in the reduced amplitudes of BAER waves (Jiang et al., 2009e). The preliminary data showed that the reduced amplitudes tended to increase and returned to normal or near normal. Apparently, following prompt treatment the impaired auditory function due to the ototoxicity of hyperbilirubinemia recovers quickly. Therefore, the bilirubin toxicity to the amplitudes of BAER waves is largely transient, which is in agreement with our findings in BAER wave latencies and intervals (Jiang et al., 2007a).

BAER Abnormalities and the Level of TSB

Bilirubin toxicity to the neonatal brains, particularly the auditory system, is often related to the level of bilirubin. In neonates with hyperbilirubinemia, the BAER abnormalities tended to occur more frequently in neonates with a higher level of TSB than in those with a lower level of TSB. The author and his colleagues found that the differences in these BAER variables between neonates with hyperbilirubinemia and age-matched normal neonates tended to be greater with the increase in the level of TSB (Jiang et al., 2007a). The latencies of waves I, III, and V and the I-V interval generally increased with the increase in TSB level. The latencies of waves I, III, and V, and the I-V interval were all slightly longer in neonates with TSB at 16-20 mg/dL than in those with TSB at 11-15 mg/dL. Similarly, all BAER wave latencies and the I-V interval in neonates with TSB greater than 20 mg/dL tended to be longer than in those with TSB at 16-20 mg/dL. The wave latencies and I-V interval in neonates with

TSB greater than 20 mg/dL were all longer than in those with TSB at 11-15 mg/dL. These results indicate that the acute toxic effect of hyperbilirubinemia on brainstem auditory function is more significant at a higher level of TSB than at a lower level. However, only some of these BAER variables showed statistical significance between different groups. Correlation analysis revealed that all latencies of waves I, III, and V were correlated significantly with the level of TSB. The I-V interval was also correlated with the level of TSB.

Similar to the above findings in BAER wave latencies and interpeak intervals, the abnormalities in BAER wave amplitudes tended to be more significant at higher levels of TSB than at lower levels. The author and colleagues found that in the neonates with TSB 11–15 mg/dL, the amplitudes of waves I and III was slightly smaller than those in age-matched normal controls, which did not reach statistical significance (Jiang et al., 2009e). Wave V amplitude and the V/I and V/III amplitude ratios were all significantly smaller than those in the controls. In the neonates with TSB 16–20 mg/dL, all the amplitudes of waves I, III and V were significantly smaller than those in the controls. Both the V/I and V/III amplitude ratios were significantly smaller than in the controls. In the neonates with TSB >20 mg/dL, the amplitudes of waves III and V were significantly smaller than those in the controls. The V/I and V/III amplitude ratios were both significantly smaller than in the controls. Distribution of data points of the amplitudes of BAER waves I, III and V at different levels of TSB in individual neonates with hyperbilirubinemia demonstrated a trend that the amplitudes of BAER waves were lower at a higher level of TSB than at a lower level (Jiang et al., 2009e). The differences in BAER amplitude variables between the neonates with hyperbilirubinemia and the normal neonates tended to be greater with the increase in the level of TSB. Thus, acute bilirubin ototoxicity is more significant at higher levels of TSB than at lower levels. However, there was only a weak correlation between these BAER abnormalities and the level of TSB.

These findings indicate that although the acute ototoxic effect of hyperbilirubinemia is generally more significant at a higher level of TSB than at a lower level, auditory impairment does not increase linearly with the increase in the level of TSB. In jaundiced Gunn rat, Rice and Shapiro (2006) found that injection of biliverdin resulted in a decreased amplitude of BAER wave III, which is closely related to the level of bilirubin. There was a significant correlation between total bilirubin and wave III amplitude. In our human neonates with hyperbilirubinemia, the abnormalities in BAER were more significant at higher levels of TSB than at lower levels, but the magnitude of the abnormalities were not always closely related to the level of TSB. It appears that the severity of auditory brainstem impairment is not closely related to the severity of hyperbilirubinemia. Why the results in our human neonates with hyperbilirubinemia are different from those found in animal models with experimental hyperbilirubinemia reported by other investigators?

In animal models, experiments can be designed with and carried out under optimal conditions to achieve optimal results. In clinical human studies, however, there are many practical issues and limitations that make it difficult to conduct experiments under optimal conditions. Consequently, it is often difficult to achieve optimal results. For instance, there are many perinatal risk factors for neurological and auditory impairment that may occur in some of those with hyperbilirubinemia (Boo et al., 1994; Gupta and Mann 1998). The perinatal conditions associated with hyperbilirubinemia were not unified in individual neonates, i.e. varying with individual infants, although none of our neonates had any major perinatal risk factors for neurological and auditory impairment except hyperbilirubinemia. It cannot be completely excluded that less severe perinatal conditions happened in some of the neonates with hyperbilirubinemia. For example, a small number of them may have mild or moderate degree of acidosis, hypoxia, or sepsis, which may facilitate bilirubin toxicity (Gupta and Mann 1998), whereas others did not. There may be also individual variation in the sensitivity to bilirubin neurotoxicity. Therefore, unlike in animal experiments, the increase in BAER wave latencies and interpeak intervals and the reduction in BAER wave amplitudes in neonates with hyperbilirubinemia are not closely related to the level of TSB. That is, the severity of brainstem auditory impairment in neonates with hyperbilirubinemia is not closely related with the severity of hyperbilirubinemia.

These BAER findings in neonates with hyperbilirubinemia demonstrate that an increase in BAER wave latencies and interpeak intervals is a useful indicator of bilirubin toxicity. The same is true of the reduction in wave III and V amplitudes and the decrease in the V/I and V/III amplitude ratio. The degrees of these abnormalities reflect, to a certain degree, the severity of neonatal hyperbilirubinemia, though may not precisely indicate the severity.

Improved Detection of Bilirubin Neurotoxicity with Maximum Length Sequence BAER

Previous studies of functional integrity of the auditory brainstem in infants with hyperbilirubinemia have used conventional BAER, i.e. the BAER that is recorded and processed using conventional averaging techniques. More recently, the MLS has been introduced to studies of auditory evoked responses, particularly BAER (Jiang, 2012; Jirsa, 2001; Lasky, 1997; Lasky et al., 1998; Lina-Granada et al., 1994; Picton et al., 1992; Wilkinson and Jiang, 2006). The author found that this relatively new technique can enhance the detectability of BAER for auditory dysfunction, brain damage, and neurological impairment in some neonatal problems, typically perinatal hypoxia or hypoxia-ischaemia and bronchopulmonary dysplasia (Jiang, 2008,2012; Jiang and Wilkinson, 2012; Jiang et al., 2000,2003,2005,2007b,2008,2009a-c,2010; Wilkinson and Jiang, 2006; Wilkinson et al., 2007). In particular, MLS BAER can detect some early or subtle degrees of neural abnormalities that cannot be shown by conventional BAER (Jiang, 2012; Jiang et al., 2007b).

In order to further our understanding of the effect of hyperbilirubinemia on the neonatal brain, specifically the auditory brainstem, the author and colleagues carried out a detailed study of MLS BAER in infants who had hyperbilirubinemia to examine functional integrity of the auditory brainstem. We found that the increase in wave V latency and I-V interval in MLS BAER was more significantly than that in conventional BAER, particularly at very high repetition rates 455 and 910/sec. Further analysis of the two sub-components of I-V interval revealed that in conventional BAER the earlier component I-III interval in infants with hyperbilirubinemia increased while the later components III-V interval did not show any abnormality. It appears that the increase in I-V interval in infants with hyperbilirubinemia is caused by the increase in I-III interval. I-III and III-V intervals reflect functional status of the more peripheral or caudal and the more central or rostral regions of the auditory brainstem, respectively (Jiang, 2010; Jiang et al., 2009d,2010). These conventional BAER findings suggest that neonatal hyperbilirubinemia affects the more peripheral regions of the auditory brainstem but does not have an appreciable effect on the more central regions of the auditory brainstem.

In MLS BAER, although I-III interval in infants with hyperbilirubinemia increased moderately at 91 and 227/s clicks, but the increase became more significant at higher rates 455 and 910/sec.

This is similar in the case of wave V latency and I-V interval, and suggests that the higher rates of clicks in MLS BAER make the abnormality in I-III interval more prominent than at lower rates and than in conventional BAER. There was a major increase, or abnormality, in the III-V interval in MLS BAER, which is in contrast to a normal III-V interval in conventional BAER. Obviously, MLS BAER can reveal some central auditory impairment due to bilirubin neurotoxicity that is not clearly shown by conventional BAER.

The results of MLS BAER indicate that, in addition to affecting functional status of the more peripheral regions of the auditory brainstem, neonatal hyperbilirubinemia also affects the more central regions. Compared with conventional BAER, MLS BAER is particularly more sensitive to bilirubin neurotoxicity to the more central regions. The MLS BAER elicited with very high rates enhances the detection of sub-optimal auditory function or sub-clinical auditory abnormalities.

The above findings of group data comparisons were clearly demonstrated in the distribution of data points in individual infants (Figures 1-3). In conventional BAER, the data points of wave V latency and I-V interval in individual infants with hyperbilirubinemia were distributed slightly higher than those in the normal controls (Figure 1A, Figure 2A). In MLS BAER, the data points of the latency and interval were distributed much higher than in the controls. About half of the data point fell in the normal range and the other half above the rage, which were true for both 455 and 910/s (Figure 1B and C, and Figure 2B and C).

In conventional BAER, the data points of III-V interval in infants with hyperbilirubinemia intermingled with those in the controls, and almost all fell in the normal range (Figure 3A). In MLS BAER, however, the data points in infants with hyperbilirubinemia were distributed much higher than the controls. At both 455 and 910/s, about 60% of the data point fell in the normal ranges (Figure 3B and C).

The distribution of data points for I-III interval showed similar, but less marked, difference between conventional and MLS BAER. Similar findings in the differences between MLS BAER and conventional BAER and between higher and lower rates of clicks were also seen in the amplitudes of BAER wave components.

The neonates with hyperbilirubinemia also demonstrated a significant reduction in the amplitudes of MLS BAER waves III and particularly V at all click rates 91-910/s. The reduction tended to be more significant at higher than lower rates. Wave I amplitude was reduced at 910/s. V/I amplitude ratio was decreased at all click rates. Therefore, the amplitudes of MLS BAER, particularly later, waves were all reduced. The reduction tended to be more significant than in conventional BAER. These abnormalities in MLS BAER wave amplitudes further indicate that brainstem auditory electrophysiology is suppressed in neonates with hyperbilirubinemia.

All MLS BAER wave latencies and interpeak intervals in the neonates with hyperbilirubinemia were generally increased with the increase in the level of TSB. These MLS BAER variables were correlated significantly with the level of TSB at some or most click rates between 91 and 910/sec.

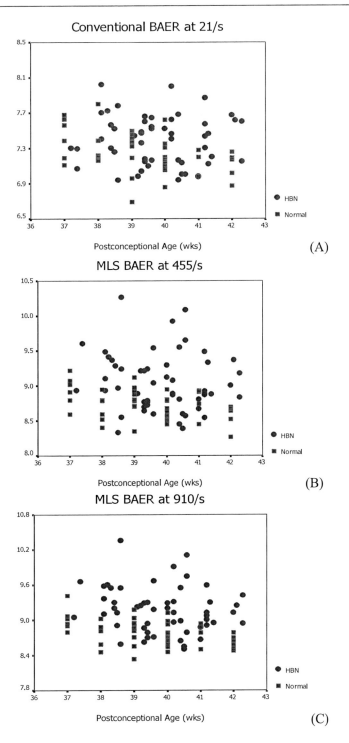

Figure 1. Distribution of data points of wave V latency in individual neonates with HBN and normal controls in conventional BAER at 21/s clicks (A) and MLS BAER at 455/s (B) and 910/s (C) (some data points are overlapped due to very similar or same values). In conventional BAER the data points in HBN are distributed only slightly higher than those of normal controls (A), with most falling in the normal range. In contrast, in MLS BAER the data points in HBN are distributed much higher than those in normal controls, with about half above the normal ranges (B and C).

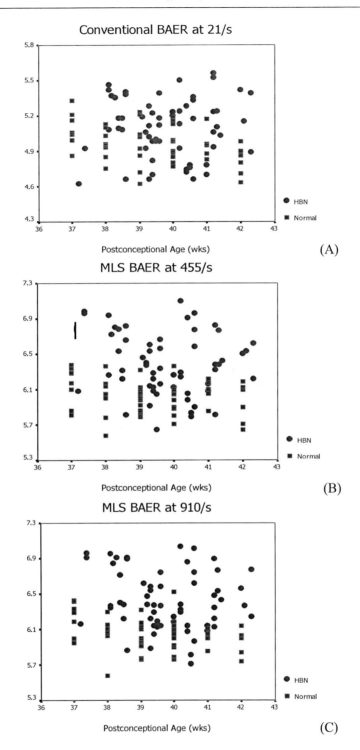

Figure 2. Distribution of data points of I-V interval in individual neonates with HBN and normal controls in conventional BAER at 21/s clicks (A) and MLS BAER at 455/s (B) and 910/s (C) (some data points are overlapped due to very similar or same values). In conventional BAER the data points in HBN are distributed only slightly higher than those of normal controls (A), with most falling in the normal range. In contrast, in MLS BAER the data points in HBN are distributed much higher than those in normal controls, with about half above the normal range (B and C).

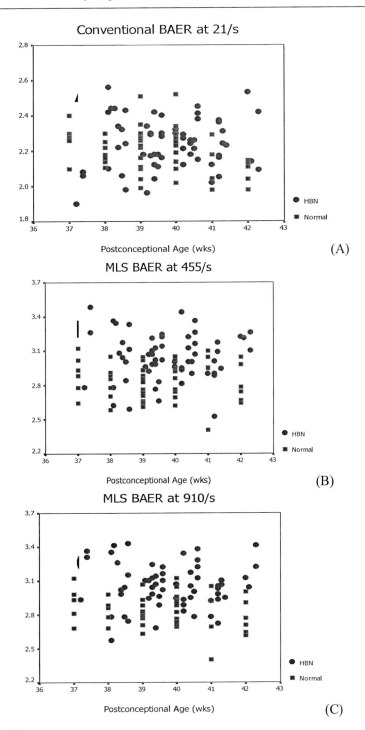

Figure 3. Distribution of data points of III-V interval in individual neonates with HBN and normal controls in conventional BAER at 21/s clicks (A) and MLS BAER at 455/s (B) and 910/s (C). In conventional BAER, the data points of the III-V interval in the neonates with HBN intermingled with those in the controls. Almost all data points fell in the normal range (A). In MLS BAER, however, the data points of the latency in neonates with HBN were distributed obviously higher than the controls. At both 455 and 910/s, about 40% of the data point in the neonates with HBN fell in the normal ranges, with the remaining intermingling with the data points in the controls (B and C).

The amplitudes of MLS BAER wave components tended to be reduced with the increase in the level of TSB. All wave amplitudes were correlated significantly with TSB level at some or most click rates, though not all rates. These findings indicate that the severity of auditory brainstem impairment in neonatal hyperbilirubinemia is related to, though not completely in parallel with, the severity of hyperbilirubinemia.

Several days and/or weeks after the first recording of MLS BAER and treatment with phototherapy and/or exchange transfusion, the MLS BAER abnormalities returned to normal or near normal in most infants. These findings are generally comparable with those reported by other investigators in their follow-up studies in conventional BAER, and in visual evoked potentials. Therefore, bilirubin toxicity to the amplitudes of evoked potentials is largely transient. Following prompt treatment, the impaired auditory function due to bilirubin neurotoxicity recovers quickly. In other words, the toxic effect of hyperbilirubinemia on the brainstem and auditory system is largely transient, provided that prompt treatment is initiated.

Conclusion

The immature brain and auditory system is sensitive to bilirubin neurotoxicity and can be damage in hyperbilirubinemia. Early detection of the damage is crucial for timing treatment to reduce the risk of occurring kernicterus and auditory impairment. The BAER has been used as an important tool to study and assess bilirubin neurotoxicity to the immature auditory system, specifically the auditory brainstem, and monitors the changes in functional integrity of the auditory brainstem in infants with hyperbilirubinemia. Most investigators found abnormalities in BAER, although a few others did not. Several days and/or weeks after the occurrence of hyperbilirubinemia and treatment with phototherapy and/or exchange transfusion, BAER abnormalities return to normal or near normal in most infants. This indicates that bilirubin neurotoxicity to the functional integrity of the auditory brainstem is largely transient, and, with prompt treatment, the impaired auditory function due to bilirubin neurotoxicity recovers quickly. Early BAER testing also has some prognostic value, although further studies are warranted.

In MLS BAER, infants with hyperbilirubinemia show significant abnormalities, particularly at very high rates of clicks, although there were only moderate abnormalities in conventional BAER. These major MLS BAER abnormalities indicate major impairment in functional integrity of the auditory brainstem in neonates with hyperbilirubinemia. MLS BAER made central auditory abnormalities due to bilirubin neurotoxicity, which were not clearly shown in conventional BAER, prominent and detectable. Furthermore, the stimulus rate-dependent changes in MLS BAER variables provided information regarding impaired efficacy of central synaptic transmission in hyperbilirubinemia. Therefore, MLS BAER enhances the detection of brainstem auditory impairment due to bilirubin neurotoxicity, particularly for the impairment at the more central regions of the brainstem that may not be shown by conventional BAER, improving early detection of bilirubin encephalopathy in neonates with hyperbilirubinemia.

The abnormalities in conventional BAER and, in particular, MLS BAER were generally more significant with the increase in the level of TSB. Most MLS BAER variables were correlated significantly with the level of TSB at some or most click rates. Apparently, the severity of auditory brainstem impairment in neonatal hyperbilirubinemia is related to, though not completely in parallel with, the severity of hyperbilirubinemia. These BAER and MLS BAER findings in neonates with hyperbilirubinemia demonstrate that an increase in BAER wave latencies and interpeak intervals and the reduction in wave amplitudes are useful indicators of bilirubin neurotoxicity to the neonatal auditory brainstem. The degrees of these abnormalities reflect, to a certain degree, the severity of neonatal hyperbilirubinemia, though may not precisely indicate the severity.

References

Ahdab-Barmada M, Moossy J. The neuropathology of kernicterus in the premature neonate: Diagnostic problems. *J Neuropath Exp Neurol* 1984;43:45-56.

Ahlfors CE, Parker AE. Unbound bilirubin concentration is associated with abnormal automated auditory brainstem response for jaundiced newborns. *Pediatrics* 2008;121:976-8.

Ahlfors CE, Amin SB, Parker AE. Unbound bilirubin predicts abnormal automated auditory brainstem response in a diverse newborn population. *J Perinatol* 2009;29:305-9.

Amin SB, Ahlfors C, Orlando MS, Dalzell LE, Merle KS, Guillet R. Bilirubin and serial auditory brainstem responses in premature infants. *Pediatrics* 2001;107:664-70.

Bergman I, Hirsch RP, Fria TJ, Shapiro SM, Holzman I, Painter MJ. Cause of hearing loss in the high-risk premature infant. *J Pediatr* 1985;106:95-101.

Boo NY, Oakes M, Lye MS, Said H. Risk factors associated with hearing loss in term neonates with hyperbilirubinaemia. *J Tropic Pediatr* 1994;40:194-7.

Chen YJ, Kang WM. Effects of bilirubin on visual evoked potentials in term infants. *Eur J Pediatr* 1995;154:662-6.

Chen WX, Wong V. Visual evoked potentials in neonatal hyperbilirubinemia. *J Child Neurol* 2006;21:58-62.

Chiappa KH. Brainstem auditory evoked potentials: Methodology. In: Chiappa KH (ed) Evoked potentials in clinical medicine. Raven Pres, New York, 1990, pp 173-221.

Chisin R, Perlman M, Sohmer H. Cochlear and brain stem responses in hearing loss following neonatal hyperbilirubinemia. *Ann Otol* 1979;88:352-7.

Dennery P A, Seidman DS, Stevenson DK. Neonatal HBN. *N Engl J Med* 2001;344:581-90.

Dublin W. Fundamentals of sensorineural auditory pathology. Springfield, IL: Charles C. Thomas, 1976.

Dublin W. Central auditory pathology. *Otolaryngol Head Neck Surg* 1986;95:363-424.

Funato M, Tamai H, Shimada S, Nakamura H. Vigintiphobia, unbound bilirubin, and auditory brainstem responses. *Pediatrics* 1994;93:50-3.

Funato M, Teraoka S, Tamai H, Shimida S. Follow-up study of auditory brainstem responses in hyperbilirubinemic newborns treated with exchange transfusion. *Acta Paediatr Jpn* 1996;38:17-21.

Gupta AC, Mann SB. Is auditory brainstem response a bilirubin neurotoxicity marker? *Am J Otolaryngol* 1998;19:232-6.

Hansen TWR, Bratlid D, Walaas SI. Bilirubin decreases phosphorylation of synapsin I, a synaptic vesicle-associated neuronal phosphoprotein, in intact synaptosomes from rat cerebral cortex. *Pediatr Res* 1988a;23:219-23

Hansen TWR, Paulsen O, Gjerstad L, Bratlid D. Short-term exposure to bilirubin reduces synaptic activation in rat transverse hippocampal slices. *Pediatr Res* 1988b;23:453–6.

Hansen TWR, Cashore WJ, Oh W. Changes in piglet auditory brainstem response amplitudes without increases in serum or cerebraospinal fluid neuron-specific enolase. *Pediatr Res* 1992;32:524-9.

Hung K L. Auditory brainstem responses in patients with neonatal hyperbilirubinaemia and bilirubin encephalopathy. *Brain Dev* 1989;11;297-301.

Ives NK, Bolas NM, Gardiner RM. The effects of bilirubin on brain energy metabolism during hyperosmolar opening of the blood-brain barrier: an in vivo study using 31P nuclear magnetic resonance spectroscopy. *Pediatr Res* 1989;26:356–61.

Jiang ZD. Brainstem electrophysiological changes after perinatal hypoxia ischemia. In: Hämäläinen E, editor. New Trends in Brain Hypoxia Ischemia Research. New York, USA: Nova Science Publishers; 2008. p. 203-20.

Jiang ZD. Damage of chronic sublethal hypoxia to the immature auditory brainstem. In: Fiedler D, Krause R, editors. Deafness, Hearing Loss, and the Auditory System. New York, USA: Nova Science Publishers; 2010. p. 159-80.

Jiang ZD. Maximum length sequence technique improves detection of brainstem abnormalities in infants. New York: Nova Science Publishers; 2012 (in press).

Jiang ZD, Wilkinson AR. Differential effects of acute severe hypoxia and chronic sublethal hypoxia on the neonatal brainstem. New York: Nova Science Publishers; 2012 (in press).

Jiang ZD, Brosi DM, Shao XM, Wilkinson AR: Maximum length sequence brainstem auditory evoked response in infants after perinatal hypoxia-ischaemia. *Pediatr Res* 2000;48:639-45.

Jiang ZD, Brosi DM, Wang J, Xu X, Chen GQ, Shao XM, Wilkinson AR: Time course of brainstem pathophysiology during first month in term infants after perinatal asphyxia, revealed by MLS BAER latencies and intervals. *Pediatr Res* 2003;54:680-7.

Jiang ZD, Brosi DM, Li ZH, Chen C, Wilkinson AR. Brainstem auditory function at term in preterm infants with and without perinatal complications. *Pediatr Res* 2005;58:1164-9.

Jiang ZD, Liu TT, Chen C, Wilkinson AR. Changes in BAER wave latencies in term neonates with hyperbilirubinaemia. *Pediatr Neurol* 2007a;37:35-41.

Jiang ZD, Xu X, Brosi DM, Shao XM, Wilkinson AR. Sub-optimal function of the auditory brainstem in term neonates with transient low Apgar scores. *Clin Neurophysiol* 2007b;118:1088-96.

Jiang ZD, Brosi DM, Shao XM, Wilkinson AR. Sustained depression of brainstem auditory electrophysiology during the first month in term infants after perinatal asphyxia. *Clin Neurophysiol* 2008;119:1496-505.

Jiang ZD, Brosi DM, Chen C, Wilkinson AR. Brainstem response amplitudes in neonatal chronic lung disease and differences from perinatal asphyxia. *Clin Neurophysiol* 2009a;120:967-73.

Jiang ZD, Brosi DM, Wilkinson AR. Impairment of perinatal hypoxia-ischaemia to the preterm brainstem. *J Neurolog Sci* 2009b;287:172-7.

Jiang ZD, Brosi DM, Wilkinson AR. Depressed brainstem auditory electrophysiology in preterm infants after perinatal hypoxia-ischemia. *J Neurolog Sci* 2009c;281:28-33.

Jiang ZD, Brosi D, Wu YY, Wilkinson AR. Relative maturation of the peripheral and central regions of the auditory brainstem from preterm to term and the influence of preterm birth. *Pediatr Res* 2009d;65:657-62.

Jiang ZD, Brosi DM, Wilkinson AR. Changes in BAER wave amplitudes in relation to total serum bilirubin level in term neonates. *Eur J Pediatr* 2009e;168:1243-50.

Jiang ZD, Brosi DM, Wilkinson AR. Differences in impaired brainstem conduction between neonatal chronic lung disease and perinatal asphyxia. *Clin Neurophysiol* 2010;121:725-33.

Jirsa RE. Maximum length sequences-auditory brainstem responses from children with auditory processing disorders. *J Am Acad Audiol* 2001;12:155-64.

Joint Committee on Infant Hearing - Year 2000 Position Statement: Principles and Guidelines for Early Hearing Detection and Intervention Programs. *Pediatrics* 2000;106:798-817.

Johnson L, Boggs TR. Bilirubin-dependent brain damage: Incidence and indications for treatment. In: Odell GB, Schaffer R, Sionpoulous AP, eds. Phototherapy in the newborn: An overview. Washington: *National Academy of Sciences* 1974:122-49.

Kaga K, Kitazumi E, Kodama K. Auditory brain stem responses of kernicterus infants. *Int J Ped Otorhinolaryngol* 1979;1:255-94.

Kotagal S, Rudd D, Rosenberg C, Horenstein S. Brain-stem auditory evoked potentials in neonatal hyperbilirubinemia. *Neurol* 1981;31:48.

Kountakis SE, Skoulas I, Phillips D, Chang CYJ. Risk factors for hearing loss in neonates: a prospective study. *Am J Otolaryngol* 2002;23:133-7.

Lasky RE. Rate and adaptation effects on the auditory evoked brainstem response in human newborns and adults. *Hear Res* 1997;111:165-76.

Lasky RE, Barry D, Veen V, Maier MM. Nonlinear functional modelling of scalp recorded auditory evoked responses to maximum length sequences. *Hear Res* 1998;120:133-42.

Levine RL, Fredericks WR, Rapoport S. Entry of bilirubin into brain due to opening of the blood-brain barrier. *Pediatrics*.1982;69 :255– 9.

Lenhardt ML, McArtor R, Bryant B. Effects of neonatal hyperbilirubinemia on the brainstem electrical response. *J Pediatr* 1984;104:281-4.

Lina-Granada G, Collet L, and Morgon A. Auditory-evoked brainstem responses elicited by maximum-length sequences in normal and sensorineural ears. *Audiology* 1994;33:218–36.

Naeye RL. Amniotic fluid infections, neonatal hyperbilirubinemia and psychomotor impairment. *Pediatrics* 1978;62:497-503.

Nakamura H, Takada S, Shimabuku R, Matsuo M, Matsuo T, Negishi H. Auditory nerve and brainstem responses in newborn infants with hyperbilirubinemia. *Pediatrics* 1985;75:703-8.

Nwaesei CG, Van Aerde J, Boyden M, Perlman M. Changes in auditory brainstem responses in hyperbilirubinemic infants before and after exchange transfusion. *Pediatrics* 1984;74:800-3.

Newman TB, Liljestrand P, Jeremy RJ, Ferriero DM, Wu YW, Hudes ES, Escobar GJ. Outcomes among newborns with total serum bilirubin levels of 25 mg per deciliter or more. *N Engl J Med* 2006;354:1889-1900.

Newton V. Adverse perinatal conditions and the inner ear. *Semin Neonatol* 2001;6:543-51.

Okumura A, Kidokoro H, Shoji H, Nakazawa T, Mimaki M, Fujii K, Oba H, Shimizu T. Kernicterus in preterm infants. *Pediatrics* 2009;123:e1052-8.

Perlman M, Fainmesser P, Sohmer H, Tamari H, Wax Y, Pevsmer B. Auditory nerve-brainstem evoked responses in hyperbilirubinemic neonates. *Pediatrics* 1983;72:658-64.

Picton TW, Champagne SC, Kellett AJC. Human auditory potentials recorded using maximum length sequences. *Electroencephalogr Clin Neurophysiol* 1992;84:90-100.

Rance G, Beer DE, Cone-Wesson B, Shepherd RK, Dowell RC, King AM, Rickards FW, Clark GM. Clinical findings for a group of infants and young children with auditory neuropathy. *Ear Hear* 1999;20:238-52.

Rhee CK, Park HM, Jang YJ. Audiologic evaluation of neonates with severe hyperbilirubinemia using transiently evoked otoacoustic emissions and auditory brainstem responses. *Laryngoscope* 1999;109:2005-8.

Rice AC, Shapiro SM. Biliverdin-induced brainstem auditory evoked potential abnormalities in the jaundiced Gunn rat. *Brain Res* 2006;1107:215-21.

Rubin RA, Balow B, Fisch RO. Neonatal serum bilirubin levels related to cognitive development at ages 4 through 7 years. *J Pediatr* 1979;94(4):601-4.

Salamy A, Eldredge L, Tooley WH. Neonatal status and hearing loss in high-risk infants. *J Pediatr* 1989;114(5):847-52.

Sano M, Kaga K, Kitazumi E, Kodama K. Sensorineural hearing loss in patients with cerebral palsy after asphyxia and hyperbilirubinemia. *Int J Pediatr Otorhinolaryngol* 2005;69:1211-7.

Shapiro SM, Nakamura H. Bilirubin and the auditory system. *J Perinatol* 2001;21(Suppl. 1):S52-5..

Shapiro SM. Bilirubin toxicity in the developing nervous system. *Pediatr Neurol* 2003;29:410-21.

Shapiro SM. Chronic bilirubin encephalopathy: diagnosis and outcome. *Semin Fetal Neonatal Med* 2010;15:157-63.

Shi HB, Kakazu, Y. Shibata S, Matsumoto N, Nakagawa T, Komune S. Bilirubin potentiates inhibitory synaptic transmission in lateral superior olive neurons of the rat. *Neurosci Res* 2006;55:161-170.

Simmons JL, Beauchaine KL. Auditory neuropathy: Case study with hyperbilirubinemia. *J Am Acad Audiol* 2000;11:337-47.

Smith CM, Barnes GP, Jacobson CA, Oelberg DG. Auditory brainstem response detects early bilirubin neurotoxicity at low indirect bilirubin values. *J Perinatol* 2004;24:730-2.

Soares I, Collet L, Delorme C, Salle B, Morgon A. Are click-evoked BAEPs useful in case of neonate hyperbilirubinaemia? *Int J Pediatr Otorhinolaryngo* 1989;117:231-7.

Soorani-Lunsing I, Woltil HA, Hadders-Algra M. Are moderate degrees of hyperbilirubinemia in healthy term neonates really safe for the brain? *Pediatr Res* 2001;50:701-5.

Streletz LJ, Graziani LJ, Branca PA, Desai HJ, Travis SF, Mikaelian DO. Brainstem auditory evoked potentials in fullterm and preterm newborns with hyperbilirubinemia and hypoxemia. *Neuropediatrics* 1986;17:66-71.

Watchko JF. Neonatal Hyperbilirubinemia - What Are the Risks? *N Engl J Med* 2006;354:1947-9.

Wennberg RP, Ahlfors CE, Bickers R, McMurtry CA, Shetter JL. Abnormal auditory brainstem response in a newborn infant with hyperbilirubinemia: Improvement with exchange transfusion. *J Pediatr* 1982;100:624-6.

Wennberg RP, Hance AJ. Experimental bilirubin encephalopathy: importance of total bilirubin, protein binding, and blood-brain barrier. *Pediatr Res* 1986;20:789–792.

Wennberg RP, Johanssen BB, Folbergrova J, Siesjo BK. Bilirubin-induced changes in brain energy metabolism after osmotic opening of the blood-brain barrier. *Pediatr Res* 1991;30:473–8.

Wennberg RP, Ahlfors CE, Bhutani VK, Johnson LH, Shapiro SM. Toward understanding kernicterus: a challenge to improve the management of jaundiced newborns. *Pediatrics* 2006;117:474-85.

Wilkinson AR, Jiang ZD. Brainstem auditory evoked response in neonatal neurology. *Semin Fet Neonatol Med* 2006;11:444-51.

Wilkinson AR, Brosi DM, Jiang ZD. Functional impairment of the brainstem in infants with bronchopulmonary dysplasia. *Pediatrics* 2007;120:362-71.

Yilmaz Y, Degirmenci S, Akdas F, Kulekci S, Ciprut A, Yuksel S, Yildiz F, Karadeniz L, Say A. Prognostic value of auditory brainstem response for neurologic outcome in patients with neonatal indirect hyperbilirubinemia. *J Child Neurol* 2001;16:772-5.

Wong V, Chen WX, Wong KY. Short- and long-term outcome of severe neonatal nonhemolytic hyperbilirubinemia. *J Child Neurol* 2006;21:309-15.

In: Bilirubin: Chemistry, Regulation and Disorder
Editors: J. F. Novotny and F. Sedlacek
ISBN: 978-1-62100-911-5
© 2012 Nova Science Publishers, Inc.

Chapter VIII

Transcriptional Regulation of Human Bilirubin UDP-Glucuronosyltransferase *UGT1A1* Gene and Implication of Defects in the *UGT1A1* Gene Promoter

Junko Sugatani[*,1,2] *and Masao Miwa*[1]

[1]Department of Pharmaco-Biochemistry and
[2]Global Center of Excellence for Innovation in Human Health Sciences,
School of Pharmaceutical Sciences, University of Shizuoka,
Yada, Surugaku, Shizuoka City, Shizuoka, Japan

Abstract

The human UDP-glucuronosyltransferase (UGT) 1A1 plays a critical role in the detoxification and excretion of endogenous compounds such as bilirubin, many drugs and other xenobiotics by conjugating them with glucuronic acid in the liver. Defective or reduced UGT1A1 activity causes unconjugated hyperbilirubinemia (Gilbert's syndrome and Crigler-Najjar syndrome). Based on the finding that phenobarbital treatment dramatically reduces hyperbilirubinemia in patients with inherently reduced UGT1A1 enzyme activity, we succeeded in determining the phenobarbital response enhancer module in the human UGT1A1 gene and its transcriptional factor.

This chapter describes the transcriptional regulation of human UGT1A1 gene through distal and proximal promoter and nuclear receptors. A 290-bp distal enhancer module at -3499/-3210 of *UGT1A1* gene fully accounts for constitutive androstane receptor (CAR)-, pregnane X receptor (PXR)-, aryl hydrocarbon receptor (AhR)-,

---

[*]Address correspondence to: Junko Sugatani, Department of Pharmaco-Biochemistry, School of Pharmaceutical Sciences, University of Shizuoka, 52-1 Yada, Surugaku, Shizuoka City, Shizuoka 422-8526, Japan. Phone: (81)-54-264-5779; Fax: (81)-54-264-5779; E-mail: sugatani@u-shizuoka-ken.ac.jp

glucocorticoid receptor (GR)-, peroxisome proliferator-activated receptor α (PPARα)- and NF-E2-related factor 2 (Nrf2)-mediated activation of the *UGT1A1* gene.

In addition, hepatocyte nuclear factor 1α (HNF1α) bound to the proximal promoter motif not only enhances the basal reporter activity of *UGT1A1*, but also influences the transcriptional regulation of *UGT1A1* by these nuclear factors.

Moreover, the TA repeat polymorphism and T-3279G mutation of the 290-bp distal enhancer module affects the transcriptional activation of *UGT1A1* by the nuclear receptors. Thus, activation by specific inducers of the *UGT1A1* gene is offered an excellent clinical trial in treating patients with unconjugated hyperbilirubinemia and in preventing side effects of drug treatment such as SN-38-induced toxicity. Therefore, we screened some UGT1A1 inducers in foods and plants and found that dietary flavones were able to induce the UGT1A1 gene expression. Taken together, these results indicate that even in subjects with inherited deficiency of the *UGT1A1* gene, it would be possible to prevent drug side effects by increasing expression of the enzyme. Screening dietary compounds affecting the expression of UGT1A1 leads to protection against drug side effects, achieving more advanced therapeutic effects by smaller dosages of drugs, and reducing the medical cost.

Introduction

Hydrophobic bilirubin is a toxic breakdown product of heme that is detoxified mainly in the liver via its conjugation with uridinediphosphateglucuronic acid [1]. The resulting hydrophilic bilirubin mono- and di-conjugate are then excreted in bile [2]. Drug-metabolizing phase II enzyme UDP-glucuronosyltransferase, UGT1A1, plays a critical role in the detoxification of bilirubin by conjugating it with glucuronic acid for excretion in bile (Figure 1), and conjugates drugs and other xenobiotics.

The UGT1A1 enzyme is encoded by five exons located at the 3' end of the UGT1A locus (Figure 2). Among all of the UGT isoforms identified to date, UGT1A1 is the only relevant bilirubin-glucuronidating enzyme (Figure 2).

The requirements for the glucuronidation of bilirubin for excretion were first elucidated on the basis of the finding that the severe form of unconjugated hyperbilirubinemia in the absence of hemolysis or other liver disease, known as Crigler–Najjar type I syndrome, can be fatal [3]. Defective or reduced UGT1A1 activity causes unconjugated hyperbilirubinemia (Crigler–Najjar syndrome and Gilbert's syndrome) [4] and decreases the glucuronidation of SN-38 (a pharmacologically active metabolite of the anticancer drug,irinotecan), leading to an increased risk forthe development of severe irinotecan-associated toxicity [5]. Bilirubin-glucuronide is secreted across the bile-canalicular membrane of the hepatocyte by ATP-binding cassette subfamily C member 2 (ABCC2, also known as MRP2) (Figure 1). Loss of ABCC2 function causes Dubin-Johnson syndrome, which is a condition characterized by jaundice due to accumulation of the glucuronidated product.

Phenobarbital is used as a therapeutic drug for patients with Crigler–Najjar type II syndrome, because it increases the expression of bilirubin glucuronosyltransferase and markedly reduces the incidence of unconjugated hyperbilirubinemia[6-10]

The 51-bp phenobarbital-responsive enhancer module (PBREM), which is conserved between species, has been identified in phenobarbital-inducible *CYP2B* genes and is regulated by the nuclear receptor known as constitutive active receptor (CAR) in response to phenobarbital induction [11].These observations prompted us to find the enhancer module in the 5'-flanking region of the phenobarbital-inducible human *UGT1A1* gene and identified UGT1A1 phenobarbital-responsive enhancer module [12].

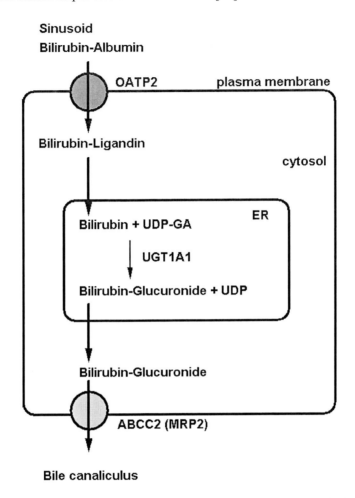

Figure 1. Uptake of bilirubin, conjugation of the bilirubin, and its excretion from the hepatocyte.

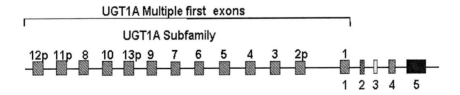

Figure 2. Schematic diagram of the human UGT1A gene structure.

Instead of uniform medical treatment with therapeutic drugs and dosage selected based on disease, tailor-made medical treatment based on the polymorphism of genes encoding drug-metabolizing enzymes is being tried.In the end of 20th century, the Human Genome Project did much work and generally finished determining the human DNA sequence. Along with other recent technological innovations, tailor-made medical treatment has become available for use. Information on defects of the drug-metabolizing enzyme/drug transporter-translating genes is useful for tailor-made medical treatment, together with gene mutations that may be causal factors of diseases. Drugs and xenobiotics are metabolized by phase I and phase II drug-metabolizing enzymes; phase I enzymes introduce a functional groups such as hydroxyl group into endogenous and exogenous substrates, and phase II enzymes conjugate the functional group with glucuronic acid, glutathione, or sulfate to yield a hydrophilic product for its excretion. Some genetic polymorphisms of drug-metabolizing enzymes in human are associated with reduction or enhancement in rates of drug metabolism. For example, CYP2C19 has been found to be a genetic polymorphism characterized by individual differences in S-mephenytoin metabolism [13]. The subjects with the alleles of CYP2C19m1 and CYP2C19m2 are the poor metabolizers (PM) for CYP2C19 substrates. On the other hand, drug-metabolizing capacity in subjects who are ill or taking drugs is considered to be different from those in normal subjects. Moreover, the levels of drug-metabolizing enzymes expressed in our body are also affected by non-geneticfactors including environmental factors. Drug-metabolizing phase I and II enzymes are up-regulated or down-regulated by the mechanism mediated through transcription factors such as nuclear receptors in response to exogenous signals such as drugs, ingested food, environmental pollutants and carcinogens. The nuclear receptors, CAR and pregnane X receptor (PXR), have been proposed as xenobiotic-responsive transcription factors that regulate multiple drug-metabolizing phase I and II enzymes [14]. Thus, we cannot cope with tailor-made medical treatment focused only on the genotypes of translating genes. In this chapter, we described the induction mechanism of human bilirubin UDP-glucuronosyltransferase UGT1A1, the association of the defect in the UGT1A1 gene promoter with hyperbilirubinemia and the enhancement of UGT1A1 expression by food components, and discussed combined effects of drugs and food on drug metabolism.

(1) Identification of the Phenobarbital Response Enhancer Module (gtPBREM) in the Human Bilirubin UDP-Glucuronosyltransferase*UGT1A1* Gene and Its Induction Mechanism

Crigler-Najjar syndrome is an autosomal recessive conditions caused by complete (type I) or incomplete deficiency (type II) of hepatic bilirubin UGT1A1 activity. Phenobarbital has been used for reducing the serum bilirubin load in Crigler-Najjar type II patients [6-10]. Since CAR has been identified as the transcription factor regulating PBREM in response to phenobarbital induction, we wondered whether CAR regulates phenobarbital-inducible transcription of the human *UGT1A1* gene.To confirm whether CAR can regulate the expression of the *UGT1A1* gene in CAR overexpressed HepG2 cells (g2car-3 cells), the cells were pretreated with a repressor of CAR, androstenol. The UGT1A1 mRNA level was decreased by treatment with androstenol, and the phenobarbital-type inducer TCPOBOP increased UGT1A1 mRNA [12].

Since the *UGT1A1* gene appeared to be directly regulated by CAR, we aimed to identify DNA enhancer elements that can be regulated by CAR. To elucidate the location of the CAR-regulated enhancer element in the *UGT1A1* gene, various DNA fragments were generated from an 11-kbp 5' flanking region of the *UGT1A1* gene, and were placed in front of the reporter luciferase gene. These deletion constructs were examined for their enhancer activity in HepG2 cells co-transfected with a human CAR-expression plasmid. Thus, we identified the phenobarbital-responsive enhancer moduleat −3499/-3210 from the transcription start site of UGT1A1, gtPBREM (Figure 3), which is regulated by the nuclear receptor known as CAR in response to phenobarbital treatment [12].

(A)
```
                         SP1
-165 AAACCTAATAAAGCTCCACCTTCTTATCTCTGAAAGTGAACTCCCTG
                                                    HNF1
     CTACCTTTGTGGACTGACAGCTTTTTATAGTCACGTGACACAGTCAAACA
                                  TATA box
     TTAACTTGGTGTATCGATTGGTTTTTGCCATATATATATATATAAGTAGGA
     GAGGGCGAACCTCTGG  -1
```

(B)
```
                           DR4 (CAR, PXR)
-3499 TACACTAGTAAAGGTCACTCAATTCCAAGGGGAAAATGATTAACCAA
      AGAACATTCTAACGTTCATAAAGGGTATTAGGTGTAATGAGGATG TGTTAT
      GRE1 (GR)            gtNR1 (CAR, PXR)
      CTCACCAGAACAAACTTCTGAGTTTATATAACCTCTAGTTACATAACCTGAA
         ARE1 (Nrf2)   ARE2    XRE (AhR)
      ACCCGGACTTGGCGCTTGGTAAGCACGCAATGACCAGTCATAGTAAGCTGGCC
      DR1 (PPARα)  DR3 (CAR, PXR)                   GRE2 (GR)
      AAGGGTAGAGTTCAGTTTGAACAAAGCAATTTGAGAACATCAAAGGAAGTTT
      GGGGAACAGCAAGGGATCCAGAATGGCTAGAGGG  -3210
```

Figure 3. Nucleotide sequence of human UGT1A1 proximal promoter gene (-165/-1) (A) and distal enhancer module (-3499/-3210) (B).

In addition to that by phenobarbital, UGT1A1 induction by rifampicin has been reported in HepG2 cells and primary rat hepatocytes (15). Therefore, we investigated the mechanism by which UGT1A1 induction in response to rifampicin is regulated in HepG2 cells. Using transient transfection studies, we characterized an upstream enhancer element that responds to rifampicin in the UGT1A1 5' flanking region. We identified and characterized a distal enhancer module at −3499/-3210 that directs the transactivation of the *UGT1A1* gene induced by PXR ligand rifampicin and aryl hydrocarbon receptor (AhR) ligand benzo[a]pyrene and chrysin [16]. Strikingly, this element is the same as the phenobarbital-responsive enhancer module gtPBREM.

The 290-bp distal enhancer module is a complex array of transcription-factor-binding sites. To define the roles of each NR motif in the CAR- or PXR-dependent activation of the 290-bp reporter gene, five motifs were singularlymutated. The mutated DNAs were incorporated into luciferase reporter gene plasmids, and subjected to transient transfection assays in HepG2 cells co-transfected with the CAR-expression plasmid or the PXR-expression plasmid. Mutation of gtNR1 resulted in a 100% decrease inboth rifampicin-induced PXR-dependent enhancer activity and CAR-dependent enhancer activity (Figure 4). The gel shift assay also indicated that the mixture of CAR with RXR or PXR with RXR bound to gtNR1 specifically. Thus, gtNR1 wasthe major binding site of the CAR: RXR or PXR:RXRheterodimer [16,17]. On the other hand, single mutations of DR4, PXRE, or DR3 retained about 50 to 80%of the original activity. These results indicate that gtNR1 plays the most significant role in the activation of 290-bp DNA by CAR and PXR. However, DR4, PXRE, and DR3 appeared to be required to confer full enhancer activity to the 290-bp reporter gene. Moreover, the induction of UGT1A1 by dexamethasone has been extensively documented both in vivo in the liver and in vitro in cultured rat hepatocytes [18,19]. The physiological concentration of dexamethasone slightly induced the expression of UGT1A1 mRNA in HepG2 cells, but the rifampicin-inducible expression of UGT1A1 mRNA was synergistically enhanced by dexamethasone. Hence, to investigate whether the induction of UGT1A1 by submicromolar concentrations of dexamethasone is caused directly by a distinct mechanism mediated by the GR, we searched for a functional glucocorticoid-responsive element (GRE) in the 11-kbp 5'-flanking region of the UGT1A1 gene. In HepG2 cells with exogenously expressed CAR and PXR, a submicromolar concentration of dexamethasone slightly but significantly enhanced the reporter activity of the 290-bp distal enhancer module.

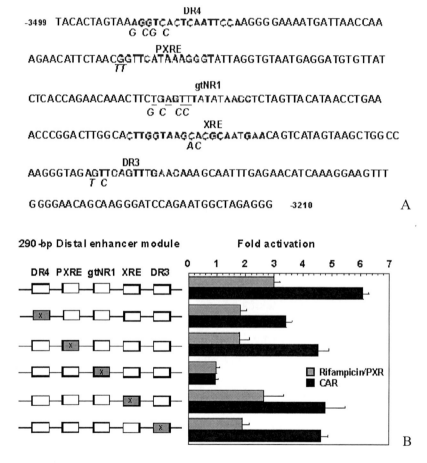

Figure 4. Roles of nuclear receptor/transcription factor binding elements in the transcriptional activation of UGT1A1 290 bp reporter gene by CAR or PXR, assessed by expression of a reporter gene encoding firefly luciferase (17). (A) Mutations of nuclear receptor/transcription factor binding elements, which are underlined, were introduced into the UGT1A1 290-bp reporter gene. Italicized characters show the bases for the mutation. (B) Each mutated DNA (indicated by the closed boxes with X) was co-transfected with expression plasmids for CAR, PXR, or the control vector pCR3 into HepG2 cells with pRL-SV40. The transfected cells were incubated with the vehicle (DMSO) or 5×10^{-6} M rifampicin for 24 h, harvested, and assayed for luciferase activity. Fold activation was calculated by dividing the activity with CAR by that without CAR or with the inducer by that without the inducer. Data are the average of 3 independent experiments ± SD.

These results suggest a possible GRE on the 290-bp distal enhancer module of UGT1A1. Using site-directed mutagenesis and electrophoretic mobility shift assays, we identified the glucocorticoid-response element. The GR-regulated enhancer elements were located at -3404/-3389 and -3251/-3236 within the gtPBREM [20]. In addition, the 165-bp region upstream of the transcriptionstart site involves the HNF1α binding site (Figure 3). We prepared a promoter sequence comprised ofthe distal 290-bp gtPBREM(-3499/-3210) linked to the proximal 165-bp region upstream of the transcriptionstart site [21]. To assess whether HNF1α contributes to the basal transcriptional activity of the distal and proximal promoters of UGT1A1, the HNF1α-binding site in the proximal promoter was mutated [22]. Mutation of the HNF1α-binding site resulted in decreased basal and CAR-mediated transcriptional activity. The basal and CAR-mediated transcriptional activities of the HNF1α-bindingsite-mutated promoter were almost the same as those of gtPBREM. Moreover, rifampicin-, dexamethasone-, and benzo[a]pyrene-stimulated, that is, PXR-, GR-, and AhR-mediated, transcriptional activities of the distal enhancer module linked to the proximal UGT1A1 promoter increased to 2.8-, 1.8-, and 1.4-fold that of gtPBREM, respectively. Mutation of the HNF1α-binding site in the UGT1A1 proximal promoter abolishes the basal promoter function. HNF1α bound to the proximal promoter motif not only enhances the basal reporter activity of UGT1A1 including the distal (-3570/-3180) and proximal (-165/-1) regions, but also influences the transcriptional regulation of UGT1A1 by CAR, PXR, GR, and AhR to markedly enhance reporter activities.

To date, in addition to CAR [12]-, PXR [17]-, AhR [23]- and GR [20]-response elements, NF-E2-related factor 2 (Nrf2) [24]- and peroxisome proliferators-activated receptor α (PPAR α) [25]-response elements have been demonstrated (Figure 3 and Table 1).Surprisingly, all these elements are located within the 290-bp distal enhancer module of human UGT1A1 gene. Nrf2-KeapI-dependent UGT1A1 induction by prooxidants has been considered to represent a key adaptive response to cellular oxidative stress that defends against a variety of environmental insults [24].

(2) Identification of a Defect in the UGT1A1 Gene Promoter and Its Association with Hyperbilirubinemia

Reduction of UGT1A1 activity causes unconjugated hyperbilirubinemia (Crigler-Najjar syndrome and Gilbert's syndrome). Many UGT1A1 mutant alleles responsible for clinical hyperbilirubinemia have been identified [4,26]. Genetic lesions causing an absence of UGT1A1 activity result in Crigler-Najjar syndrome type I, which is characterized by potentially lethal hyperbilirubinemia (serum total bilirubin level, 20 to 50 mg/dl [342 to 855 mmol/l]) in the absence of hemolysis or other liver disease. Mutations causing a severe but incomplete lack of UGT1A1 activity result in Crigler-Najjar syndrome type II, which is characterized by intermediate levels of hyperbilirubinemia (6 to 20 mg/dl [103 to 342 mmol/l]). Gilbert's syndrome is characterized by mild, chronic unconjugated hyperbilirubinemia in the absence of liver disease. An elongated TA repeat [A[TA]$_7$TAA (UGT1A1*28) instead of the more usual A[TA]$_6$TAA (UGT1A1*1)], located 41 nucleotides upstream of the translational start site of the UDP-glucuronosyltransferase gene [27-29], and a missense mutation involving G to A substitution at nucleotide number 211 in exon 1 of

UGT1A1 (G211A; also known as G71R) (UGT1A1*6) [30-32], both result in decreased bilirubin-glucuronidating activity and lead to mildly elevated serum bilirubin levels (27,33,34). Variants with the TA repeats [A(TA)$_8$TAA (UGT1A1*37) and A(TA)$_5$TAA (UGT1A1*36)] are associated with reduced [(TA)$_8$] and increased [(TA)$_5$] transcription (35), but their allele frequencies are markedly lower than that of the TA repeat [A[TA]$_7$TAA (UGT1A1*28) in all races (36). The mild hyperbilirubinemia seen in individuals with Gilbert's syndrome has been reported to be linked to homozygous A[TA]$_7$TAA and G211A mutations (27-39). Their allele frequencies can be stratified by races; A[TA]$_7$TAA (UGT1A1*28) frequencies in White (allele frequency 0.334) and in African/African American (allele frequency 0.404) are higher than that in Asian/Asian American (allele frequency 0.139). On the other hand, G211A allele frequencies in Asian/Asian American (allele frequency 0.13) are higher than those in White (allele frequency 0.005) and African/African American (allele frequency 0.00) [36]. Although G211A is the most common mutation encountered in Japanese patients with Gilbert's syndrome, the majority of these mutations are heterozygous and the transferase activity level (60% of normal) is somewhat too high to explain the development of mild hyperbilirubinemia [34]. Because the hepatic UGT1A1 activity in patients with Gilbert's syndrome is decreased to about 30% of normal [40,41], additional factors have been thought to contribute to the development of this syndrome. We identified a new type of polymorphism in the gtPBREM DR3 region of the *UGT1A1* gene. This consisted of a single nucleotide polymorphism in which the T normally present at nucleotide number −3279 was substituted by G (UGT1A1*60) [42]. The frequency of alleles carrying the T-3279G mutation was significantly higher in the adult hyperbilirubinemic group (0.55) than in the adult control group (0.22; p<0.001) (Table 2)

The T-3279G mutation, which is located in the spacer of the DR3 element of UGT1A1 gtPBREM, slightly but significantly reduces CAR-dependent transcriptional activities (Figure 5). The frequencies of alleles for the A[TA]$_7$TAA and exon 1 mutations were also significantly higher in the hyperbilirubinemic group (0.29 and 0.34, respectively) than in the control group (0.01 and 0.16, respectively). Analysis of the UGT1A1 gene promoter gtPBREM revealed that 31 of the 41 subjects (76%) in the hyperbilirubinemic group had an identical T to G mutation at nucleotide number −3279 in the gtPBREM DR3 region. Fourteen of these 31 patients were homozygous; the other 17 were heterozygous. The 14 heterozygotes for the T-3279G mutation were also heterozygous for the G to A mutation at nucleotide number 211 in exon 1 (G71R). In addition, two of the 31 patients were compound heterozygotes for mutations in the gtPBREM (T-3279G), TATA element (A[TA]$_7$TAA) and exon 1 (G211A).

In the control group, fifteen of the 38 subjects (39%) had the T to G mutation at nucleotide number −3279 in the UGT1A1 gtPBREM (13 heterozygous and 2 homozygous). One of these was a double heterozygote for T-3279G and A[TA]$_7$TAA. Ten of the 38 subjects (30%) had the exon 1 mutation G211A (8 heterozygous and 2 homozygous), of whom four were double heterozygotes for G211A and T-3279G. The remaining 23 subjects had no detectable mutations in the gtPBREM, TATA element or exons 1 through 5. Although Maruo et al. [43] have reported that all patients with the A[TA]$_7$TAA mutation possessed the T-3279G mutation, in Japanese control and hyperbilirubinemic groups we found patients with the T-3279G mutation alone (Table 2), indicating that T-3279G and A[TA]$_7$TAA mutations are not linked.

Table 2. Distribution of genotypes of phenobarbital-responsive enhancer module of UGT1A1 gtPBREM DR3, TATA element and UGT1A1 G211A mutations in control and hyperbilirubinemic groups

Table 3. UGT1A1 mutations and total bilirubin concentrations in the control group [42]

Group	Mutation	Other mutation	n	Plasma bilirubin (mg/dl)
1	None	None	14	0.45 ± 0.11
2	gtPBREM [-3263T→G] Hetero	None	4	0.61 ± 0.04[*]
3	gtPBREM [-3263T→G] Hetero	UGT1A1 codon 71 [G71R Hetero]	3	0.82 ± 0.09[***, #, +]
4	UGT1A1 codon 71 [G71R Hetero or Homo]	None	5 (4 Hetero and 1 Homo)	0.69 + 0.06[***]
5	gtPBREM [-3263T→G] Hetero	A(TA)₇TAA Hetero	1	0.86

Hetero and Homo indicate heterozygous and homozygous mutations.
Bilirubin values are given as means ± SD.
*, ***$p<0.05, 0.001$ versus group 1; #$p<0.05$ versus group 2; +$p<0.05$ versus group 4.

The plasma total bilirubin levels of the subject with G211A were significantly higher than those of the subjects with no mutations (p <0.01) (Table 3). Moreover, levels in the double heterozygotes for T-3279G and G211A were significantly higher than those in the normal subjects (p<0.001), the heterozygotes for T-3279G (p<0.01) and the subjects with G211A (p<0.05). These observations indicate that double heterozygous mutations in the UGT1A1 gene promoter (T-3279G) and exon 1 (G211A) may result in more strongly reduced transferase activity than a single heterozygous mutation.

Sai et al. [44] also have conducted a comprehensive haplotype analysis of *UGT1A1* gene, including the enhancer, the promoter, and all 5 exons and their flanking regions, from 195 Japanese subjects and investigated the association of these haplotypes with area under the concentration – time curve (AUC) ratios (7-ethyl-10-hydroxycamptothecin glucuronide [SN-38G]/irinotecan metabolite 7-ethyl-10-hydroxycamptothecin[SN-38] and pretreatment levels of serum total bilirubin in 85 cancer patients who received irinotecan.

They have demonstrated highly significant associations between the haplotypes of UGT1A1*28 (A[TA]₇TAA) and both a reduced AUC ratio and an increased total bilirubin level and between the haplotypes of UGT1A1*60 (T-3279G) and an increased total bilirubin level, and that the reduction in the AUC was remarkable in combination with *6/*60 or *6/*28. These observations are basically consistent with our results.

As described above, in adult Japanese patients with Gilbert's syndrome, the T-3279G mutation is a risk factor. Neonatal hyperbilirubinemia is frequent and severe in Japanese newborns, but the T-3279G mutation has been reported not to be associated with the neonatal hyperbilirubinemia in Japaneses [45]. The G211A mutation is likely to be a risk factor for developing neonatal hyperbilirubinemia in Asians, but not Caucasians [46].

To assess the influence of the T-3279G and A[TA]₇TAA mutations on CAR-, PXR-, GR-, and AhR-mediated transactivation of the human *UGT1A1* gene, we mutated T to G at nucleotide number -3279 and/or the TATA element to A[TA]₇TAA in the UGT1A1 3570-bp fragment (-3570/-1) (21).The wild-type, T to G at nucleotide number -3279, and/or A[TA]₇TAA UGT1A1 3570-bp promoterwere cotransfected intoHepG2 cells with plasmids of expressing CAR(Figure 5A), PXR(Figure 5B) or no expression plasmid (Figure 5C) and then the cells were cultured with ligands [no ligand (Figure 5A), 5 μM rifampicin or 5 μM rifampicin plus 0.4 μM dexamethasone (Figure 5B), or 1 μM benzo[a]pyrene (Figure 5C)].

The T-3279G mutation reduced the CAR-, PXR- and AhR-dependent transcriptional activities to 87% (p<0.05), 93% and 89% of wild-type, respectively, whereas the A[TA]$_7$TAA mutation reduced these activities to 63%, 67% and 42% of wild-type, respectively (Figure 5).

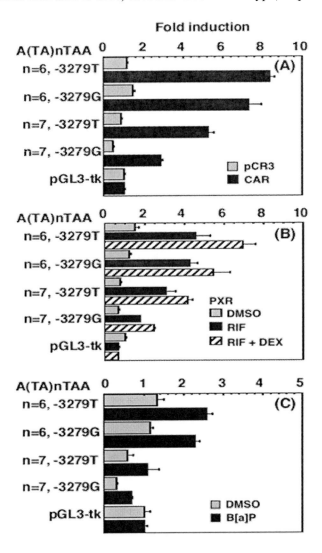

Figure5. Effect of the TA repeat polymorphism and gtPBREM variation on transcriptional activity mediated by CAR (A), PXR (B), or AhR (C), assessed by expression of a reporter gene encoding firefly luciferase (21). Effect of the TA repeat polymorphism and gtPBREM variation on transcriptional activity mediated by CAR (A), PXR (B), or AhR (C), assessed by expression of a reporter gene encoding firefly luciferase (21). HepG2 cells were cotransfected with expression plasmids for CAR, PXR, or the control vector pCR3, as well as the distal enhancer module linked to proximal promoter gene-tk-luciferase plasmid. The transfected cells were treated with vehicle (DMSO, dimethylsulfoxide), 5 x 10^{-6} M rifampicin (RIF), 5 x 10^{-6} M rifampicin plus 4 x 10^{-7} M dexamethasone (DEX) (B) or 1 x 10^{-6} M benzo[a]pyrene (B[a]P) (C) for 24 h, harvested, and assayed for luciferase activity. Luciferase activity was measured using the dual-luciferase reporter assay system. Fold activation was calculated for each reporter construct by dividing the activity with CAR by that without CAR or with inducer by that without inducer. Data presented are the average of 3 independent experiments ± SD. Transcriptional activity of the pGL3-tk reporter gene without CAR or with vehicle was calculated as one.

As shown in Figure 5, the CAR-, PXR-, and AhR-dependent transcriptional activities of the *UGT1A1* promoter exhibited the following order of activity; the promoter containing A[TA]$_6$TAA and -3279T (highest), the promoter containing A[TA]$_6$TAA and -3279G, the promoter containing A[TA]$_7$TAA and -3279T, and the promoter containing A[TA]$_7$TAA and -3279G. The transcriptional activities of 4 types of the UGT1A1 3570-bp promoter by GR were rather low and hard to be compared. Since activated GR by dexamethasone enhances CAR- and PXR-mediated UGT1A1 transactivation [20], we compared the transcriptional activities of 4 types of the *UGT1A1* 3570-bp promoter by rifampicin and dexamethasone.

The T-3279G and A[TA]$_7$TAA mutations more strongly reduced the transcriptional activity by rifampicin and dexamethasone than by rifampicin alone (Figure 5B). The PXR plus GR-mediated transcriptional activities of the promoters also exhibited the following order of activity: the promoter containing A[TA]$_6$TAA and -3279T (highest), the promoter containing A[TA]$_6$TAA and -3279G, the promoter containing A[TA]$_7$TAA and -3279T, and the promoter containing A[TA]$_7$TAA and -3279G (Figure 5B). These observations suggest that the A[TA]$_7$TAA mutation plays a more important role in the transcription of the *UGT1A1* gene than the T-3279G mutation. In conclusion, the manifestation of Gilbert's syndrome observed among Japanese may be attributable to mutations in both the UGT1A1 gene promoter (T-3279G and A[TA]$_7$TAA) and exon 1 (G211A), particularly when these arise as compound mutations.Furthermore, haplotype analysis of the UGT1A1 promoter has been conducted in different ethnic groups. Innocent et al. [47] have reported that the T-3279G (UGT1A1*60) and A-3156G (UGT1A1*93) variants are found to be common (allele frequency 0.39 and 0.30, respectively), and that T-3279G is more common in Caucasians than African Americans (P=0.001). In Caucasians, the linkage disequilibrium is highly significant between sites -3279, -3156, and the (TA)n polymorphism (P<0.0001), and in African Americans, only marginal levels of significance are observed between (TA)n and -3279 (P=0.02) and between -3279 and -3156 (P=0.04). The haplotype structure of promoter is considered to be probably different between Caucasians and African Americans.

To date, more than 113 mutations causing familial non-hemolytic unconjugated hyperbilirubinemia (Crigler-Najjar syndrome and Gilbert syndrome) have been identified [26]. Almost all mutations first demonstrated were *UGT1A1* gene polymorphisms associated with Crigler-Najjar syndrome and predicted the substitution of a single amino acid resulting in the reduction of UGT1A1 activity. Not only the allelic homogeneity of Crigler-Najjar syndrome type II, but also the allelic heterogeneity is demonstrated [39, 48-52]. In addition to the structural mutations,recent study demonstrates that the patient with Crigler-Najjar syndrome type II is a compound heterozygote for A[TA]$_7$TAA and the G923A mutation in exon 2 of UGT1A1(known as G308E) (UGT1A1*11) [53]. Thus, mutations of both the enhancer/promoter region and the coding region including the intron-exon junctions should be determined in patients with Crigler-Najjar syndrome and Gilbert syndrome.

(3) Induction of Human UGT1A1 Gene Expression by Dietary Flavonoids and Its Clinical Application

Mutations found in patients with Crigler-Najjar syndrome type II and Gilbert syndrome do not completely abolish the enzyme activity. Phenobarbital can be an effective therapeutic drug for treating patients with Crigler-Najjar syndrome type II and Gilbert's syndrome and increasing the expression of partially active UGT1A1 enzyme. However, phenobarbital has the serious disadvantage of inducing sleep. Moreover, patients with reduced enzyme activity possess a higher risk for severe hematological toxicity and/or diarrhea by the anticancer drug irinotecan. Bilirubin levels have been demonstrated to be influenced by not only genetic factors but also nongenetic factors such as hemoglobin levels, body mass index, and fasting time [54]. Thus, the modulation of UGT1A1 expression by dietary factors is of considerable interest. Expression of UGT1A1 has been found to be increased by the flavonoid chrysin and quercetin in HepG2 and Caco2 cells (55-58). Accordingly, we investigated the effects of dietary flavonoids on human UGT1A1 induction, utilizing a screening system for assessing UGT1A1 induction with the 290-bp distal enhancer module, and the characterization of UGT1A1 induction by flavonoids [16]. We examined the effects of 26 flavonoids on UGT1A1 expression by measuring the reporter activity of the 290-bp distal enhancer moduleafter 48 h-incubation of human hepatoma cell line HepG2 cells and pig proximal tube-like cell line LLC-PK1 cells with those compounds. Flavone had the most potent induction; 5,7-dihydroxyflavones with varying substituents in the B-ring, such as luteolin, apigenin, chrysin and baicalein, 5-hydroxyflavone, 7-hydroxyflavone and α-naphthoflavone at a concentration of 25μM elevated the reporter activities to 306% to848% in HepG2 cells and 205% to567% in LLC-PK1 cells (Figure 6) [16].

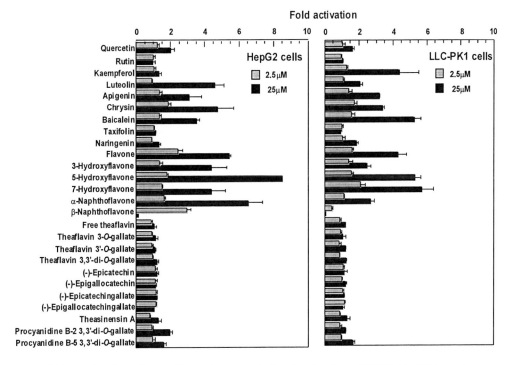

Figure 6. Effects of various flavonoids on the transcriptional activation of UGT1A1 290 bp enhancer module in HepG2 cells and LLC-PK1 cells (16). The 290 bp-tk-luciferase plasmid was cotransfected with pRL-SV40 into the cells. The transfected cells were incubated with various flavonoids at concentrations of 2.5 and 25 μM for 48 h, harvested, and assayed for luciferase activity. Fold activation was calculated by dividing the activity with the inducer by that without the inducer. Data presented are the average of 3 independent experiments ± SD.

While black tea components such as theaflavin, theaflavin 3-*O*-gallate, theaflavin 3'-*O*-gallate and theaflavin 3,3'-di-*O*-gallate and green tea polyphenol components such as (-)-epicatechin, (-)-epigallocatechin, (-)-epicatechingallate and (-)-epigallocatechingallate at 2.5 to 25 µM had no effect on luciferase activity, procyanidine B-2 3,3'-di-*O*-gallate and procyanidine B-5 3,3'-di-*O*-gallate (gallocatechin dimers) at 25 µM potently enhanced the reporter activities in both HepG2 cells and LLC-PK1 cells (121% to196% of the control activity in DMSO-treated cells). A dose-dependent effect was produced by these flavones and gallocatechin dimers, except apigenin, which had maximal activation at 2.5 µM.

While maximum induction by apigenin and luteolinat 2.5 µM and procyanidine B-5 3,3'-di-*O*-gallate at12.5 µM occurred at 24 h, induction by chrysin at 5 µM and procyanidine B-2 3,3'-di-*O*-gallate at12.5 µM reached near the plateau at 24 h. These observations indicate that the reporter gene assay is quite sensitive within a short time period, compared with the catalytic activity assay. Moreover, Petri et al. [59] have determined the *in vivo* enhancement in gene expression of glutathione-S-transferase A1 (GSTA1) and UGT1A1 in the jejunal segment perfused an onion (quercetin-3,4'-diglucoside) and brocoli (sulforaphane) extract, that was performed in six volunteers using a Loc-I-Gut perfusion tube. These results indicate that food components induce drug-metabolizing enzymes such as UGT1A1, i.e., that they have the potential to modify drug-metabolizing capacity.

Next, we investigated the effect of T–3279G mutation on CAR- and PXR-mediated UGT1A1 induction and enhancement by chrysin. In the presence of exogenously expressed CAR, treatment of the cells with chrysin (5 µM) resulted in increases in luciferase activity[1.84 times that of the DMSO control]. On the other hand, treatment of PXR-transfected cells with chrysin (5 µM) elicited increases in the luciferase activity 1.32 times that in the cells transfected with the control vector pCR3 and treated with chrysin (5 µM).The activation extent of the mutated 290-bp reporter gene by 5 µM chrysin (2.17 and 2.99 times that of the DMSO control) was greater than that of the wild-type 290-bp reporter gene (1.84 and 1.57 times that of the DMSO control) in the exogenously CAR- and PXR-expressed cells, respectively. Similarly, rifampicin (5 µM) increased the PXR-mediated reporter gene activities of the wild-type and mutated genes (2.47 and 3.66 times that of the DMSO control, respectively), indicating that flavones and xenobiotics are effective in the transactivation of not only the wild-type reporter gene but also the mutant gene. Moreover, citrus fruit intake has been found to be associated with lower serum bilirubin concentrations in subjects with the UGT1A1*28 polymorphism [60]. Serum bilirubin levels in both men and women decrease in response to isothiocyanates from cruciferous vegetables feeding. Cruciferous vegetable supplementation lowers serum bilirubin concentrations in a dose-dependent manner, and decreases in serum bilirubin concentrations occur with a lower cruciferous dose and to a greater extent among subjects with the *28/*28 genotype compared with those among subjects with one or more *1 allele [60,61].

Taken together, these results indicate that even in subjects with inherited deficiency of the UGT1A1 gene, it would be possible to improve glucuronidation and prevent drug side effects by increasing expression of the enzyme through dietary intervention. Screening dietary compounds affecting the expression of drug-metabolizing enzymes/drug transporters leads to protection against drug side effects, achieving more advanced therapeutic effects by smaller dosages of drugs, and reducing the medical cost. Studies on the combined effects of drugs and food on drug metabolism in human subjects with different genetic backgrounds are expected for tailor-made medical treatment.

References

[1] Ostrow, J.D. and Murphy, N.H. (1970) Isolation and properties of conjugated bilirubin and bile. *Biochem. J.* 120, 311-327.

[2] Ritter, J.K., Crawford, J.M., and Owens, I.S. (1991) Cloning of two human liver bilirubin UDP-glucuronosyltransferasecDNA with expression in COS1 cell. *J. Biol. Chem.* 266, 1043-1047.

[3] Crigler, J.F.Jr. and Najjar, V.A. (1952) Congenital familial non-hemolytic jaundice with kernicterus. *Pediatrics* 10, 169-180.

[4] Mackenzie, P.I., Owens, I.S., Burchell, B., Bock, K.W., Bairoch, A., Belanger, A., Fournel-Gigleux, S., Green, M., Hum, D.W., Iyanagi, T., Lancet, D., Louisot, P., Magdalou, J., Chowdhury, J.R., Ritter, J.K., Schachter, H., Tephly, T.R., Tipton, K.F., and Nebert, D.W. (1997) The UDP glycosyltransferase gene superfamily: recommended nomenclature update based on evolutionary divergence. *Pharmacogenetics* 7, 255-269.

[5] Tukey, R.H., Strassburg, C.P., and Mackenzie, P.I. (2002) Pharmacogenomics of human UDP-glucuronosyltransferases and irinotecan toxicity. *Mol. Pharmacol.* 62, 446-450.

[6] Jansen, P.L.M. (1999) Diagnosis and management of Crigler-Najjar syndrome. *Eur. J. Pediatr.* 158 [Supp. 2], S89-S94.

[7] Rubaltelli, F.F. and Griffith, P.F. (1992) Management of neonatalhyperbilirubinaemia and prevention of kernicterus. *Drugs* 43, 864-872.

[8] Ritter, J.K., Kessier, F.K., Thompson, M.T., Grove, A.D., Auveung, D.J., and Fisher, R.A. (1999) Expression and inducibility of the human bilirubin UDP-glucuronosyltransferase UGT1A1 in liver and cultured primary hepatocytes: evidence for both genetic and environmental influences. *Hepatology* 30, 476-484.

[9] Dennery, P.A. (2002) Pharmacological interventions for the treatment of neonatal jaundice. *Semin. Neonatol.* 7, 111-119.

[10] Yaffe, S.J.,Levym, G., Matsuzawa, T., andBaliah, T. (2003) Enhancement of glucuronide-conjugating capacity in a hyperbilirubinemic infant due to apparentenzyme induction by phenobarbital. *N. Engl. J. Med.*275, 1461-6.

[11] Sueyoshi, T., Kawamoto, T., Zelko, I., Honkakoski, P., andNegishi, M. (1999) The repressed nuclear receptor CAR responds to phenobarbital in activating the human *CYP2B6*gene. *J. Biol. Chem.* 274, 6043-6046.

[12] Sugatani, J., Kojima, H., Ueda, A., Kakizaki, S., Yoshinari, K., Gong, Q.H., Owens, I.S., Negishi, M., andSueyoshi, T. (2001) The phenobarbital response enhancer module in the human bilirubin UDP-glucuronosyltransferase*UGT1A1* gene and regulation by the nuclear receptor CAR. *Hepatology* 33, 1232-1238.

[13] Wedlund, P.J., Aslanian, W.S., Jacqz, E., McAllister, C.B., Branch, R.A., and Wilkinson, G.R. (1985) Phenotypic differences in mephenytoinpharmacokinetics in nornal subjects. *J. Pharmacol. Exp. Ther.* 234, 662-669.

[14] Pascussi, J.M., Gerbal-Chaloin, S., Drocourt, L., Maurel, P., and Vilarem,. MJ. (2003) The expression of CYP2B6, CYP2C9 and CYP3A4 genes: a tangle of networks of nuclear and steroid receptors. *Biochim. Biophys. Acta* 1619, 243-253.

[15] Moore, L.B., Parks, D.J., Jones, S.A., Bledsoe, R.K.,Consler, T.G.,Stimmel, J.B, Goodwin, B.,Liddle, C,. Blanchard, S.G.,Willson, T.M., Collins, J.L.,and Kliewer, S.A. (2000) Orphan nuclear receptors constitutive androstane receptor and pregnane X receptor share xenobiotic and steroid ligands. *J. Biol.Chem.*275, 15122-7.

[16] Sugatani, J., Yamakawa, K., Tonda, E., Nishitani, S., Yoshinari, K., Degawa, M., Abe, I., Noguchi, H., and Miwa, M. (2004)The induction of human UDP-glucuronosyltransferase 1A1 mediated through a distal enhancer module by flavonoids and xenobiotics. *Biohem. Pharmacol.* 67, 989-1000.

[17] Sugatani, J., Sueyoshi, T., Negishi, M.,and Miwa, M. (2005) Regulation of the human *UGT1A1* gene by nuclear receptors CAR, PXR and GR. In: Sies H, Packer L (eds) Conjugation Enzymes. *Methods in Enzymology*, Elsevier, Philadelphia, pp.92-104.

[18] Emi, Y., Ikushiro, S., and Iyanagi, T. (1995) Drug-responsive and tissue-specific alternative expression of multiple first exons in rat UDP-glucuronosyltransferase family 1 (UGT1) gene complex. *J Biochem* 117: 392-399.

[19] Jemnitz, K., Lengyel, G., and Vereczkey, L. (2002) In vitro induction of bilirubin conjugation in primary rat hepatocyte culture. *Biochem.Biophys. Res. Commun.* 291, 29-33.

[20] Sugatani, J., Nishitani, S., Yamakawa, K., Yoshinari, K., Sueyoshi, T., Negishi, M., and Miwa, M. (2005) Transcriptional regulation of human UGT1A1 gene expression: Activated glucocorticoid receptor enhances constitutive androstane receptor/pregnane X receptor-mediated UDP-glucuronosyltransferase 1A1 regulation with glucocorticoid receptor-interacting protein 1. *Mol. Pharmacol.* 67, 845-855.

[21] Sugatani, J., Mizushima, K., Osabe, M., Yamakawa, K., Kakizaki, S., Takagi, H., Mori, M., Ikari, A., and Miwa, M. (2008) Transcriptional regulation of human *UGT1A1* gene expression through distal and proximal promoter motifs: implication of defects in the *UGT1A1* gene promoter. *NaunynSchmiedebergs Arch. Pharmacol.* 377 (4-6), 597-605.

[22] Bernard, P., Goudonnet, H., Artur, Y., Desvergne, B., and Wahli, W. (1999) Activation of the mouse TATA-less and human TATA-containing UDP-glucuronosyltransferase 1A1 promoters by hepatocyte nuclear factor 1. *Mol. Pharmacol.* 56, 526-536.

[23] Yueh, M.-F., Huang, Y.-H., Chen, S., Nguyen, N. andTukey, R.H. (2003) Involvement of the xenobotic response element (XRE) in Ah-receptor mediated induction of human *UDP-glucuronosyltransferase1A1*. *J. Biol. Chem.* 278, 15001-15006.

[24] Yueh, M.-F. andTukey, R.H. (2007) Nrf2-Keap1 signaling pathway regulates human UGT1A1 expression *in vitro* and in transgenic *UGT1* mice. *J. Biol. Chem.* 282, 8749-8756.

[25] Senekeo-Effenberger, K., Chen, S., Brace-Sinnokrak, E., Bonzo, J.A., Yueh, M.-F., Argikar, U., Kaeding, J., Trottier, J., Remmei, R.P., Ritter, J.K., Barbier, O., and Tukey, R.H. (2007) Expression of the human *UGT1* locus in transgenic mice by 4-chloro-6-(2,3-xylidino)-2-pyrimidinylthioacetic acid (WY-14643) and implications on drug metabolism through peroxisome proliferators-activated receptor α activation. *Drug Metab. Disos.* 35, 419-427.

[26] UDP-glucuronosyltransferasealleles nomenclature committee. (2010) UGT1A1 and common exons allele nomenclature. http://www.pharmacogenomics.pha.ulaval.ca/cms/ugt_alleles/

[27] Bosma, P.J., Chowdhury, J.R, Bakker, C., Gantla, S., de Boer, A., Oostra, B.A., Lindhout, D., Tytgat, G.N., Jansen P.L., Oude Elferink, R.P., and Chowdhury, N.R. (1995) The genetic basis of the reducedexpression of bilirubin UDP-glucuronosyltransferase 1 in Gilbert's syndrome.*New Eng. J. Med.* 333, 1171-1175.

[28] Monaghan, G., Ryan, M.,Seddon, R., Hume, R., and Burchell, B. (1996) Genetic variation in bilirubin UPD-glucuronosyltransferase gene promoter and Gilbert's syndrome. *Lancet* 347, 578-581.

[29] Monaghan, G.,McLellan, A., McGeehan, A., Li Volti, S.,Mollica, F.,Salemi, I., Din, Z. Cassidy, A., Hume, R., and Burchell, B. (1999) Gilbert's syndrome is a contributory factor inprolonged unconjugated hyperbilirubinemia of the newborn. *J. Pediatr*. 134, 441-446.

[30] Aono, S., Adachi, Y.,Uyama, E., Yamada, Y.,Keino, H.,Nanno, T.,Koiwai, O., and Sato, H. (1995) Analysis of genes for bilirubin UDP-glucuronosyltransferase in Gilbert's syndrome. *Lancet* 345, 958-959.

[31] Maruo, Y.,Nishizawa, K., Sato, H.,Sawa, H., and Shimada, M. (2000) Prolonged unconjugated hyperbilirubinemia associated with breast milk and mutations of the bilirubin uridinediphosphate- glucuronosyltransferase gene. *Pediatr.* 106, E59.

[32] Maruo, Y.,Nishizawa, K., Sato, H.,Doida, Y., and Shimada, M. (1999) Association of neonatal hyperbilirubinemia with bilirubin UDP-glucuronosyltransferase polymorphism. *Pediatr.* 103, 1224-1227.

[33] Raijmakers, M.T., Jansen, P.L.,Steegers, E.A., and Peters, W.H. (2000) Association of human liver bilirubin UDP-glucuronyltransferase activity with a polymorphism in thepromoter region of the UGT1A1 gene. *J. Hepatol*. 33, 348-351.

[34] Yamamoto, K., Sato, H., Fujiyama, Y.,Doida, Y., and Bamba, T. (1998) Contribution of two missense mutations (G71R and Y486D) of the bilirubin UDP glycosyltransferase(UGT1A1) gene to phenotypes of Gilbert's syndrome and Crigler-Najjar syndrome type II. *Biochim. Biophys. Acta*. 1406, 267-273.

[35] Beutler, E., Gelbary, T., and Demina, A. (1998) Racial variability in the UDP-glucuronosyltransferase 1 (UGT1A1) promoter: A balanced polymorphism for regulation of bilirubin metabolism? *Proc.Natl. Acad. Sci. USA* 95, 8170-8174.

[36] Palomaki, G.E., Bradley, L.A., Douglas, M.P, Kolor, K., and Dotson, W.D. (2009) Can *UGT1A1* genotyping reduce morbidity and mortality in patients with metastatic colorectal cancer treated with irinotecan? An evidence-based review. *Genet Med* 11, 21-34.

[37] Clarke, D.J.,Moghrabi, N., Monaghan, G., Cassidy, A., Boxer, M., Hume, R., and Burchell, B. (1997) Genetic defects of the UDP-glucuronosyltransferase-1 (UGT1) gene that causefamilial non-haemolytic unconjugated hyperbilirubinaemias. *Clin. Chim. Acta*. 266,63-74.

[38] Burchell, B., Soars, M., Monaghan, G., Cassidy, A., Smith, D., and Ethell, B. (2000) Drug-mediated toxicity caused by genetic deficiency of UDP-glucuronosyltransferases. *Toxicol. Letters*. 112-113, 333-340.

[39] Kadakol, A.,Ghosh, S.S.,Sappal, B.S., Sharma, G.,Chowdhury, J.R., and Chowdhury, N.R. (2000) Genetic lesions of bilirubin uridine-diphosphoglucuronate glucuronosyl

[40] transferase (UGT1A1) causing Crigler-Najjar and Gilbert syndromes: correlation of genotype to phenotype. *Human Mut.* 16, 297-306.
[40] Black, M. and Billing, B.H. (1969) Hepatic bilirubin udp-glucuronyltransferase activity in liver disease and gilbert's syndrome. *New Engl. J. Med.* 280, 1266-1271.
[41] Adachi, Y. and Yamamoto, T. (1982) Hepatic bilirubin-conjugating enzymes of man in the normal state and in liver disease. *Gastroenterol. Jpn.* 17, 235-240.
[42] Sugatani, J., Yamakawa, K., Yoshinari, K., Machida, T., Takagi, H., Mori, M., Kakizaki, S., Sueyoshi, T., Negishi, M., and Miwa, M. (2002) Identification of adefect in the *UGT1A1* gene promoter and its association with hyperbilirubinemia. *Biochem.Biophys. Res.Commun.*292, 492-7.
[43] Maruo, Y., Addario, C.D., Mori, A., Iwai, M., Takahashi, H., Sato, H., and Takeuchi, Y. (2004) Two linked polymorphic mutations (A(TA)7TAA and T-3279G) of *UGT1A1* as the principal cause of Gilbert syndrome. *Hum. Genet.* 115, 525-526.
[44] Sai, K., Saeki, M., Saito, Y., Ozawa, S., Katori, N., Jinno, H., Hasegawa, R., Kaniwa, N., Sawada, J., Komamura, K., Ueno, K., Kamakura, S., Kitakaze, M., Kitamura, Y., Kamatani, N., Minami, H., Ohtsu, A., Shirao, K., Yoshida, T., and Saijo, N. (2004) Pharmacogenetics and genomics: UGT1A1 haplotypes associated with reduced glucuronidation and increased serum bilirubin in irinotecan-administered Japanese patients with cancer. *Clin. Pharmacol. Ther.* 75, 501-515.
[45] Kanai, M., Kijima, K., Shirahata, E., Sasaki, A., Akaba, K., Umetsu, K., Tezuka, N., Kurachi, H., Aikawa, S., and Hayasaka, K. (2005) Neonatal hyperbilirubinemia and the bilirubin uridinediphosphate-glucuronosyltransferase gene: The common -3263T>G mutation of Phenobarbital response enhancer module is not associated with the neonatal hyperbilirubinemia in Japanese. *Pediatr. Int.* 47, 137-141.
[46] Long, J., Zhang, S., Fang, X., Luc, Y., and Liu, J. (2011) Association of neonatal hyperbilirubinemia with uridinediphosphate-glucuronosyltransferase 1A1 gene polymorphisms: a meta-analysis. *Pediatr. Int.*, in press.
[47] Innocenti, F., Grimsley, C., Das, S., Ramirez, J., Cheng, C., Kuttab-Boulos, H., Ratain, M.J., and Di Rienzo, A. (2002) Haplotype structure of the UDP-glucuronosyltransferase 1A1 promoter in different ethnic groups. *Pharmacogenetics* 12, 725-733.
[48] Aono, S., Yamada, Y., Keino, H., Hanada, N., Nakagawa,T., Sasaoka, Y., Yazawa, T., Sato, H., and Koiwai, O. (1993) Identification of defect in the genes for bilirubin UDP-glucuronosyl-transferase in a patient with Crigler-Najjar syndrome type II. *Biochem. Biophys. Res. Commun.* 197, 1239-1244.
[49] Moghrabi, N., Clarke, D.J., Boxer, M., and Burchell, B. (1993) Identificationof an A-to-G missense mutation in exon 2 of the UGT1 gene complex that causes Crigler-Najjar syndrome type 2. *Genomics* 18, 171-173.
[50] Seppen, J., Bosma, P.J., Goldhoorn, B.G., Bakker, C.T., Chowdhury, J.R., Chowdhury, N.R., Jansen, P.L., Oude Elferrink R.P. (1994) Discrimination between Crigler-Najjar type I and II by expression of mutant bilirubin uridinediphosphate-glucuronosyltransferase. *J. Clin. Invest.* 94, 2385-2391.
[51] Seppen, J., Steenken, E., Lindhout, D., Bosma, P.J., and Elferink, R.P. (1996) A mutation which disrupts the hydrophobic core of the signal peptide of bilirubin UDP-glucuronosyltransferase, an endoplasmic reticulum membrane protein, causes Crigler-Najjar type II. *FEBS Lett.* 390, 294-298.

[52] Kadakol, A., Sappal, B.S., Ghosh, S.S., Lowenheim, M., Chowdhury, A., Chowdhury, S., Santra, A., Arias, I.M., Chowdhury, J.R., and Chowdhury, N.R. (2001) Interaction of coding region mutations and the Gilbert-type promoter abnormality of the UGT1A1 gene causes moderate degrees of unconjugated hyperbilirubinaemia and may lead to neonatal kernicterus. *J. Med. Genet.* 38, 244-249.

[53] Costa, E., Vieira, E., Martins, M., Saraiva, J., Cancela, E., Costa, M., Bauerle, R., Freitas, T., Carvalho, J.R., Santos-Silva, E., Barbot, J., and dosSantos, R. (2006) Analysis of the UDP-glucuronosyltransferase gene in Portuguese patients with a clinical diagnosis of Gilbert and Crigler-Najjar syndromes. *Blood Cells Mol. Dis.* 36, 91-97.

[54] Rodrigues, C., Costa, E., Vieira, E., De Carvalho, J., Santos, R., Rocha-Pereira, P., Santos-Silva, A., and Bronze-da-Rocha, E. (2011) Bilirubin is mainly dependent on UGT1A1 polymorphisms, hemoglobin, fasting time and body mass index. *Am. J. Med. Sci.* in press.

[55] Galijatovic, A., Otake, Y., Walle, U.K., and Walle, T. (2001) Induction of UDP-glucuronosyltransferase UGT1A1 by the flavonoid chrysin in Caco-2cells-potential role in carcinogen bioinactivation. *Pharm. Res.* 18, 374-379.

[56] Galijatovic, A., Walle, U.K., and Walle, T. (2000) Induction of UDP-glucuronosyltransferase by the flavonoid chrysinand quercetinin Caco-2cells.*Pharm. Res.* 17, 21-26.

[57] Walle, T., Otake, Y., Galijatovic, A., Ritter, J.K., and Walle, U.K. (2000) Induction of UDP-glucuronosyltransferase UGT1A1 by the flavonoid chrysin in the human hepatoma cell line HepG2. *Drug Metab.Dispos.* 28, 1077-1082.

[58] Walle, U.K. andWalle, T. (2002) Induction of human UDP-glucuronosyltransferase UGT1A1 by flavonoids-structural requirements. *Drug Metab.Dispos.* 30, 564-569.

[59] Petri, N., Tannergren, C., Holst, B., Mellon, F.A., Bao, Y., Plumb, G.W., Bacon, J., O'Leary, K.N., Kroon, P.A., Knutson, L., Forsell, P., Eriksson, T., Lennernas, H., and Williamson, G. (2003) Absorption/metabolism of sulforaphane and quercetin, and regulation of phase II enzymes, in human jejunum in vivo. *Drug Metab. Dispos.* 31, 805-813.

[60] Saracino, M.R., Bigler, J., Schwarz, Y., Chang, J.-L., Li, S., Li, L., White, E., Potter, J.D., and Lampe, J.W. (2009) Citrus fruit intake is associated with lower serum bilirubin concentration among women with the *UGT1A1*28* polymorphism. *J. Nutr.* 139, 555-560.

[61] Navarro, S.L., Peterson, S., Chen, C., Makar, K.W., Schwarz, Y., King, I.B., Li, S.S., Li, L., Kestin, M., and Lampe, J.W. (2009) Cruciferous vegetable feeding alters UGT1A1 activity: diet- and genotype-dependent changes in serum bilirubin in a controlled feeding trial. *Cancer Prev. Res.* 2, 345-352.

In: Bilirubin: Chemistry, Regulation and Disorder
Editors: J. F. Novotny and F. Sedlacek
ISBN: 978-1-62100-911-5
© 2012 Nova Science Publishers, Inc.

Chapter IX

Drug-Induced Cholestatic Liver Injury

Karel Urbánek[1], Ondřej Krystyník[2] and Vlastimil Procházka[2]
[1]Department of Pharmacology
[2]Department of Internal Medicine II – Gastroenterology and Hepatology
Faculty of Medicine, Palacký University and University Hospital,
Olomouc, Czech Republic

Abstract

The review deals with the probable etiology, diagnostics, classification, most likely causative drugs, risk factors and disease course of drug-induced cholestasis. Cholestatic and mixed forms of drug-induced liver injury (DILI) account for nearly half of all reported cases. Medications are probably responsible for 2 – 5 % of cases of jaundice requiring hospital admission; moreover, all forms of DILI are currently the most common adverse drug reaction resulting in withdrawal of new drugs from clinical research. Cholestatic syndromes caused by drugs can be divided into acute (bland cholestasis, cholestatic hepatitis and cholangiolitis) and, less frequent, chronic (vanishing bile duct syndrome and extrahepatic biliary obstruction). The etiology seems to be mostly idiosyncratic, with a supposed genetic predisposition. Bile salt export pump (BSEP) is known to be subject to drug inhibition in susceptible patients. Besides rare mutations that have been linked to drug-induced cholestasis, the common p.V444A polymorphism of BSEP, DRB1*1501 HLA class II haplotype and homozygosity for GSTM1 null and/or GSTT1 null alleles have been identified as a potential risk factors. No specific tests are available for establishing drug etiology of cholestasis; therefore causality assessment is performed in the same way as in other adverse drug reactions. Several structured causality assessment methods for DILI have also been proposed, e.g. RUCAM, CIOMS score etc. Drugs known to cause cholestatic syndromes include various antibiotics, oral contraceptives, oral antidiabetics and numerous other drugs including herbal medicines.

Keywords: Cholestasis, drug-induced liver injury, cholestatic hepatitis, adverse drug reaction

Introduction

Being the main metabolizing organ for xenobiotics, the liver is also a primary target organ for their adverse effects. Hepatic damage is a relatively rare adverse drug reaction (ADR), but at the same time one of the most feared. The occurrence of drug-induced liver injury (DILI) forces clinicians to change the medication, often to a less effective alternative; moreover, it is a frequent cause of withdrawal of new, commonly potent and promising, drugs from clinical use. It seems that such cases are found in an increasing number of newly introduced drugs, with ximelagatran and troglitazone being probably the most notorious examples. Cholestasis, defined as abnormal bile flow from the liver to the duodenum, may be one of several symptoms of such damage. Cholestatic and mixed forms of DILI account for nearly half of all reported cases. Their clinical manifestations can mimic all the other forms of cholestatic liver disease; a failure to recognize medication as a cause of the disease can lead to prolonged exposure to the causative factor with more serious outcomes. For all these reasons, an extensive effort to recognize the causes and course of this condition has been made.

Classification of Adverse Drug Reactions

For a proper understanding of the etiology, course and treatment of the disease, drug-induced cholestatic liver injury must be regarded not simply as one of many other types of cholestasis, but primarily as an adverse drug reaction (ADR). A recent definition describes an ADR as "an appreciably harmful or unpleasant reaction, resulting from an intervention related to the use of a medicinal product, which predicts hazard from future administration and warrants prevention or specific treatment, or alteration of the dosage regimen, or withdrawal of the product" [1].

The vast majority of the DILI cases seem to have an idiosyncratic character (type B ADR; see Table 1). These adverse effects occur with an incidence lower than 1:10,000 drug users. Therefore, it may be supposed that patient predispositions are at least of the same importance for the etiology of injury as are the characteristics of the causative drug. This also means that it is almost impossible to detect them in pre-registration studies or later to obtain sufficiently large cohorts with similar patient characteristics.

Table 1. Types of adverse drug reactions

Type of ADR	Mnemonic	Dose dependence / Predictability	Characteristics
A	Augmented	Dose-related, predictable	Most usual; frequent
B	Bizarre	Dose-unrelated, unpredictable	Immune-mediated reactions or idiosyncrasies; rare
C	Continuous	Related to a cumulative dose and time	Effect of chronic use, late toxicity
D	Delayed	Related to time	Long time from drug cessation
E	End of use	Related to drug withdrawal	Immediately after withdrawal

From that point of view, every case of drug-induced cholestatic injury should be documented as well as possible, and also reported to a pharmacovigilance system for further evaluation.

Drug-Induced Liver Injury

Drug-induced liver injury (DILI) can be defined as an abnormality in the liver function tests related to medication intake [2, 3]. The estimated incidence of DILI for individual medications varies from 1:10,000 to 1:100,000. According to the criteria established by the Council for International Organizations of Medical Sciences (CIOMS), it has been classified into three types [4], characterized by changes in the biochemical liver function tests:

1. Hepatocellular, in which alanine transaminase (ALT) serum levels are increased at least two-fold the upper limit of normal values (ULN) and serum alkaline phosphatase (AP) levels are within normal ranges or ALT/AP ratio is higher than 5;
2. Cholestatic, in which ALP serum levels are increased two-fold the ULN and ALT/AP ratio is lower than 2;
3. Mixed, with both AP and ALT >2xULN and ALT/AP = 2 - 5.

According to etiology, DILI may be both idiosyncratic and intrinsic (direct toxic). Nevertheless, only two commonly used drugs cause dose-dependent liver injury: paracetamol (acetaminophen) and methotrexate. Moreover, these drugs have a toxicity threshold, which ensures the safety of low doses. All other drugs associated with DILI cause idiosyncratic reactions (i.e. immune-mediated or genetically determined), which are dose-independent and unpredictable under the current state of knowledge.

Drug-Induced Cholestatic Syndromes

Biliary tract injury caused by drugs may exhibit several patterns, differing by the pathologic manifestation and partially also by clinical symptoms. The basic patterns are intrahepatic (further being divided into acute and chronic) and extrahepatic. A more detailed commonly used division is shown below, as well as the drugs typically associated with a certain pattern of damage; however, it must be remembered that any single drug can exhibit more than one type of injury [5, 6].

Acute Syndromes

Bland cholestasis, also called pure, simple or canalicular is not associated with hepatocellular damage; it is manifested histologically only by pure canalicular cholestasis. Bile plugs are seen in hepatocytes or canaliculi and are most prominent in zone 3. It may be caused by estrogens (e.g. oral contraceptives) and anabolic steroids.

Mixed cholestatic hepatitis, also known as hepatocanalicular hepatitis, is characterized by cholestasis and portal inflammation with concomitant varying degrees of hepatocyte damage or necrosis. It is probably more frequent than other syndromes, being typically caused by co-amoxiclav, sulfonylureas and macrolide antibiotics.

Ductular cholestasis or cholangiolytic cholestasis is a syndrome of cholestasis with bile duct injury (cholangiolitis), without significant parenchymal liver damage. It may again be caused by co-amoxiclav, but also by carbamazepine, co-trimoxazole and other drugs.

Chronic Syndromes

Drug-induced cholestatic liver injury is considered chronic if symptoms persist longer than six months. According to the histological pattern of injury, three types of chronic syndromes are usually distinguished: *mild bile duct disarray* or ductopenia, *vanishing bile duct syndrome* and *sclerosing cholangitis-like syndrome*. Whilst the former two types of chronic damage are associated with a number of systemic drugs (e.g. flucloxacillin, phenothiazines or ibuprofen), the latter has been described to be caused by 2-deoxy-5-fluorouridine applied intra-arterially for the treatment of hepatic metastasis of colorectal carcinoma as well as by some other topically administered agents [6].

Extrahepatic Syndromes

Extrahepatic drug-induced cholestasis may be caused by the sclerosing cholangitis-like syndrome with predominance of extrahepatic bile duct damage or by cholelithiasis induced by medication. Octreotide and estrogens (when used as hormone replacement therapy) are probably the best-known examples of therapeutically used agents that increase the incidence of gallstones [7, 8].

Epidemiology

The real incidence of DILI cannot be considered to be exactly known; it is estimated to be 1 to 2 cases per 100,000 person-years [9]. The highest incidence found in any epidemiologic study was that of Sgro et al: in an area with 81,000 inhabitants the total crude incidence was 14 per 100,000 per year [10]. Undoubtedly, the total incidence of ADR should be different in different countries and dependent on the utilization of drugs with a potential to cause the adverse reaction. The incidence of cholestatic form of DILI is even more difficult to be estimated. The majority of older data was obtained from countries with very high rates of paracetamol-induced liver disease (very specific by its etiology and being more an overdose than an ADR), which falsely decreases the proportion of cholestatic cases. On the basis of these data, cholestatic form of DILI was considered to be of much lower importance than hepatocellular damage. More recent studies have yielded somewhat different results concerning this issue.

The proportion of cholestatic injury on the total number of drug-induced liver injuries reported to the Danish Committee on Adverse Drug Reactions during a ten-year period was published in the study of Friis and Andreasen [11]. Among 1,100 reported cases, 16.2% were recognized as acute cholestatic DILI. Björnsson and Olsson reviewed 784 cases of DILI received by the Swedish Adverse Drug Reactions Advisory Committee between 1970 and 2004; 26% were found to be cholestatic and 22% to be a mixed form of injury [12]. Of course, an estimate of the incidence of any adverse effect from the number of spontaneously reported cases must only be indicative. In a prospective study performed by Drug-Induced Liver Injury Network (DILIN), Chalasani et al. found 23% of cholestatic and 20% of mixed form of DILI among 298 cases [13]. The proportion even increased when the calculation was made from laboratory test values taken at the peak of serum bilirubin elevations; thus, at an international clinical research workshop on DILI in 2008 it was emphasized that the cases are more likely to be designated as mixed or cholestatic when laboratory tests are taken later or when the patient is first seen later in the course of disease [14]. Therefore, it can be assumed that cholestasis occurs in more than one-third and less than half of all cases of liver damage caused by medications.

Etiology

Despite a significant progress made in recent years, the etiology of drug-induced cholestatic injury remains unknown. As in all B-type adverse drug reactions, two main pathophysiological pathways may be involved: genetic predisposition and immune-mediated reaction, with possible overlap of these mechanisms. Extensive studies of hepatobiliary transporters and their genetic variants are discussed, but it should be stressed that the rarity of the disease and its dose-independence exclude determination by any common genetic variant.

Predisposing Factors

Female gender and age over 50 have been generally considered to be risk factors for the DILI occurrence. However, in the cholestatic liver injury these factors may have been less pronounced. In the DILIN prospective study [13], a somewhat higher mean age (54 years) and a higher proportion of men (50%) were found in the cholestatic form of DILI in comparison to hepatocellular type of injury (48 years and 40% of men).

Pre-existing liver disease does not seem to be a factor increasing the susceptibility to cholestatic liver injury, which is consistent with our knowledge on idiosyncratic ADRs. Unfortunately, little is known about the possible association of a history of gallstones and the risk of drug-induced cholestasis.

On the other hand, any type of DILI in patients with a pre-existing liver disease would probably result in a more severe disease course and worse outcomes. Critically ill patients, at risk of impaired hepatic excretion due to many causative factors, may at least be more susceptible to cholestasis induced by penicillins and carbapenem antibiotics [15].

Mechanisms of Injury

Cholestasis may result from both a functional defect in bile formation in the hepatocyte and an impairment of bile secretion and flow in the bile duct. Bland cholestasis may be caused by the drug effect on sinusoidal uptake, intracellular transport, or canalicular secretion of bile. A dose-dependent inhibition of bilirubin uptake in an animal experiment has been described in several drugs [16].

Rarity and dose-independence in the majority of drug-induced bland cholestasis cases implicates the presence of an intermediate step (possibly immune-mediated) also in the pathophysiology of this form of injury. Cholestatic hepatitis has been supposed to be caused mainly by inflammation and hepatocyte injury.

One of the supposed mechanisms causing this type of DILI may be a genetically determined deficient variant of cellular transporters responsible for drug excretion. This can result in accumulation of drug metabolites damaging the cytoskeletal elements responsible for bile export.

The pathophysiology of ductular cholestasis has not yet been well understood. It has been hypothesized that drug-induced impairment of phospholipid bile transporter (MDR3) can cause bile salt-induced cholangiocyte damage [5]. The emergence of chronic drug-induced cholestatic syndromes, especially the vanishing bile duct syndrome, has been usually attributed to immune-mediated mechanisms. This theory is supported by a repeatedly reported simultaneous occurrence of Stevens-Johnson syndrome in patients (often children) suffering from drug-induced VBD syndrome [17].

Role of Biliary Transporters

As a result of recent research, the knowledge on the role of hepatocellular transporters in different types of acquired cholestasis has been continuously increasing [18]. The key bile salt transporters, possibly involved in acquired cholestasis, are bile salt export pump (BSEP), sodium-taurocholate co-transporting polypeptide (NTCP), organic anion transporting polypeptides (OATPs) and the phosphatidylcholine translocator known as multi-drug resistance protein 3 (MDR3). Inhibition of these transporters by drugs or their metabolites may be one of the pathophysiological mechanisms of DILI, of course in the presence of a rare immunologically or genetically conditioned co-factor.

A dose-dependent interference with the uptake of bilirubin by binding to the BSEP in an experiment in rat liver has been described in rifampicin and ciclosporin, glibenclamide and an estrogen metabolite estradiol-17-beta-glucuronide [16].

Pro-inflammatory cytokines, namely IL-1, IL-6 and TNF-alpha, have been identified to act as suppressors of bile salt transporters as well as sinusoidal solute carrier transporters [19], thereby probably lowering the threshold for liver injury.

In experiments on the human NTCP- and human BSEP-expressing LLC-PK1 cells monolayer, drugs causing cholestatic liver damage, such as rifampicin, glibenclamide and ciclosporin, reduced the basal-to-apical transport and the apical efflux clearance of taurocholate by inhibiting both NTCP and BSEP [20].

Genetic Predispositions

Among possible genetic determinants of drug-induced cholestasis, gene mutations and polymorphisms in the biliary transporters BSEP and MDR 3 have obviously been studied, but till now with only limited results. The common p.V444A polymorphism of the BSEP has been identified as a potential risk factor for drug-induced cholestasis [21] although its common occurrence in the population excludes that it could be the main or even the sole cause. Also, the heterozygous mutation p.D676Y of the BSEP might possibly be associated with hepatocellular cholestasis, as might be the heterozygous mutation p.1764L of MDR3 [22].

The supposed mechanism of cholestatic DILI, involving an immune-mediated damage of the biliary system, has led to the research into a possible association of HLA class I and II genes with this type of liver injury, in particular. In a study by O'Donohue et al. [23], an odds ratio of 9.25 was found for the DRB1*1501 HLA class II haplotype in patients who developed cholestasis induced by co-amoxiclav, so the authors hypothesized that co-amoxiclav associated hepatotoxicity may have a genetic basis. A significant genetic risk for HLA-DRB1*15 was confirmed by a recent study which has also provided evidence of a protective effect of the HLA-DRB1*07 family of alleles [24]. A genetic basis for the susceptibility to flucloxacillin-induced liver injury has also been extensively studied. The HLA-B*5701 genotype was found to be a major immunogenetic determinant of this type of DILI [25].

Although all these results for individual drugs are very interesting, the assumed general susceptibility genes for drug-induced cholestasis require further investigation. Promising findings have been brought by the studies of a possible association of DILI and a relatively common absence of glutathione S-transferases M1 and T1 due to gene deletions. Homozygosity for GSTM1 null and/or GSTT1 null alleles has been found to be a risk factor for DILI, with associations described independently for several drugs (e.g. isoniazid, co-amoxiclav and troglitazone) [26].

Causative Medications

Cholestatic form of drug-induced liver injury can be caused by many xenobiotics, including medications, herbal preparations, dietary supplements, illicit agents and toxins. Therapeutically used drugs reported as causative in drug-induced cholestatic liver injury are listed in Table 2. At least some of them, representing the causes of different types of cholestatic injury, also deserve more detailed mention, which can be found in the following paragraphs. The clinical importance of a specific drug risk depends not only on its ability to cause an adverse effect, but also on the frequency of its use in the population and on the underlying diseases or other predispositions of patients treated with a specific drug. For these reasons, a list of the drugs most frequently causing drug-induced cholestasis may vary in different countries or populations. Generally, the causative agents most likely encountered in clinical medicine belong to the drug class of antimicrobials and oral contraceptives. Although it is the medication effects that are stressed here, the considerable risks of dietary supplements must not be forgotten.

Table 2. Therapeutically used drugs associated with cholestatic liver injury

Type of syndrome	Syndrome	Causative agents
Acute intrahepatic	Bland cholestasis	anabolic steroids, azathioprine, cetirizine, ciclosporin, estrogens, fosinopril, glimepiride, infliximab, nevirapine, tamoxifen, warfarin
	Mixed cholestatic hepatitis	acitretin, atorvastatin, azathioprine, benzodiazepines, bupropion, captopril, carbamazepine, cefalexin, celecoxib, chlorambucil, chlorpromazine, co-amoxiclav, danazol, dapsone, dextromethorphan, fenofibrate, gabapentin, gemcitabine, glimepiride, griseofulvin, halothane, hydrochlorothiazide, isoniazid, isoflurane, itraconazole, macrolide antibiotics, mesalazine, metformin, methimazole, methyldopa, NSAIDs, orlistat, oxacillin, pioglitazone, propafenone, propylthiouracil, pyritinol, repaglinide, risperidone, TCADs, terbinafine, ticlopidine, troglitazone
	Ductular cholestasis	carmustine, co-amoxiclav, dextropropxyphene, flucloxacillin, gold compounds, pioglitazone, tenoxicam
Chronic intrahepatic	Mild bile duct disarray	as for the VBD syndrome
	Vanishing bile duct syndrome	aceprometazine, ajmaline, amineptine, amitriptyline, ampicillin, anabolic steroids, azathioprine, barbiturates, carbamazepine, carbutamide, chlorothiazide, chlorpromazine, cimetidine, ciprofloxacin, clindamycin, co-amoxiclav co-trimoxazole, cromolyn sodium, cyamemazine, cyclohexyl propionate, cyproheptadine, penicillamine, diazepam, erythromycin, estradiol, flucloxacillin, glibenclamide, haloperidol, ibuprofen, imipramine, phenylbutazone, phenytoin, prochlorperazine, terbinafine, tetracyclines, thiabendazole, tiopronin, trifluoperazine, tolbutamide, troleandomycin, xenalamine
	Sclerosing cholangitis-like syndrome	absolute ethanol, hypertonic saline, iodine solution, floxuridine, formaldehyde, silver nitrate (administered topically)
Extrahepatic	Cholelithiasis	octreotide, estrogens in hormone replacement therapy

Oral Contraceptives

In combined oral contraceptives, the estrogen component (most commonly ethinylestradiol) is considered to be responsible for the known cholestatic effect. Cholestasis occurs in women taking oral contraceptives with an estimated prevalence of 1:10,000 to 1:4,000. Liver injury typically presents as bland cholestasis: mild elevation in liver enzymes occurs typically 2–3 months after starting the drug, sometimes with jaundice and pruritus. The prognosis of this type of cholestatic liver injury is good, with clinical symptoms and laboratory abnormalities resolving within several days to weeks after drug withdrawal. Cholestatic hepatitis can develop in up to 15% of affected women [5]. As mentioned above, molecular studies have revealed that estrogen metabolite estradiol-17-beta-glucuronide has the capacity to cause cholestasis through a dose-dependent inhibition of the BSEP.

Co-Amoxiclav

Co-amoxiclav is a combination of amoxicillin, a broad-spectrum penicillin antibiotic, and clavulanic acid, a beta-lactamase inhibitor. It is among the most popular antibiotics in many countries for its excellent antibacterial effects and safety. The prevalence of liver injury caused by this combination is estimated to be 1:100,000 exposed patients. Males over 60 years of age seem to be at the highest risk, while a pre-existing liver disease does not seem to be a significant risk factor. Symptoms such as fatigue, nausea, abdominal pain and sometimes jaundice with pruritus occur 4–7 weeks after treatment initiation, which means that liver injury is usually diagnosed several weeks after treatment cessation. Liver biopsy findings are usually consistent with the pattern of cholestatic hepatitis. The time to recovery after drug withdrawal varies widely and it can take up to 4 months; prolonged cholestasis, vanishing bile duct syndrome and even death have also been reported. An immune-mediated reaction to clavulanic acid is usually supposed to be the trigger of liver injury; however, amoxicillin may play a role here as well. Genetic predispositions for this type of DILI in the main histocompatibility complex are mentioned above. Recently, the results of a genome-wide association study using 822,927 single nucleotide polymorphism (SNP) markers from 201 cases of DILI following co-amoxiclav administration and 532 population controls have confirmed this relationship. The study has proved that several class I and II HLA genotypes (e.g. HLA-DRB1*1501-DQB1*0602) affect the susceptibility to co-amoxiclav-induced DILI, indicating the importance of the adaptive immune response in the pathogenesis [27].

Flucloxacillin

Flucloxacillin is a narrow-spectrum penicillin antibiotic with activity against some beta-lactamase producing bacteria. Approximately 1 in 15,000 users of flucloxacillin will develop cholestatic liver injury. Jaundice and pruritus may first appear several weeks after administration of the drug has ceased and are typically severe and protracted, whilst death is uncommon. Liver tests abnormalities may persist for months after symptomatic recovery [28]. Liver biopsy shows a pattern of ductular cholestasis: centrizonal bile stasis with portal tract inflammation and variable loss of bile ducts. An experimental study has shown that hepatocytes metabolize flucloxacillin via CYP3A4 to several metabolites, including 5'-hydroxymethylflucloxacillin that acts as a cytotoxic agent in susceptible biliary epithelial cells. This may at least contribute to the pathogenesis of flucloxacillin-induced cholestasis [29].

Octreotide

Octreotide is a somatostatin analog, a more potent inhibitor of growth hormone, glucagon, and insulin. It is administered parenterally for the treatment of acromegaly, diarrhea and flushing episodes in carcinoid syndrome, and diarrhea in patients with vasoactive intestinal peptide-secreting tumors. Long-term octreotide administration in acromegalic patients leads to the formation of small, cholesterol gallstones that are typically asymptomatic. The mechanism of octreotide-associated gallstone formation may involve a decrease in gallbladder emptying, hepatic bile secretion, and sphincter of Oddi motility, as well as modification of bile composition. Gallbladder stasis may sequentially lead to increased bile concentration, precipitation of cholesterol and calcium salts, retention of biliary precipitates, and maturation of gallstones [30]. A recent retrospective study in 459 acromegalic patients has revealed a prevalence of gallstones in acromegaly of 8% at the time of diagnosis, with an additional 35% developing during somatostatin analog treatment. Obesity, dyslipidemia, and somatostatin analog treatment were independent predictors of onset of gallstones, whereas gender and age were not [31].

Clinical Manifestation

Drug-induced liver injury may occur as an acute or, more rarely, chronic disease. As mentioned earlier, any single drug can exhibit more than one type of liver injury. The clinical pattern may vary from mild forms with symptoms which can be easily overlooked, to fulminant or chronic forms with typical clinical and laboratory symptoms of liver failure. It should be noted that there are no specific manifestations which would distinguish drug-induced cholestatic liver injury from other forms of such a disease. The clinical pattern in acute cholestatic syndromes highly depends on the amount of liver parenchyma impairment. In simple cholestasis, where the liver parenchyma is intact, pruritus is usually the first and often the only obvious symptom. Jaundice may follow although cholestatic liver impairment without jaundice is also commonly described [5, 6]. The most common form of drug-induced cholestatic liver injury, cholestatic hepatitis, is histologically characterized by cholestasis and portal inflammation with concomitant varying degrees of hepatocyte damage or necrosis. This syndrome has usually been characterized by two different clinical phases. The prodromal phase is manifested by the emergence of nausea, vomiting, loss of appetite, anorexia, malaise, fatigue, abdominal pain or discomfort, and fever. The second, icteric phase manifested by jaundice usually follows [6, 32]. Once cholestasis is present, the risk of secondary biliary infection due to impaired bile flow increases. Therefore, cholangitis is another possible manifestation of drug-induced cholestasis. Early clinical symptoms of incipient cholangitis are usually jaundice, fever and right upper abdominal pain. The three symptoms are known as Charcot's triad. Dark urine and pale stools may confirm the interruption of bile flow. In addition to the already mentioned acute cholestatic syndromes, a chronic course of disease with slow progression and loss of hepatic function may also occur. Although the role of immunologic mechanisms involved in drug-induced cholestatic liver injury remains largely unknown, symptoms of systemic hypersensitivity may sometimes be seen. Fever, rush, facial edema, eosinophilia or atypical lymphocytosis are among the early features [2, 32].

Diagnosis and Causality Assessment

The international criteria for drug-induced liver toxicity, established by the CIOMS, are mentioned above. Serum AP activity (usually more than three times the upper limit of normal) is the most important laboratory finding for the diagnosis of drug-induced cholestasis. Serum ALT levels indicate the level of hepatocellular injury, while serum bilirubin levels predict the severity of the injury and also the development of chronic liver disease [6]. Liver biopsy is useful not only for establishing the diagnosis, but also for predicting the prognosis. Unfortunately, in many cases it is not performed for various reasons. It is also considered the gold standard to distinguish cholestatic from hepatocellular form of injury and, subsequently, to differentiate between the cholestatic syndromes. However, the results may be biased by sampling error, and may also vary with the timing: hepatocellular injury predominates during initial days or weeks of injury, whereas cholestatic forms are more frequently diagnosed later [14].

As is usual with the vast majority of idiosyncratic adverse drug reactions, no specific clinical or biochemical test for establishing the drug-induced etiology of cholestatic liver injury is available. Therefore, establishing the diagnosis in clinical practice requires a high degree of clinical suspicion and has been based on the following criteria:

1. Cholestatic liver injury occurs in association with the administration of a drug;
2. All other common causes of the disease have been excluded;
3. Symptoms of cholestatic liver injury disappear after drug withdrawal;
4. Symptoms recur after a re-challenge of the suspected drug.

These criteria are associated with certain problems in all ADRs, not only in such a difficult one as DILI. The first criterion seems to be easy to meet until we remember that monotherapy in our patients becomes increasingly scarce. When DILI occurs later in the course of pharmacotherapy, it is not easy to determine which drug is most suspicious. It is necessary to evaluate the possible involvement of all drugs used by the patient within the last 3–6 months. For that reason, an effort to obtain an accurate causality assessment tool has been made from the beginning of the research into this type of adverse drug reactions.

Currently, several structured scoring systems, helping to establish the causal relationship between an offending drug and liver injury, are used. Nevertheless, none of them has been universally adopted. The majority of causality assessment methods focus primarily on the time dependence of the drug use and injury onset, secondarily on the exclusion of possible non-drug-related causes. Three existing well-established scoring systems deserve a mention here. The RUCAM (Roussel Uclaf Causality Assessment Method) was the first widely accepted method designed to assess causality in drug-induced liver injury [33]. In this scoring system, the suspicion of drug-induced injury has been divided into highly probable (score >8), probable (score 6–8), possible (score 3–5), unlikely (score 1–2), and excluded (score $\leq$ 0). Low test-retest and inter-rater reliability as well as the qualitative character of clinical criteria have been identified as the main disadvantages of this test [14]. The CIOMS (Council of International Organizations of Medical Sciences) scoring scale represents a modification of RUCAM with a higher sensitivity (86%) and specificity (89%) of the test. The parameters are partly quantitative and, therefore, well validated [34]. Probably the most suitable test battery based on the CIOMS scale seems to be the structured causality assessment system recently published by Teschke et al. [35]. It evaluates drug-induced cholestatic liver injury by pre-test, main test and post-test. Pre-test parameters are qualitative and targeted to exclude unrelated or unevaluable causality. In the case of a positive signal for causality, the main test follows. This part of evaluation is based on quantitative parameters. The total score is categorized into ranges of causality: excluded (≤ 0), unlikely (1–2), possible (3–5), probable (6–8) and highly probable (>8). Post-test parameters have a qualitative character and exclude other hepatic and extrahepatic diseases possibly involved in the case of liver injury undergoing investigation. Causality scoring systems are useful tools to increase the accuracy in diagnosing DILI and should be demanded for use not only in major trials, but also in published case reports which still comprise the base of knowledge on the risk of individual agents.

Course of the Disease and Treatment

The period between the initiation of the causative drug and the onset of symptoms varies widely. It may be short (several hours to days), intermediate or delayed (1–8 weeks), or sometimes long (1–12 months), depending on the specific agent [6]. Drug-induced cholestatic liver injury is usually manifested as a mild disorder and the liver and biliary tract function usually recovers spontaneously without any consequences. In patients with cholestatic or mixed liver injury enrolled in the DILIN prospective study, the mean time to the resolution of jaundice was 38 days, which was insignificantly more than in those with hepatocellular injury [13]. Rarely, chronic hepatitis, ductopenia, cirrhosis, or even death may also occur [2]. Persistent use of the causative agent seems to be an important predictor of the disease transition to chronicity. Bad prognosis is associated mainly with a fulminant form of mixed liver injury, the most frequent cause of death in DILI.

The management of drug-induced cholestatic liver injury consists of causative and symptomatic components. Withdrawal of the drug possibly implicated as causative is a fundamental step in the therapy because only a rapid discontinuation of the offending agent may prevent irreversible liver injury. As mentioned above, polypharmacy has been encountered more and more frequently; therefore, determination of the correct drug causing liver injury and its withdrawal remains problematic. Relief of clinical and laboratory symptoms may immediately follow the withdrawal of the causative agent. On the other hand, cholestatic and parenchymal liver damage as well as abnormal laboratory tests may persist several weeks after withdrawal of the causative drug because the time to a complete recovery of liver function may vary substantially [14]. Among the therapeutic options, greatest attention has been paid to the use of ursodeoxycholic acid. Its choleretic effect is well-known, but cytoprotective, antiapoptotic and immunomodulatory effects of this compound have also been described. Ursodeoxycholic acid has recently been shown to increase the expression of MDR3 transporter in humans, which may also partly explain its beneficial effect in cholestatic diseases [36]. Therefore, it remains to be the first-line pharmacotherapy in drug-induced cholestatic liver injury. The use of corticosteroids, no matter how logical due to the supposed involvement of immune mechanisms in the pathophysiology of drug-induced syndromes, has not yet been proved to be beneficial.

In the case of extrahepatic drug-induced cholelithiasis or symptomatic biliary strictures due to sclerosing cholangitis-like damage, it is necessary to immediately restore the bile flow to avoid secondary complications. Therefore, endoscopic retrograde cholangiopancreatography (ERCP) should be the first-line therapeutic approach in these cases. Percutaneous transhepatic biliary drainage may also be used, if necessary. Liver transplantation remains reserved for the rare cases of injury resulting in vanishing bile duct syndrome, cirrhosis, or fulminant liver failure.

The second component of the disease management is focused on the symptoms present as a consequence of ongoing liver injury. Patients often complain of pruritus. Severe forms of pruritus may lead to sleep deprivation, which can result in psychological abnormalities [6]. First-line therapeutic agents used in pruritus are bile acid resins (e.g. colestyramine). These agents exchange chloride anions with organic anions, thus sequestering bile acids from enterohepatic circulation and reducing their reabsorption from the intestine. Another agent sometimes used to manage pruritus is rifampicin. It can be used in moderate-to-severe itching as well as in patients with contraindications or intolerance to resins. Rifampicin induces expression of CYP isoforms involved in detoxification processes of bile salts. Hepatic microsomal enzymes may also be activated by phenobarbital; thus, where available, it can also be used for pruritus treatment. Finally, opiate antagonists (naloxone, naltrexone) have been recommended in some cases. Symptomatic therapy could also be focused on the supplementation of fat-soluble vitamins (A, E, D and K), the deficiency of which is caused by prolonged cholestasis leading to impaired fat emulgation and absorption.

Conclusion

In recent years, significant advances have been made in the research into drug-induced cholestatic liver injury. It has been found that this form of drug-induced liver injury is more frequent than previously accepted. Important findings have been made in identifying genetic predispositions of the disease and in the study of subcellular structures involved in the pathophysiology of bile flow impairment by xenobiotics. In the field of clinical medicine, the most important steps forward have been the formation of the networks for studying DILI and the acceptance of improved causality scoring systems. Future research will be targeted at genomic and proteomic studies on the one hand, and at collecting and processing the data from large clinical networks on the other. However, only a close cooperation between experimental and clinical research may provide a better insight into the causes and mechanisms of this adverse drug reaction and allow its prevention.

Acknowledgements

Supported by grants IGA UPOL LF_2011_005 and P303/12/G163.

References

[1] Edwards IR, Aronson JK. Adverse drug reactions: definitions, diagnosis, and management. *Lancet* 2000; 356(9237):1255-9.
[2] Bleibel W, Kim S, D'Silva K, Lemmer ER. Drug-induced liver injury: review article. *Dig. Dis. Sci.* 2007; 52(10):2463-71.
[3] Chang CY, Schiano TD. Review article: drug hepatotoxicity. *Aliment Pharmacol. Ther.* 2007; 25(10):1135-51.
[4] Watkins PB, Seeff LB. Drug-induced liver injury: summary of a single topic clinical research conference. *Hepatology* 2006; 43:618–31.
[5] Hamilton JP, Laurin JM. Drug-Induced Cholestasis. In: Lindor KD, Talwalkar JA. (Eds.) *Cholestatic liver disease.* Humana Press, New Jersey, 2008, pp. 21-43.
[6] Padda MS, Sanchez M, Akhtar AJ, Boyer JL. Drug-induced cholestasis. *Hepatology* 2011; 53(4):1377-87.

[7] Hofmann AF. Increased deoxycholic acid absorption and gall stones in acromegalic patients treated with octreotide: more evidence for a connection between slow transit constipation and gall stones. *Gut* 2005; 54(5):575-8.

[8] Hart AR, Luben R, Welch A, Bingham S, Khaw KT. Hormone replacement therapy and symptomatic gallstones - a prospective population study in the EPIC-Norfolk cohort. *Digestion* 2008; 77(1):4-9.

[9] de Abajo FJ, Montero D, Madurga M, Garcia Rodriguez LA. Acute and clinically relevant drug-induced liver injury: a population based case-control study. *Br. J. Clin. Pharmacol.* 2004; 58:71-80.

[10] Sgro C, Clinard F, Ouazir K et al. Incidence of drug-induced hepatic injuries: a French population-based study. *Hepatology* 2002; 36(2):451-5.

[11] Friis H, Andreasen PB. Drug-induced hepatic injury: an analysis of 1100 cases reported to the Danish Committee on Adverse Drug Reactions between 1978 and 1987. *J. Intern. Med.* 1992; 232(2):133-8.

[12] Björnsson E, Olsson R. Outcome and prognostic markers in severe drug-induced liver disease. *Hepatology* 2005; 42(2):481-9.

[13] Chalasani N, Fontana RJ, Bonkovsky HL, et al. Causes, clinical features, and outcomes from a prospective study of drug induced liver injury in the United States. *Gastroenterology* 2008; 135:1924-1934.

[14] Fontana RJ, Seeff LB, Andrade RJ et al. Standardization of nomenclature and causality assessment in drug-induced liver injury: summary of a clinical research workshop. *Hepatology* 2010; 52(2):730-42.

[15] Frenette AJ, Dufresne ME, Bonhomme V, Albert M, Williamson DR. Drug-induced hepatic cholestasis in the critically ill. *Intensive Care Med.* 2011; 37(7):1225-6.

[16] Stieger B, Fattinger K, Madon J, Kullak-Ublick GA, Meier PJ. Drug- and estrogen-induced cholestasis through inhibition of the hepatocellular bile salt export pump (BSEP) of rat liver. *Gastroenterology* 2000; 118(2):422-30.

[17] Okan G, Yaylaci S, Peker O, Kaymakoglu S, Saruc M. Vanishing bile duct and Stevens-Johnson syndrome associated with ciprofloxacin treated with tacrolimus. *World J. Gastroenterol.* 2008; 14(29):4697-700.

[18] Pauli-Magnus C, Meier PJ. Hepatobiliary transporters and drug-induced cholestasis. *Hepatology* 2006; 44(4):778-87.

[19] Fardel O, Le Vée M. Regulation of human hepatic drug transporter expression by pro-inflammatory cytokines. *Expert Opin. Drug Metab. Toxicol.* 2009; 5(12):1469-81.

[20] Mita S, Suzuki H, Akita H et al. Inhibition of bile acid transport across Na+/taurocholate cotransporting polypeptide (SLC10A1) and bile salt export pump (ABCB 11)-coexpressing LLC-PK1 cells by cholestasis-inducing drugs. *Drug Metab. Dispos.* 2006; 34(9):1575-81.

[21] Pauli-Magnus C, Meier PJ, Stieger B. Genetic determinants of drug-induced cholestasis and intrahepatic cholestasis of pregnancy. *Semin. Liver Dis.* 2010; 30(2):147-59.

[22] Lang C, Meier Y, Stieger B et al. Mutations and polymorphisms in the bile salt export pump and the multidrug resistance protein 3 associated with drug-induced liver injury. *Pharmacogenet Genomics* 2007; 17(1):47-60.

[23] O'Donohue J, Oien KA, Donaldson P et al. Co-amoxiclav jaundice: clinical and histological features and HLA class II association. *Gut* 2000; 47(5):717-20.

[24] Donaldson PT, Daly AK, Henderson J et al. Human leucocyte antigen class II genotype in susceptibility and resistance to co-amoxiclav-induced liver injury. *J. Hepatol.* 2010; 53(6):1049-53.

[25] Daly AK, Donaldson PT, Bhatnagar P et al. HLA-B*5701 genotype is a major determinant of drug-induced liver injury due to flucloxacillin. *Nat. Genet.* 2009; 41(7):816-9.

[26] Lucena MI, Andrade RJ, Martínez C et al. Glutathione S-transferase m1 and t1 null genotypes increase susceptibility to idiosyncratic drug-induced liver injury. *Hepatology* 2008; 48(2):588-96.

[27] Lucena MI, Molokhia M, Shen Y et al. Susceptibility to amoxicillin-clavulanate-induced liver injury is influenced by multiple HLA Class I and II alleles. *Gastroenterology* 2011; 141(1): 338-47.

[28] Devereaux BM, Crawford DH, Purcell P, Powell LW, Roeser HP. Flucloxacillin associated cholestatic hepatitis. An Australian and Swedish epidemic? *Eur. J. Clin. Pharmacol.* 1995; 49(1-2):81-5.

[29] Lakehal F, Dansette PM, Becquemont L et al. Indirect cytotoxicity of flucloxacillin toward human biliary epithelium via metabolite formation in hepatocytes. *Chem. Res. Toxicol.* 2001; 14(6):694-701.

[30] Redfern JS, Fortuner WJ 2nd. Octreotide-associated biliary tract dysfunction and gallstone formation: pathophysiology and management. *Am. J. Gastroenterol.* 1995; 90(7):1042-52.

[31] Attanasio R, Mainolfi A, Grimaldi F et al. Somatostatin analogs and gallstones: a retrospective survey on a large series of acromegalic patients. *J. Endocrinol. Invest.* 2008; 31(8):704-10.

[32] Geubel AP, Sempoux CL. Drug and toxin-induced bile duct disorders. *J Gastroenterol Hepatol* 2000; 15(11):1232-8.

[33] Danan G. Causality assessment of drug-induced liver injury. Hepatology Working Group. *J. Hepatol.* 1988; 7(1):132-6.

[34] Danan G, Bénichou C. Causality assessment of adverse reactions to drugs – I. A novel method based on the conclusions of international consensus meetings: application to drug-induced liver injuries. *J. Clin. Epidemiol.* 1993; 46: 1323–30.

[35] Teschke R, Schwarzenboeck A, Hennermann KH. Causality assessment in hepatotoxicity by drugs and dietary supplements. *Br. J. Clin. Pharmacol.* 2008; 66(6): 758-66.

[36] Marschall HU, Wagner M, Zollner G, Trauner M. Clinical hepatotoxicity. Regulation and treatment with inducers of transport and cofactors. *Mol. Pharm.* 2007; 4(6): 895-910.

In: Bilirubin: Chemistry, Regulation and Disorder ISBN: 978-1-62100-911-5
Editors: J. F. Novotny and F. Sedlacek, pp. 293-302 © 2012 Nova Science Publishers, Inc.

Chapter X

Genetic Service, Counseling and Research for Hyperbilirubinemic Patients in Taiwan

Ching-Shan Huang

Abstract

Neonatal hyperbilirubinemia was a serious problem in Taiwan before 1980 and glucose-6-phosphate dehydrogenase (G6PD) deficiency was found being an important factor for the development of such a disease. For example, among male newborn infants suffering from neonatal hyperbilirubinemia, bilirubin concentration; number of bilirubin concentration ≥342 µM (20.0 mg/dL) and mortality rate were significantly higher in G6PD-deficient subjects than in G6PD-normal analogs. The results of further analysis showed that, in G6PD-deficient newborn infants, contact of oxidant chemical or drugs; such as naphthalene and some herbals; was the main cause of severe neonatal hyperbilirubinemia. Therefore, the Cathay General Hospital (CGH) in Taipei established a quantitative G6PD screening test for every neonate born at the CGH in 1981. The incidence of G6PD deficiency detected by this quantitative method was approximate 3%. At the CGH, the medical staffs were aware of any G6PD deficiency within 3 days of birth and health education was provided to parents before the infants' discharge from the hospital. Such a service resulted in that severity of neonatal hyperbilirubinemia caused by G6PD deficiency had improved. The Department of Health, Taiwan (DOH) has decided G6PD being one item of nationwide neonatal screening tests since 1987. At present, dry blood spots of neonates are sent to three neonatal screening institutions, which were certified by the DOH, and the turn around time of G6PD report is restrict less than 10 days of birth in order to health education being given as earlier as possible.

In Taiwan, study of molecular biology for G6PD was started in 1992. Up to now, it is known that at least 15 different types of single-point mutations responsible for G6PD deficiency among Taiwanese. Among the 15 types, the single-point mutation at nucleotide (nt) 1376 (G>T, Arg459Leu) is predominant, accounts for about 50% of

G6PD deficiency in Taiwanese. Further studies of DNA-based conformation for G6PD deficiency offered useful medical information for Taiwanese.

Although the number of Taiwanese neonates suffering from severe hyperbilirubinemia was reduced after nationwide-neonatal G6PD screening test being performed, incidence of neonatal hyperbilirubinemia was found to be 7.8% and 15.6% in G6PD-normal and G6PD-deficienct neonates, respectively, with 14.6% of the neonates with hyperbilirubinemia having peak bilirubin levels $\geqq 342$ µM. It was concerned that other genetic issues may be associated with hyperbilirubinemia in Taiwanese. Serum bilirubin levels are dependent on both bilirubin production and elimination. The relationship between G6PD deficiency and hyperbilirubinemia may be attributable to that life span of G6PD-deficienct erythrocytes is shorter than G6PD-normal erythrocytes and thus more amount of bilirubin is produced by G6PD-deficienct subjects. Uridine-diphospho-glucuronosyl transferase (UGT) 1A1 had been known being the sole enzyme responsible for glucuronidation for bilirubin. After glucuronidation, bilirubin is more water-soluble and feasible for elimination. However, the variation status of UGT gene was never studied for Taiwanese until 2000. The first article concerning UGT 1A1 gene in Taiwanese indicated that among the four variant sites found within the coding region, 211 G>A (Gly71Arg) was the predominate one, 1091 C>T (Pro364Leu) was a novel variation and 686 C>A (Pro229Gln) was associated with A(TA)$_6$TAA>A(TA)$_7$TAA at nt −53 in the promoter area. Thereafter, further studies revealed that variation in UGT1A1 gene was an important risk factor for the development of hyperbilirubinemia in Taiwanese.

In conclusion, G6PD deficiency and variation of UGT1A1 gene are the main genetic issues associated with hyperbilirubinemia in Taiwanese. However, in Taiwanese, mutation type of G6PD and variation status of UGT1A1 gene are different from those in other ethnic groups. Understandings of such genetic problems are useful for health counseling for our citizens.

Neonatal Hyperbilirubinemia and Glucose-6-Phosphate Dehydrogenase Deficiency

In full-term Asian (Chinese, Japanese and Korean) and American Indian neonates, the peak serum levels of unconjugated bilirubin are double those in Caucasian and black populations [1]. Maternal race (Chinese, Japanese, Korean and Native American) is one of the risk factor for causing neonatal hyperbilirubinemia [2]. Moreover, the incidence of kernicterus is also higher among Asian newborn infants [3]. These findings suggest that genetic factors are involved in the development of neonatal hyperbilirubinemia. Glucose-6-phosphate dehydrogenase (G6PD) deficiency was first concerned in the assessment of such genetic factors, sine it is an X-linked (Xq28) abnormality [4].

It is known that G6PD deficiency is the most-commonly seen genetic defect, affecting over 400 million individuals worldwide [4]. As a consequence of a G6PD deficiency, the mean life span of erythrocytes may be shortened somewhat, and/or low-grade hemolysis may occur, both of which situations may lead to the increased production of bilirubin and the occurrence of unconjugated hyperbilirubinemia [4]. G6PD deficiency may also cause moderate to severe acute hemolysis and unconjugated hyperbilirubinemia when neonates expose to certain oxidative materials [4].

Neonatal hyperbilirubinemia was a serious problem in Taiwan before 1980 and G6PD deficiency was found being an important factor for the development of such a disease. [5,6]. For example, among male newborn infants suffering from neonatal hyperbilirubinemia [peak bilirubin concentration ≥256.5 μM (15.0 mg/dL) within 10 days after birth], bilirubin concentration, number of bilirubin concentration ≥342 μM (20.0 mg/dL) and mortality rate were significantly different between G6PD-deficient and G6PD-normal subjects [Table 1]. The results of further analysis showed that, in G6PD-deficient newborn infants, contact of oxidant chemical or drugs; such as naphthalene and some herbals; was the main cause of severe neonatal hyperbilirubinemia [5-7].

The Cathay General Hospital (CGH) in Taipei established a quantitative G6PD screening test for every neonate born at the CGH in 1981, by using umbilical cord blood as specimen [5]. The incidence of G6PD deficiency detected by this quantitative method was approximate 3% [5,7,8]. For male G6PD-deficient neonates, erythrocyte G6PD activity was almost undetectable [5,7,8].

At the CGH, the medical staffs were aware of any G6PD deficiency within 3 days of birth, the administration of any oxidative chemical or drugs is avoided and health education was provided to parents before the infants' discharge from the hospital [7,8]. Such a service resulted in that severity of neonatal hyperbilirubinemia caused by G6PD deficiency had improved [9].

The Department of Health, Taiwan (DOH) has decided G6PD being one item of nationwide neonatal screening tests since July 1, 1987 [6,10]. A follow-up system comprising of referral hospitals, including outlying islands, was organized for confirmatory test, medical care and genetic counseling [10]. For such a neonatal screening test, the effective collection rate has reached more than 96% of all newborns since 1993 [11] and the coverage rate was 99% in 1997 [10]. Moreover, to assess the reliability of the confirmatory test and screening tests, an external quality assurance program for G6PD assay was developed [10-12]. Dried blood spots and lyophilized quality control materials were prepared from whole blood and erythrocytes for screening and confirmatory tests, respectively [12]. Such an external quality assurance program has been useful for monitoring the performance of the screening laboratories and referral hospitals and helpful for the participating laboratories to improve their test quality [12].

Table 1. Neonatal hyperbilirubinemia in male newborn infants at CGH (year 1980)[a]

	G6PD deficiency (N = 27)	G6PD normal (N = 81)	P
Bilirubin μM, mean (SD)	415.5 (160.7)	289.0 (75.2)	< 0.001[b]
Bilirubin≥342 μM, n (%)	18 (66.7)	17 (21.0)	< 0.001[c]
Mortality, n (%)	5 (18.5)	0 (0.0)	0.00072[d]

[a]Studied by Huang CS.
[b]Calculated by Student's *t* test, [c]Calculated by chi-square test, [d]Calculated by Fisher's exact probability test.

At present, dry blood spots of neonates are sent to three neonatal screening institutions, which were certified by the DOH, and the turn around time of G6PD report is restrict less than 10 days of birth in order to health education being given as earlier as possible [13]. There are current 20 referral hospitals available for G6PD deficiency in Taiwan [13].

Molecular Biologic Studies for G6PD Deficiency

In Taiwan, study of molecular biology for G6PD deficiency was started in 1992 [14-17]. Up to now, it is known that at least 15 different types of single-point mutation, causing a change among the 515 amino acids, responsible for G6PD deficiency among Taiwanese [7-10,14-24]. Among the 15 types, the single-point mutation at nucleotide (nt) 1376 (G>T, Arg459Leu) is predominant, accounts for about 50% of G6PD deficiency in Taiwanese [7-10,14-24].

Table 2 illustrates the nucleotide substitutions and change of amino acids in G6PD-deficient Taiwanese.

Table 2. Nucleotide substitutions and change of amino acids in G6PD-deficient Taiwanese

Nucleotide substitution	Change of amino acid
95 A>G	His 32 Arg
392 G>T	Gly 131 Val
487 G>A	Gly 163 Ser
493 A>G	Asn 165 Asp
517 T>C	Phe 173 Leu
519 C>G	Phe 173 Leu
592 C>T	Arg 198 Cys
835 A>T	Thr 279 Ser
871 G>A	Val 291 Met
1004 C>A	Ala 335 Asp
1024 C>T	Leu 342 Phe
1360 C>T	Arg 454 Lys
1376 G>T	Arg 459 Leu
1387 C>T	Arg 463 Cys
1388 G>A	Arg 463 His

Further studies of DNA-based conformation for G6PD deficiency in Taiwanese indicate those:

1) Polymerase chain reaction (PCR)-restriction enzyme digestion method [8] and denaturing high-performance liquid chromatography [25] are suitable for diagnosis of G6PD deficiency compared with the PCR-sequencing method.
2) In human G6PD, Arginine 198 may be an important amino acid to bind the substrate G6P since it is possibly located within the putative G6P binding domain [26].
3) In Taiwan aboriginal Tribes, the predominant G6PD mutation is at nt 493 (A>G, Asn165Asp) for Saisiat subjects and nt 592 (C>T, Arg198Cys) for Ami subjects, respectively [20].

4) Among the G6PD-deficient newborn infants subsequently suffering from neonatal hyperbilirubinemia, most (78.3%) bring the nt 1376 mutation and the outcome of phototherapy is worse in subjects with nt 1376 mutation than subjects with the non-nt 1376 mutation [7]. Among the G6PD-deficient infants requiring blood exchange transfusion therapy after birth, most (64.7%) carry the mutation at nt 1376 [18].

5) G6PD mutations are associated with F8C/G6PD haplotypes and there is a strong linkage disequilibrium between some of the G6PD mutations and F8C/G6PD haplotypes [22,27].

6) Compared with normal donors, concentration of reduced glutathione in the stored blood of G6PD deficient donors is 67%. However, in adult patients transfused with 1 U G6PD-deficient erythrocytes, hemolysis is not observed [28].

7) A greater levels of infection with dengue virus (DENV-2 New Guinea C strain) is found in the monocytes of 12 G6PD-deficient patients (seven subjects carrying single-point mutation at nt 1376, four at nt 1388 and one at nt 493) than in 24 healthy controls (mean+/-SD: 33.6%+/-3.5 vs 20.3%+/-6.2, P<0.01). Similar results are observed in infection with the DENV-2 16681 strain (40.9%+/-3.9 vs 27.4%+/-7.1, P<0.01) [29].

8) When human coronavirus 229E is inoculated to G6PD-knockdown and G6PD-deficient cells, gene expression and particle production of virus are much higher than G6PD-normal cells [30]. Enterovirus 71 replicates more efficiently and produces greater cytopathic effect and loss of viability in G6PD-deficient (single-point mutation at nt 1376) cells than in G6PD-normal cells [31].

Uridine-Diphospho-Glucuronosyl Transferase and Hyperbilirubinemia

Although the number of Taiwanese neonates suffering from severe hyperbilirubinemia was reduced after nationwide-neonatal G6PD screening test being performed, incidence of neonatal hyperbilirubinemia was found to be 7.8% and 15.6% in G6PD-normal and G6PD-deficienct neonates, respectively [8], with 14.6% of the neonates with hyperbilirubinemia having peak bilirubin levels ≥ 342 µM [32]. It was concerned that other genetic issues may be associated with hyperbilirubinemia in Taiwanese.

Serum bilirubin levels are dependent on both bilirubin production and elimination. The relationship between G6PD deficiency and hyperbilirubinemia may be attributable to that life span of G6PD-deficienct erythrocytes is shorter than G6PD-normal erythrocytes and thus more amount of bilirubin is produced by G6PD-deficient subjects [4]. Uridine-diphospho-glucuronosyl transferase (UGT) 1A1 had been known being the sole enzyme responsible for glucuronidation for bilirubin [33]. After glucuronidation, bilirubin is more water-soluble and feasible for elimination. UGT1A1 gene is located at chromosome 2q37 and encodes 533 amino acids [34]. However, the variation status of UGT gene was never studied for Taiwanese until 2000 [35]. The first article concerning UGT 1A1 gene in Taiwanese [35] indicated that the four variant sites found within the coding region, 211 G>A (Gly71Arg) was the predominate one, 1091 C>T (Pro364Leu) was a novel variation and 686 C>A

(Pro229Gln) was associated with A(TA)$_6$TAA>A(TA)$_7$TAA at nt −53 in the promoter area. Thereafter, studies of UGT1A1 gene in Taiwanese reveal those:

1) A(TA)$_6$TAA>A(TA)$_7$TAA and 211 G>A are the major risk factors for Gilbert's syndrome. Coinheritance of those two variations usually causes a synergistic effect on hyperbilirubinemia [36].
2) Variation at nt 211 of the UGT1A1 gene is a main cause for neonatal hyperbilirubinemia [9,32].
3) For G6PD-deficient male neonates who subsequently suffer from neonatal hyperbilirubinemia, homozygous variation at nt 211 in the UGT1A1 gene is an additive risk factor [8].
4) Homozygous variation in the UGT1A1 gene may cause pronounced hyperbilirubinemia in G6PD-deficient male adults [23].
5) Neonates are at high risk to suffer from severe hyperbilirubinemia if they carry variations at nt 211 in the UGT1A1 gene and at nt 388 in the organic anion transporting polypeptide 2 (OATP 2) gene, as well as feed with breast milk [9].
6) Carriage of homozygous variation [37] or A(TA)$_6$TAA>A(TA)$_7$TAA [38] in the UGT1A1 gene is one of causes for cholelithiasis in adults.
7) A high risk to suffer from unconjugated hyperbilirubinemia exists in adults who carry certain haplotypes in UGT1A1, OATP2 and G6PD genes [24].
8) Increased risk of Gilbert's syndrome is found in adults who carry variant UGT1A7 alleles plus variant UGT1A1 alleles [such as A(TA)$_6$TAA>A(TA)$_7$TAA (UGT1A1*28) and 211 G>A (UGT1A1*6)] (odds ratio = 13.96 for subjects with combined genotype bringing at least one variant allele of UGT1A7 and UGT1A1) [39].
9) The −3279T>G variant of UGT1A1 gene (UGT1A1*60) is another factor for causing hyperbilirubinemia. Among hyperbilirubinemic patients, coinheritance of *60/*60 and *28/*28 in the UGT1A1 gene is a determinant of relatively higher bilirubin values [40].
10) Variation at nt 211 in the UGT1A1 gene is a risk factor for causing early-onset neonatal breastfeeding hyperbilirubinemia [41] and causing prolonged hyperbilirubinemia in male breast-fed neonates [42]. Variation at nt 211 in the UGT1A1 gene, G6PD deficiency and vaginal delivery are risk factors for hyperbilirubinemia in breast-fed neonates [43].
11) Bilirubin concentrations in both α- and β-thalassemia heterozygotes are influenced by variation status of the UGT1A1 gene. More bilirubin being produced in β-thalassemia heterozygotes than in α-thalassemia heterozygotes leads that different bilirubin concentrations are observed between α- and β-thalassemia heterozygotes. [44].
12) A heterozygous C>T substitution at nt 625 in exon 1 (Arg209Trp) plus a novel heterozygous G deletion at nt 1186 in exon 4 (frameshift), a novel homozygous T>A mutation at nt 479 in exon 1 (Val160Glu), a novel homozygous A>G mutation at nt 610 in exon 1 (Met204Val) and a homozygous T>G variation at nt 1456 in exon 5 (Tyr486Asp) plus a heterozygous G>A variation at nt 211 in exon 1 have been observed in the patients suffering from Crigler-Najjar syndrome type 2, respectively [45].

13) Variant 211 G>A in the UGT1A1 gene is a protecting factor for nonalcoholic fatty liver disease in obese children [odds ratio = 0.31 (95% confidence interval: 0.11-0.91; P = 0.033)] [46].

In conclusion, G6PD deficiency and variation of UGT1A1 gene are the main genetic issues associated with hyperbilirubinemia in Taiwanese. However, in Taiwanese, mutation types of G6PD and variation status of UGT1A1 gene are different from those in other ethnic groups. Understandings of such genetic problems are useful for health counseling for our citizens.

References

[1] Halamek LP, Stevenson D: Diseases of the fetus and infants. In: Fanaroff AA, Martin RJ (eds) Neonatal-Perinatal Medicine. Mosby Co, St. Louis, 1997 pp 1345-1389.

[2] Maisels MJ: Jaundice. In: Avery GB, Fletechen MA, MacDonald MG (eds) Neonatology, Pathophysiology and Management of the Newborn. 4th ed. JB Lippincott Co, New York, 1994 pp 630-675.

[3] Anthony JP, Barbara JS: Digestive system disorder. In: Robert MK, Richard EB, Hal BJ, Bonita FS (eds) Nelson textbook of pediatrics. 18th ed. Saunders Elsevier Co, Philadelphia 2007 pp 753-766.

[4] Beutler E: G6PD deficiency. *Blood* 1994; 84: 3616-3636.

[5] Huang CS, Chen TH, Wei C, Chein TY, Jang JF: The clinical application of glucose-6-phosphate dehydrogenase quantitative test. *J. Formos Med. Assoc.* (Taiwan Yi Xue Hui Za Zhi) 1982; 81: 938-944.

[6] Yu MW, Hsiao KJ, Wuu KD, Chen CJ: Association between glucose-6-phosphate dehydrogenase deficiency and neonatal jaundice. *Int. J. Epidemiol.* 1992; 21: 947-952.

[7] Huang CS, Hung KL, Huang MJ, Li YC, Liu TH, Tang TK: Neonatal jaundice and molecular mutations in glucose-6-phosphate dehydrogenase deficient newborn infants. *Am. J. Hematol.* 1996; 51: 19-25.

[8] Huang CS, Chang PF, Huang MJ, Chen ES, Chen WC: Glucose-6-phosphate dehydrogenase deficiency, the UDP-glucuronosyltransferase 1A1 gene and neonatal hyperbilirubinemia. *Gastroenterology* 2002; 123: 127-133.

[9] Huang MJ, Kua KE, Teng HC, Tang KS, Weng HW, Huang CS: Risk factors for severe hyperbilirubinemia in neonates. *Pediatr Res.* 2004; 56: 682-689.

[10] Chiang SH, Wu SJ, Wu KF, Hsiao KJ: Neonatal screening for glucose-6-phosphate dehydrogenase deficiency in Taiwan. *Southeast Asian J. Trop. Med. Public Health* 1999; 30 Suppl 2: 72-74.

[11] Chiang SH, Wu KF, Liu TT, Wu SJ, Hsiao KJ: Quality assurance program for neonatal screening of glucose-6-phosphate dehydrogenase deficiency. *Southeast Asian J. Trop. Med. Public Health* 2003; 34 Suppl 3: 130-134.

[12] Chiang SH, Fan ML, Hsiao KJ: External quality assurance programme for newborn screening of glucose-6-phosphate dehydrogenase deficiency. *Ann. Acad. Med. Singapore* 2008; 37 (12 Suppl): 84.

[13] Neonatal screening tests for prevention of genetic diseases (http://www.bhp.doh.gov.tw/BHPnet/Portal/).

[14] Tang TK, Huang CS, Huang MJ, Tam KB, Yen CH, Tang CJ: Diverse point mutations result in glucose-6-phosphate dehydrogenase (G6PD) polymorphism in Taiwan. *Blood* 1992; 79: 2135-2140.

[15] Huang CS, Tang CJ, Huang MJ, Tang TK: Diagnosis of glucose-6-phosphate dehydrogenase (G6PD) mutations by DNA amplification and allele-specific oligonucleotide probes. *Acta Haematol.* 1992; 88: 92-95.

[16] Chang JG, Chiou SS, Perng LI, Chen TC, Liu TC, Lee LS, et al: Molecular characterization of glucose-6-phosphate dehydrogenase (G6PD) deficiency by natural and amplification created restriction sites: five mutations account for most G6PD deficiency cases in Taiwan. *Blood* 1992; 79: 2135-2140.

[17] Perng LI, Chiou SS, Liu TC, Chang JG: A novel C to T substitution at nucleotide 1360 of cDNA which abolishes a natural Hha I site accounts for a new G6PD deficiency gene in Chinese. *Hum. Mol. Genet.* 1992; 1: 205.

[18] Lo YS, Lu CC, Chiou SS, Chen BH, Chang JG: Molecular characterization of glucose-6-phosphate dehydrogenase deficiency in Chinese infants with or without severe neonatal hyperbilirubinemia. *Br. J. Hematol.* 1993; 86: 858-862.

[19] Chiu DT, Zuo L, Chao L, Chen E, Louie E, Lubin B, et al: Molecular characterization of glucose-6-phosphate dehydrogenase (G6PD) deficiency in patients of Chinese descent and identification of new base substitutions in the human G6PD gene. *Blood* 1993; 81: 2150-2154.

[20] Tang TK, Huang WY, Tang CJ, Hsu M, Cheng TA, Chen KH: Molecular basis of glucose-6-phosphate dehydrogenase (G6PD) deficiency in three Taiwan aboriginal tribes. *Hum. Genet.* 1995; 95: 630-632.

[21] Chen HL, Huang MJ, Huang CS, Tang TK: G6PD NanKang (517 T-->C; 173 Phe-->Leu): a new Chinese G6PD variant associated with neonatal jaundice. *Hum. Hered.* 1996; 46: 201-204.

[22] Chen HL, Huang MJ, Huang CS, Tang TK: Two novel glucose-6-phosphate dehydrogenase deficiency mutations and association of such mutations with F8C/G6PD haplotype in Chinese. *J. Formosan. Med. Assoc* 1997; 96: 779-783.

[23] Huang MJ, Yang YC, Yang SS, Lin MS, Chen ES, Huang CS: Coinheritance of variant UDP-glucuronosyltransferase 1A1 gene and glucose-6-phosphate dehydrogenase deficiency in adults with hyperbilirubinemia. *Pharmacogenetics* 2002; 12: 663-666.

[24] Huang CS, Huang MJ, Lin MS, Yang SS, Teng HC, Tang KS: Genetic factors related to unconjugated hyperbilirubinemia amongst adults. *Pharmacogenet Genomics* 2005; 15: 43-50.

[25] Tseng CP, Huang CL, Chong KY, Hung IJ, Chiu DT: Rapid detection of glucose-6-phosphate dehydrogenase gene mutations by denaturing high-performance liquid chromatography. *Clin Biochem* 2005; 38: 973-980.

[26] Tang TK, Yeh CH, Huang CS, Huang MJ: Expression and biochemical characterization of human glucose-6-phosphate dehydrogenase in Escherichia coli: a system to analyze normal and mutant enzymes. *Blood* 1994; 83: 1436-1441.

[27] Tang TK, Liu TH, Tang CJ, Tam KB: Glucose-6-phosphate dehydrogenase (G6PD) mutations associated with F8C/G6PD haplotypes in Chinese. *Blood* 1995; 85: 3767-3768.

[28] Huang CS, Sung YC, Huang MJ, Yang CS, Shei WS, Tang TK: Content of reduced glutathione and consequences in recipients of glucose-6-phosphate dehydrogenase deficient red blood cells. *Am. J. Hematol.* 1998; 57: 187-192.

[29] Chao YC, Huang CS, Lee CN, Chang SY, King CC, Kao CL: Higher infection of dengue virus serotype 2 in human monocytes of patients with G6PD deficiency. *PLoS One* 2008; 3: e1557.

[30] Wu YH, Tseng CP, Cheng ML, Ho HY, Shih SR, Chiu DT: Glucose-6-phosphate dehydrogenase deficiency enhances human coronavirus 229E infection. *J. Infect Dis.* 2008; 197: 812-816.

[31] Ho HY, Cheng ML, Weng SF, Chang L, Yeh TT, Shih SR, et al: Glucose-6-phosphate dehydrogenase deficiency enhances enterovirus 71 infection. *J. Gen. Virol.* 2008; 89: 2080-2089.

[32] Huang CS, Chang PF, Huang MJ, Chen ES, Hung KL, Tsou KI: Relationship between bilirubin UDP-glucuronosyl transferase 1A1 gene and neonatal hyperbilirubinemia. *Pediatr Res.* 2002; 52: 601-605.

[33] Bosma PJ, Seppen J, Goldhoorn BG, Bakker CT, Chowdhury JR, Chowdhury NR, et al: Bilirubin UDP-glucuronosyl transferase 1 is the only relevant bilirubin glucuronidating isoform in man. *J. Biol. Chem.* 1994; 269: 17960-17964.

[34] Ritter JK, Crawford JM, Owens IS: Cloning of two human liver bilirubin UDP-glucuronosyltransferase cDNAs with expression in COS-1 cells. *J. Biol. Chem.* 1991; 266: 1043-1047.

[35] Huang CS, Luo GA, Huang ML, Yu SC, Yang SS: Variations of the bilirubin uridine-diphosphoglucuronosyl transferase 1A1 gene in healthy Taiwanese. *Pharmacogenetics* 2000; 10: 539-544.

[36] Hsieh SY, Wu YH, Lin DY, Chu CM, Wu M, Liaw YF: Correlation of mutational analysis to clinical features in Taiwanese patients with Gilbert's syndrome. *Am. J. Gastroenterol.* 2001; 96: 1188-1193.

[37] Huang YY, Huang CS, Yang SS, Lin MS, Huang MJ, Huang CS: Effects of variant UDP-glucuronosyltransferase 1A1 gene, glucose-6-phosphate dehydrogenase deficiency and thalassemia on cholelithiasis. *World J. Gastroenterol.* 2005; 11: 5710-5713.

[38] Chu CH, Yang AM, Kao JH, Liu CY, Chang WH, Yang WS: Uridine diphosphate glucuronosyl transferase 1A1 promoter polymorphism is associated with choledocholithiasis in Taiwanese patients. *J. Gastroenterol. Hepatol.* 2009; 24: 1559-1561.

[39] Teng HC, Huang MJ, Tang KS, Yang SS, Tseng CS, Huang CS: Combined UGT1A1 and UGT1A7 variant alleles are associated with increased risk of Gilbert's syndrome in Taiwanese adults. *Clin. Genet.* 2007; 72: 321-328.

[40] Huang YY, Huang MJ, Yang SS, Teng HC, Huang CS: Variations in the UDP-glucuronosyltransferase 1A1 gene for the development of unconjugated hyperbilirubinemia in Taiwanese. *Pharmacogenomics* 2008; 9: 1229-1235.

[41] Chou HC, Chen MH, Yang HI, Su YN, Hsieh WS, Chen CY, et al: 211 G to a variation of UDP-glucuronosyl transferase 1A1 gene and neonatal breastfeeding jaundice. *Pediatr. Res.* 2011; 69:170-174.

[42] Chang PF, Lin YC, Liu K, Yeh SJ, Ni YH: Prolonged unconjugated hyperbiliriubinemia in breast-fed male infants with a mutation of uridine diphosphate-glucuronosyl transferase. *J. Pediatr.* 2009;155: 860-863.

[43] Chang PF, Lin YC, Liu K, Yeh SJ, Ni YH: Risk of Hyperbilirubinemia in breast-fed infants. *J. Pediatr.* 2011; 159: 561-565.

[44] Huang YY, Huang MJ, Wang HL, Chan CC, Huang CS: Bilirubin concentrations in thalassemia heterozygotes in university students. *Eur. J. Haematol.* 2011; 86: 317-323.

[45] Huang CS, Tan N, Yang SS, Sung YC, Huang MJ: Crigler-Najjar syndrome type 2. *J. Formos Med. Assoc.* 2006; 105: 950-953.

[46] Lin YC, Chang PF, Hu FC, Chang MH, Ni YH: Variants in the UGT1A1 gene and the risk of pediatric nonalcoholic fatty liver disease. *Pediatrics* 2009; 124: e1221-1227.

Index

A

Abraham, 96
absorption spectra, 5, 15, 16, 20, 25, 30, 34, 37, 43, 49
abstraction, 121, 122
accelerator, 69, 138
access, 219
accounting, 187, 195
acetaminophen, 182, 279
acetic acid, 127, 131
acid, viii, xi, 7, 15, 25, 26, 56, 68, 95, 104, 112, 115, 119, 121, 125, 127, 131, 133, 138, 139, 143, 148, 153, 157, 158, 159, 161, 165, 173, 174, 176, 179, 184, 195, 211, 214, 217, 218, 220, 257, 258, 260, 272, 284, 288, 290
acidic, 125, 213, 214, 215, 216, 220
acidity, 120, 149
acidosis, 156, 164, 245
acromegaly, 285
acrylic acid, 218
active site, x, 115, 116, 133, 143, 144, 145, 211, 212, 215, 216, 224
active transport, 113, 114, 132, 136
activity level, 264
acute alcoholic hepatitis, 180
acute renal failure, 155, 177, 178
AD, 87, 92, 95, 99, 108, 204, 206, 208, 228
adaptation, ix, 153, 157, 253
additives, 33
adenine, 32, 69, 184
adenocarcinoma, 201, 207
adenosine, 221
ADH, 160, 161, 164
adhesion, 96, 202
adjustment, 53, 78, 79, 89, 93
adolescents, 205
ADP, 136, 137

ADR, 278, 280
adsorption, 170, 171, 172, 181
adulthood, 190
adults, 72, 73, 79, 100, 141, 186, 187, 202, 205, 240, 253, 298, 300, 301
adverse effects, 47, 48, 49, 54, 199, 278
adverse event, 49, 88, 106, 206
Africa, 188, 192
African Americans, 77, 208, 268
African-American, 74, 100, 103, 197, 198, 207
age, 78, 79, 88, 89, 90, 93, 155, 190, 196, 240, 241, 243, 244, 281, 285, 286
aggregation, 128, 129, 130, 148
agonist, 133
alanine, 279
albumin, x, 6, 7, 18, 19, 20, 22, 23, 24, 28, 29, 33, 34, 36, 39, 51, 52, 53, 56, 58, 59, 61, 62, 68, 69, 78, 114, 132, 139, 142, 158, 159, 170, 171, 181, 182, 184, 201, 211, 221, 224, 237
alcohol use, 197
aldosterone, 160, 161, 162, 169
algae, 212
Algeria, 188
algorithm, 151
alkaline media, 61
allele, 71, 73, 74, 75, 76, 77, 80, 81, 83, 88, 89, 90, 93, 101, 102, 107, 186, 194, 195, 199, 200, 203, 264, 268, 270, 273, 298, 300
ALS, 181
ALT, 279, 286
alters, 275
amiloride, viii, 153, 158, 159
amines, 176
amino, 29, 30, 36, 37, 38, 40, 62, 73, 114, 143, 189, 190, 213, 214, 215, 216, 217, 218, 220, 268, 296, 297
amino acid, 29, 30, 36, 37, 38, 40, 73, 114, 143, 189, 190, 213, 214, 215, 216, 218, 220, 268, 296, 297

amino acids, 29, 213, 214, 215, 216, 218, 296, 297
aminoglycosides, 168
ammonia, 171
amplitude, 241, 242, 243, 244, 245, 246
anabolic steroids, 279, 284
anaphylaxis, 70
anastomosis, 156
anatomic site, 201
anemia, 166
aneurysm, 88, 105, 109
angina, 87, 92
angiography, 80, 93
angioplasty, 85, 105, 206
angiotensin II, 166
aniline, 224
annealing, 141, 151
anorexia, 286
ANS, 62
anthocyanin, 133
antibiotic, 284, 285
anticancer drug, 198, 258, 269
anticholinergic, 161
antidiuretic hormone, 63, 160, 176
antigen, 291
anti-inflammatory drugs, 168, 208
antioxidant, vii, ix, 67, 68, 69, 70, 71, 95, 96, 97, 104, 113, 117, 147, 183, 184, 185, 201, 202, 212, 224, 237
anuria, 164
aphasia, 239
apoptosis, 98
aqueous solutions, 5, 22
arginine, 150
argon, 7, 8, 32, 34, 35, 51, 55
arrhythmia, 92
arsenic, 83, 87, 88, 106
arteries, 80
artery, 78, 79, 81, 87, 92, 103, 105, 106, 107, 194, 206
aryl hydrocarbon receptor, xi, 257, 261
ascites, ix, 154, 174
Asia, 188
asphyxia, 252, 253, 254
assessment, xii, 41, 147, 161, 173, 174, 181, 277, 287, 290, 291, 294
asymmetry, 16, 28, 53
asymptomatic, 80, 105, 285
ATF, 71
atherosclerosis, x, 69, 70, 77, 79, 83, 87, 88, 90, 96, 97, 103, 104, 106, 184, 185, 194, 206
atherosclerotic plaque, 69
atoms, 126, 127, 129, 145, 219

ATP, 68, 113, 132, 136, 137, 138, 186, 189, 193, 258
atrium, 159
auditory evoked potentials, 251, 253, 254
auditory nerve, 236, 239
Austria, 197
autooxidation, 216
autopsy, 164
autosomal dominant, 194
autosomal recessive, 190, 193, 260
azotemia, 156

B

Bacillus subtilis, x, 211, 212, 225
bacteria, x, 124, 211, 212, 237, 285
balloon angioplasty, 81, 85, 88, 105, 206
barbiturates, 284
basal ganglia, 239
base, 5, 7, 131, 147, 205, 287, 300
basic research, 224
basicity, 120
batteries, 232
BD, 96, 202
Belarus, 1
beneficial effect, 81, 167, 288
benefits, 125, 168, 172, 202
benign, 157
bias, 196
bile acids, 88, 112, 175, 176, 179, 288
bile duct, xii, 155, 156, 157, 159, 163, 173, 176, 178, 179, 193, 194, 277, 280, 282, 284, 285, 288, 290, 291
biliary atresia, 176
biliary obstruction, xii, 113, 155, 157, 164, 165, 174, 193, 277
biliary stricture, 288
biliary tract, 154, 155, 158, 160, 177, 180, 288, 291
biliverdin, ix, x, 3, 36, 68, 69, 70, 71, 72, 94, 96, 97, 99, 100, 113, 116, 117, 118, 141, 143, 144, 145, 148, 151, 183, 184, 186, 189, 202, 211, 212, 242, 244
biliverdin reductase, ix, 68, 69, 96, 97, 100, 113, 116, 117, 141, 143, 144, 145, 148, 183, 184, 186, 202, 211
biochemical processes, 113
biochemistry, 96, 102, 150
biodegradation, 230
biofuel, x, 211, 212, 219, 221, 223, 224, 229, 231, 232
biological activity, 142
biomarkers, 99, 106, 200
biomolecules, 123
biopsy, 285, 286

biosensors, 232
black tea, 270
Blacks, 73
bleaching, 31, 33, 34, 36, 39
blood circulation, 73
blood flow, 167, 168, 174, 179, 181
blood plasma, 29, 193
blood pressure, 82, 88, 106, 160, 170
blood stream, 113, 189
blood transfusion, 167
blood vessels, 42, 53, 69
blood-brain barrier, 252, 253, 255
bloodstream, 237
body composition, 159
body fluid, 173, 175
body mass index, 269, 275
body weight, 159
bonding, 57, 115, 116, 129, 130, 132, 145, 215
bonds, 24, 118, 131, 132
bone, 70
bone marrow, 70
bradycardia, 161
brain, x, 70, 108, 235, 236, 237, 238, 239, 242, 245, 250, 251, 252, 253, 254, 255
brain damage, 245, 253
brain stem, 251, 253
brainstem, x, 235, 236, 237, 238, 239, 242, 243, 244, 245, 246, 250, 251, 252, 253, 254, 255
Brazil, 51, 210
breakdown, vii, 29, 184, 186, 237, 258
breast cancer, 197, 201, 207, 208
breast milk, 72, 273, 298
breastfeeding, 102, 298, 301
Britain, 188
bronchopulmonary dysplasia, 245, 255
brothers, 205
business partners, 125

C

CAD, 71, 77, 78, 80, 81, 82, 83, 84, 87, 88, 90, 91, 92, 93, 94, 194, 195, 196
cadmium, 7, 32, 33, 34, 35
calcification, 78, 103, 107
calcium, 48, 163, 285
Cameroon, 188
cancer, ix, 96, 101, 183, 184, 196, 197, 198, 203, 207, 208, 209, 266, 274
capillary, 161, 231
carbamazepine, 280, 284
carbohydrates, 217, 219
carbon, 68, 96, 97, 113, 116, 127, 129, 131, 132, 149, 184, 222, 231, 232
carbon dioxide, 184

carbon film, 231
carbon monoxide, 68, 96, 97, 113, 116
carboxyl, 132
carboxylic acid, 112, 114, 126, 128, 129, 135, 146, 147, 148, 149
carboxylic acids, 112, 126, 128, 129, 135, 146, 147, 148, 149
carboxylic groups, 113, 119, 129, 132, 135
carcinogen, 275
carcinogenesis, 196
carcinoid syndrome, 285
carcinoma, 280
cardiac output, 161
cardiogenic shock, 164
cardiovascular disease, vii, ix, 67, 68, 83, 87, 88, 92, 96, 98, 103, 183, 184, 200, 206
cardiovascular disease XE "cardiovascular disease" s, 96
cardiovascular function, ix, 70, 97, 154
cardiovascular risk, 78, 79, 104
cardiovascular system, 68, 70, 80, 94
carotid endarterectomy, 87
case study, 176
catabolism, vii, 112, 113, 115, 118, 211
catalysis, 148
catalyst, x, 139, 211, 212, 222, 224
catalytic activity, 31, 226, 227, 270
cation, 114, 134, 135
Caucasian population, 72
Caucasians, 72, 73, 74, 75, 76, 77, 79, 80, 81, 82, 83, 89, 94, 207, 266, 268
causal relationship, 287
causality, xii, 277, 287, 289, 290
cDNA, 225, 300
cell death, 61
cell line, 71, 82, 98, 269, 275
cell lines, 71, 82, 98
cell membranes, 29, 69
cellular signaling pathway, 70
central nervous system, 4, 238
cerebellum, 238
cerebral cortex, 242, 252
cerebral palsy, 238, 241, 254
cerebrospinal fluid, 237
cerebrovascular disease, 83, 87
ceruloplasmin, x, 211, 212, 214
charge density, 123, 124
chemical, xii, 6, 14, 15, 33, 60, 117, 118, 119, 121, 123, 125, 129, 141, 142, 143, 145, 146, 147, 151, 160, 220, 221, 232, 293, 295
chemical modeling, 151
chemical properties, 142
chemical reactions, 6, 123, 143

chemical reactivity, 142
chemical structures, 117
chemicals, 141
children, 2, 4, 20, 42, 47, 48, 50, 63, 88, 104, 240, 241, 253, 254, 282, 299
China, 67, 95, 192, 197
chiral recognition, 59
chloride anion, 288
chloroform, 4, 5, 6, 8, 9, 11, 12, 13, 15, 18, 24, 26, 58, 59, 149, 218
chlorophyll, 224
cholangiocarcinoma, ix, 154, 156, 157, 158, 163, 166, 174
cholangitis, 164, 166, 169, 177, 280, 284, 286, 288
choledocholithiasis, 301
cholelithiasis, 98, 162, 204, 280, 288, 298, 301
cholemia, 156, 160
cholemic nephrosis, 177
cholestasis, xi, 171, 176, 178, 193, 194, 277, 278, 279, 280, 281, 282, 283, 284, 285, 286, 289, 290
cholesterol, 69, 83, 89, 112, 154, 175, 185, 221, 228, 285
cholic acid, 175, 176
chromatography, 36
chromosome, 93, 94, 97, 108, 186, 187, 189, 193, 194, 203, 297
chromosome 10, 193
chronic diseases, ix, 184, 198, 200
cimetidine, 284
circulation, 2, 112, 165, 237, 288
cirrhosis, 155, 168, 174, 288
City, 257
classes, 71, 121, 139
classification, xi, 277
claudication, 79, 92, 102, 201
cleavage, 33, 227
clinical application, 202, 224, 299
clinical diagnosis, 275
clinical symptoms, 279, 284, 286
clinical trials, 171
cloning, 173, 225, 226
closure, 137
clusters, 227
cochlea, 239
coding, xiii, 186, 187, 190, 192, 193, 199, 268, 275, 294, 297
codon, 204, 213, 266
cognitive development, 254
colon, 201, 207
colon cancer, 201, 207
color, 2, 112, 116, 216, 224
colorectal cancer, 185, 196, 197, 198, 201, 205, 207, 208, 209, 273

combined effect, 260, 270
combustion, 222
common bile duct, 155, 174, 175, 176
communication, 221, 222
communities, ix, 183, 190
community, 79
compensation, 49, 146, 157
competitive process, 133
complement, 70, 96
compliance, 79
complications, 4, 47, 90, 170, 177, 178, 180, 252, 288
composition, 3, 25, 37, 52, 99, 285
compounds, xi, 3, 5, 36, 48, 49, 88, 116, 142, 154, 224, 257, 258, 269, 270
computer, 142
computing, 149
concentration ratios, 6, 29
conduction, x, 235, 236, 237, 253
conductivity, 133, 135
conductor, 145
conference, 289
configuration, vii, 1, 53, 141
conformational analysis, 147
congestive heart failure, 92
congress, 61
Congress, iv
conjugated bilirubin, 68, 69, 96, 113, 138, 139, 150, 155, 156, 193, 194, 211, 212, 217, 219, 220, 229, 230, 237, 238, 271
conjugation, 14, 113, 125, 258, 259, 272
consensus, 51, 291
Consensus, 172
conservation, ix, 154
constipation, 290
constituents, 69, 175
construction, 143, 215, 222, 233
consumption, 7, 70, 93
contour, 80
contraceptives, xii, 206, 277, 279, 283, 284
control group, 48, 90, 264, 266
controlled studies, 168
controlled trials, 167, 170
controversial, viii, 67, 73, 94, 165
convergence, 143, 145, 151
COOH, 14, 115
cooling, 9
cooperation, 289
coordination, 226
copper, x, 63, 211, 212, 213, 214, 215, 216, 217, 218, 219, 220, 225, 226, 227, 229, 230, 231
coronary artery bypass graft, 83, 87

Index

coronary artery disease, 71, 82, 87, 92, 95, 97, 98, 99, 103, 104, 105, 107, 108, 109, 200, 201, 202, 206, 207
coronary heart disease, 101, 102, 107, 109, 194, 200, 203
coronavirus, 297, 301
correlation, 134, 142, 151, 155, 173, 178, 201, 244, 274
cortex, 158
corticosteroids, 288
cost, xi, 130, 180, 224, 258, 270
counseling, xiii, 294, 295, 299
country of origin, 188
covalent bond, 146
Cox regression, 90
creatinine, 164, 165, 221, 228
Crigler-Najjar syndrome, xi, 187, 190, 204, 205, 209, 257, 260, 263, 268, 271, 273, 274, 275, 298, 302
crises, 190
cross-sectional study, 78, 79
CRP, 78, 81, 85, 87
crystal structure, 212, 215, 216, 224, 225
crystallization, 115
crystals, 129, 216
CT, 12, 102, 106, 204, 205, 301
cultivation, 231
culture, 29, 62, 225, 272
currency, 119, 121, 125, 140
CVD, vii, ix, 67, 68, 71, 78, 80, 82, 84, 85, 86, 87, 88, 89, 90, 91, 92, 94, 95, 184, 194, 195, 196, 197, 198
cycles, 220
cycling, 69
cysteine, 150
cytochrome, 100, 116, 148, 154, 233
cytochrome p450, 100, 148
cytokines, 70, 170, 282, 290
cytoplasm, 131, 133, 136, 137
cytoprotectant, 69, 96
cytotoxicity, 61, 291
Czech Republic, 277

D

damages, iv, xi, 221, 236, 238
damping, 24
danger, 141, 142
deaths, 87
decay, 6, 8, 9, 10, 13, 27, 219
deciliter, 253
defects, 162, 204, 221, 259, 272, 273
deficiency, xi, xii, 103, 163, 166, 176, 187, 193, 204, 258, 260, 270, 273, 289, 293, 294, 295, 296, 297, 298, 299, 300, 301
deformation, 219
degradation, x, 36, 38, 42, 63, 68, 70, 112, 116, 117, 150, 186, 196, 211, 212, 221, 229
Delta, 85
demyelination, 236
dengue, 297, 301
depolarization, 8, 13, 23, 24, 25
deposition, 69, 178
depression, 252
depth, 42, 53
derivatives, 161, 213, 218
destruction, 70
destructive process, 49
desynchronization, 236
detectable, 83, 219, 220, 250, 264
detection, x, 6, 9, 10, 11, 12, 13, 17, 18, 37, 130, 205, 224, 229, 235, 236, 237, 246, 250, 252, 300
detoxification, xi, 96, 154, 169, 199, 221, 228, 229, 257, 258, 289
deviation, 7
DFT, 131, 143, 146
diabetes, 79, 80, 81, 84, 194
diabetic patients, 81, 98, 99, 195, 206
diacylglycerol, 226
dialysis, 169, 170, 171, 181, 182
diarrhea, 199, 269, 285
diastolic blood pressure, 82, 88
dielectric constant, 131, 145
dielectric method, 149
diet, 275
diffraction, 216
diffusion, 40, 42, 49, 113, 114, 132, 146, 231
diffusion process, 146
digestion, 112, 296
dilation, 195
dimethylsulfoxide, 267
diodes, 63, 65
dipole moments, 26
dipoles, 26, 122, 123, 124, 125, 128, 146, 147
direct bilirubin, 69, 113, 139, 140, 217, 220, 221, 224, 230, 231
discomfort, 286
discrimination, 78, 219
diseases, ix, 83, 96, 97, 102, 104, 108, 113, 157, 183, 184, 189, 193, 200, 221, 260, 283, 287, 288
disequilibrium, 74, 75, 199, 268, 297
disorder, iv, 140, 190, 192, 193, 194, 288, 299
displacement, 21, 23, 25, 26, 52, 53
disposition, 100, 198
dissociation, 62, 114, 121, 132, 148

distribution, 53, 58, 71, 77, 104, 122, 225, 246
diuretic, 167
divergence, 271
diversity, 192
DMF, 130
DNA, xii, 29, 61, 97, 196, 223, 229, 236, 259, 260, 261, 262, 294, 296, 300
dogs, 158, 159, 160, 161, 165, 166, 167, 174, 176, 178
donors, 143, 297
dopamine, 167, 179
dosage, 41, 90, 199, 259, 278
double bonds, 131, 212
drainage, 156, 161, 168, 169, 172, 174, 177, 178, 179, 180, 212, 288
drawing, 133
drug design, 142, 147
drug discovery, 151
drug interaction, 198, 208
drug metabolism, 101, 185, 198, 260, 271, 272
drug reactions, xii, 277, 278, 281, 287, 289
drug resistance, 282
drug safety, 198
drug side effects, xi, 198, 258, 270
drug testing, 192
drug toxicity, 199, 200
drug treatment, xi, 258
drug withdrawal, 278, 284, 285, 287
drugs, viii, xi, xii, 73, 111, 133, 140, 160, 185, 198, 199, 257, 258, 259, 270, 277, 278, 279, 280, 282, 283, 284, 287, 290, 291, 293, 295
Dubin-Johnson syndrome, 76, 103, 193, 206, 258
duodenum, 159, 160, 278
dyes, 60, 224
dyslipidemia, 105, 106, 286
dysplasia, 238

E

East Asia, 76, 187, 192
economics, 121
edema, 286
education, xii, 293
electric current, 134
electric field, 26, 134, 139
electricity, 134, 229
electrocatalyst, 231
electrochemistry, 221, 224, 227
electrode surface, 219, 221, 233
electrodes, 134, 135, 219, 221, 222, 227, 229, 231, 232, 233, 236
electrolysis, 227
electrolyte, viii, 134, 135, 153, 154, 155, 163, 164, 167, 170

electrolyte imbalance, viii, 153, 154, 155, 163, 164
electron, x, 15, 22, 24, 25, 33, 116, 123, 131, 145, 149, 151, 211, 212, 215, 216, 218, 219, 220, 221, 227, 229, 231, 232, 233
electron pairs, 131
electron paramagnetic resonance, 149, 216
electrons, 116, 216, 221, 222
electrophoresis, 231
emergency, 2
emission, vii, 1, 7, 8, 9, 10, 11, 12, 13, 17, 18, 19, 21, 22, 23, 24, 25, 28, 30, 31, 34, 41, 42, 43, 44, 46, 47, 48, 49, 50, 51, 53, 59
emitters, 42, 46
emphysema, 98
enamel, 238
enantiomers, 16, 132
encephalopathy, x, 69, 149, 190, 235, 236, 237, 238, 240, 250, 252, 254, 255
encoding, 107, 214, 259, 262, 267
endocrine, 175
endoscopic retrograde cholangiopancreatography, 288
endothelial cells, 70
endothelium, 97
endotoxemia, 165, 168, 178, 179
endotoxins, 178
energy, 7, 9, 12, 14, 15, 17, 18, 19, 21, 22, 23, 24, 25, 26, 27, 28, 29, 39, 40, 41, 42, 44, 47, 49, 51, 53, 60, 112, 113, 115, 121, 122, 125, 126, 128, 131, 132, 136, 137, 141, 143, 145, 146, 151, 212, 229, 252, 255
energy parameters, 42
energy transfer, 17, 18, 19, 23, 24, 26, 28, 29, 39, 53, 60
England, 61
enrollment, 172
enterovirus, 301
entropy, 124
environment, 17, 24, 28, 38, 53, 56, 116, 119, 122, 125, 130, 131, 133, 135, 137, 140, 145, 146, 202
environmental conditions, 186
environmental factors, 90, 186, 260
environmental influences, 131, 271
environments, 117, 119, 123, 140, 192
enzymatic activity, 57, 187
enzyme, xi, xiii, 6, 7, 29, 31, 32, 33, 34, 37, 38, 39, 40, 49, 57, 71, 72, 74, 106, 113, 115, 116, 119, 140, 143, 145, 187, 207, 208, 212, 217, 223, 224, 225, 226, 228, 229, 230, 231, 232, 257, 258, 259, 268, 270, 294, 297
enzyme immobilization, 232

enzymes, 39, 61, 68, 71, 94, 115, 116, 140, 141, 148, 184, 208, 224, 225, 231, 236, 259, 270, 274, 275, 289, 300
eosinophilia, 286
epidemic, 291
epidemiologic, 280
epidemiology, ix, 104, 109, 183, 184, 198, 200
epidermis, 52
epithelial cells, 136, 285
epithelium, 291
EPR, 145, 216, 217, 218, 219
equilibrium, 8, 17, 18, 19, 35, 37, 112, 113, 119, 121, 128, 132, 146, 216, 237
equipment, 4, 41, 42, 50, 55
erythrocyte membranes, 61, 149
erythrocytes, xiii, 2, 68, 95, 112, 184, 187, 294, 295, 297
ESR, 145
estriol, 78
estrogen, 78, 104, 197, 208, 282, 284, 290
ethanol, 123, 138, 284
ethnic groups, xiii, 77, 190, 192, 194, 210, 268, 274, 294, 299
ethnicity, 80, 188
ethylene, 13, 56, 128
ethylene glycol, 13, 128
etiology, x, xi, 235, 236, 277, 278, 279, 280, 281, 287
eukaryotic, 62
eukaryotic cell, 62
Europe, 188
evidence, 3, 25, 28, 32, 41, 48, 62, 63, 68, 69, 70, 94, 107, 146, 150, 155, 166, 188, 196, 198, 199, 200, 204, 239, 271, 273, 283, 290
evoked potential, 236, 241, 242, 250, 251, 254
evolution, 215, 226, 229
exchange transfusion, 2, 4, 240, 241, 243, 250, 251, 253, 254, 297
excision, 174
excitation, 5, 6, 7, 8, 9, 10, 11, 12, 13, 14, 16, 17, 18, 19, 21, 22, 23, 24, 25, 26, 27, 28, 37, 38, 53, 58, 60
exciton, 18, 19, 25, 26, 39, 52, 53, 59, 60
exclusion, 287
excretion, vii, viii, x, xi, 1, 3, 40, 42, 49, 53, 69, 76, 112, 113, 150, 153, 154, 155, 157, 162, 163, 169, 175, 184, 193, 194, 215, 221, 229, 257, 258, 259, 260, 281, 282
exons, 92, 190, 192, 204, 258, 264, 266, 272, 273
experimental condition, 16
exposure, 6, 7, 29, 30, 31, 32, 33, 34, 35, 36, 37, 38, 39, 40, 41, 46, 47, 48, 49, 55, 71, 83, 87, 88, 106, 154, 238, 252, 278

extinction, 5, 7, 20, 34, 35, 36, 39, 52

F

Fabrication, 232
factories, 140
FAD, 148
families, ix, 184
family members, 68
fasting, 190, 192, 269, 275
fat, 2, 53, 112, 154, 163, 289
fat soluble, 163
fatty acids, 52, 53, 148
FDA, 51, 170, 185, 192, 198
feces, 112
ferritin, 70, 99
fetus, 141, 299
fever, 166, 238, 286
fiber, 46, 170, 236
fibrin, 178
fibrinogen, 82, 85, 105
films, 232
filters, 52
filtration, 157, 167, 168, 169, 170, 172, 181
flame, 215
flavonoids, 150, 269, 272, 275
flexibility, 112, 116, 126
fluctuations, 189
fluid, 48, 49, 154, 158, 159, 161, 166, 167, 169, 170, 172, 237, 252, 253
fluorescence, 5, 6, 8, 9, 10, 11, 12, 13, 16, 17, 18, 19, 21, 22, 23, 24, 25, 26, 29, 30, 37, 38, 47, 56, 57, 58, 61
fluorescence decay, 13, 18
fluorimeter, 5
folic acid, 29, 61
food, 260, 270, 271
Food and Drug Administration, 51, 185
force, 145, 167
formaldehyde, 284
formamide, 130, 147
formation, vii, x, 1, 3, 8, 10, 12, 14, 16, 18, 19, 22, 27, 28, 29, 34, 35, 36, 38, 39, 51, 53, 60, 61, 70, 114, 116, 125, 126, 131, 132, 138, 161, 178, 184, 193, 211, 212, 218, 222, 233, 282, 285, 289, 291
formula, 151
fragments, 26, 28, 60, 260
fraud, 125
free energy, 122, 123, 124, 125, 126, 145, 146, 149
freedom, 23
fresh frozen plasma, 170
fructose, 232
FTIR, 228
fuel cell, 229, 232

functional changes, 174
fungi, 227
fungus, 212, 229
fusion, 71, 239

G

gallbladder, 285
gallstones, 280, 281, 285, 290, 291
gel, 261
gene expression, xi, 258, 270, 272, 297
gene promoter, 70, 74, 87, 98, 99, 103, 105, 106, 201, 202, 203, 206, 207, 209, 210, 260, 264, 266, 268, 272, 273, 274
genes, viii, ix, 67, 71, 74, 77, 94, 98, 100, 106, 183, 184, 185, 186, 187, 189, 194, 197, 198, 199, 200, 202, 203, 204, 207, 258, 259, 270, 272, 273, 274, 283, 298
genetic alteration, 204
genetic background, 271
genetic components, ix, 184, 188, 189
genetic defect, 294
genetic disease, 300
genetic disorders, 150, 190, 192
genetic factors, ix, 183, 184, 199, 269, 294
genetic mutations, 190, 193
genetic predisposition, xii, 277, 281, 289
genetic testing, 185, 192, 198
genetics, x, 97, 98, 99, 102, 103, 107, 108, 109, 184, 188, 192, 193, 203
genitals, 40
genome, viii, 67, 72, 73, 74, 75, 82, 88, 89, 93, 97, 99, 107, 189, 200, 201, 285
genomics, 274
genotype, 71, 72, 81, 82, 83, 87, 90, 93, 94, 99, 105, 106, 107, 195, 196, 198, 202, 205, 206, 209, 210, 270, 274, 275, 283, 291, 298
genotyping, 102, 273
geometry, 15, 122, 143, 215
Germany, 4, 5, 7, 43
gestation, 54, 238
gestational age, 41
Giraffe, 45
globus, 238
glomerulus, 155
glucagon, 285
glucocorticoid, xi, 257, 262, 263, 272
glucocorticoid receptor, xi, 257, 272
glucose, xii, 71, 163, 167, 186, 203, 212, 221, 222, 223, 228, 229, 231, 232, 293, 299, 300, 301
glucose oxidase, 222, 229, 231
glutathione, 69, 96, 185, 202, 207, 260, 270, 283, 297, 301
glycine, 112

glycol, 221, 228
glycosylation, 217, 218
gold compound, 284
governments, 140
grades, 205
grants, 95, 289
graphite, 229
growth, 47, 239, 285
growth hormone, 285
guanine, 71
Guinea, 297

H

hair, 224, 239
hair cells, 239
half-life, 56
halogen, vii, 1, 42, 45, 46, 47, 48, 49, 53, 63, 64
haplotypes, 72, 82, 100, 101, 102, 202, 203, 205, 209, 266, 274, 297, 298, 300
Hartree-Fock, 126, 131
hazards, 62
HE, 108, 109
head and neck cancer, 197, 207
healing, 116
health, xii, 293, 294, 295, 296, 299
health education, xii, 293, 295, 296
hearing loss, 238, 239, 251, 253, 254
heart disease, 96, 98, 104, 200
heart failure, 92, 106
heart rate, 176
heart transplantation, 82, 88
heavy metals, 70
height, 121
helium, 7, 32, 33, 34, 35
hematocrit, 167
heme, vii, viii, ix, 67, 68, 70, 71, 96, 97, 98, 99, 100, 105, 106, 112, 113, 115, 116, 148, 183, 184, 186, 196, 201, 202, 206, 207, 211, 258
heme degradation, x, 71, 148, 184, 196
heme oxygenase, viii, ix, 67, 68, 96, 97, 98, 99, 100, 105, 106, 113, 116, 148, 183, 184, 201, 202, 206, 207
heme toxicity, 96
hemodialysis, 93, 170
hemoglobin, vii, 2, 46, 47, 48, 49, 52, 63, 68, 70, 77, 78, 99, 103, 112, 115, 117, 139, 184, 237, 269, 275
hemoglobinopathies, 102
hemolytic anemia, 212
hemorrhage, 167
hepatic coma, 180
hepatic encephalopathy, 170, 171
hepatic failure, 154, 170, 180, 181

hepatic injury, 290
hepatitis, xii, 180, 277, 280, 282, 284, 285, 286, 288, 291
hepatitis a, xii, 180, 277
hepatitis c, 284
hepatocytes, 68, 73, 165, 169, 171, 184, 186, 212, 261, 262, 271, 279, 285, 291
hepatoma, 171, 269, 275
hepatorenal syndrome, 155, 171
hepatotoxicity, 283, 289, 291
herbal medicine, xii, 277
heritability, 71
heterogeneity, 16, 17, 18, 28, 53, 189, 196, 204, 268
heterozygote, 264, 268
high blood pressure, 159
high density lipoprotein, 87
hippocampus, 238
histology, 165
histone, 151
history, 54, 79, 80, 83, 120, 281
HIV, 209
HLA, xii, 277, 283, 285, 290, 291
HM, 207, 254
HO-1, 98, 116, 201, 207
HO-2, 116
homeostasis, 159, 160
homogeneity, 268
homozygote, 195
hormone, 63, 159, 208, 280, 284
hormone levels, 63, 208
hormones, 88, 160
host, 213
human body, 130, 140, 143, 184
human condition, 156, 158
human subjects, 168, 271
hybrid, 123, 182
hydrogen, 14, 38, 59, 69, 114, 115, 120, 121, 125, 126, 127, 129, 130, 132, 142, 148, 216, 219, 220, 222, 224, 228
hydrogen atoms, 126
hydrogen bonds, 14, 59, 115, 129, 130, 132
hydrogen gas, 222
hydrogen peroxide, 38, 69, 224, 228
hydrolysis, 136, 137
hydroperoxides, 33
hydrophilicity, 53, 137
hydrophobicity, 137
hydroxide, 215, 216
hydroxyl, 114, 221, 260
hydroxyl groups, 114
hypercholesterolemia, 81, 83, 87
hyperlipidemia, 81, 83, 86
hypernatremia, 157

hypersensitivity, 286
hypertension, 80, 82, 86, 88, 89, 97, 98, 105, 106, 107, 160, 174, 176, 195, 207
hypertonic saline, 284
hypertriglyceridemia, 83
hyperuricemia, 163
hypokalemia, viii, 153, 156, 161, 162, 163, 176
hypomagnesemia, viii, 153, 163, 177
hyponatremia, viii, 153, 155, 157, 158, 160, 161, 162, 163, 165
hypoparathyroidism, 176
hypophosphatemia, viii, 153, 155, 163
hypoplasia, 238
hypotension, 160, 161, 164, 166, 173, 176
hypothesis, 36, 167, 168, 185
hypovolemia, ix, 154, 158, 159, 160
hypoxemia, 254
hypoxia, 70, 245, 252

I

ibuprofen, 198, 280, 284
ICAM, 70, 185
icterus, 138
ID, 206
identification, ix, 184, 200, 204, 300
identity, 28, 53, 73, 189
idiopathic, 95
idiosyncratic, xii, 277, 278, 279, 281, 287, 291
illumination, 6, 41
imbalances, ix
immune response, 69, 285
immune system, 48
immunity, 169
immunohistochemistry, 159
immunomodulatory, 288
IMO, 7
improvements, 170, 171
impurities, 13, 23
in utero, 63
in vitro, 7, 52, 53, 61, 70, 78, 114, 132, 158, 176, 185, 205, 229, 262, 272
in vivo, 52, 53, 55, 58, 64, 114, 132, 176, 185, 229, 252, 262, 270, 275
incidence, xii, 4, 82, 90, 98, 104, 105, 163, 164, 167, 168, 201, 258, 278, 279, 280, 281, 293, 294, 295, 297
incubator, 30, 46
independence, 281, 282
India, 188, 203
Indians, ix, 74, 102, 183, 187, 203
indirect bilirubin, 69, 139, 217, 220, 221, 224, 237, 254
indirect effect, 189

individual differences, 260
individuals, 72, 79, 80, 83, 89, 90, 93, 104, 107, 176, 185, 186, 187, 189, 192, 194, 195, 197, 198, 199, 200, 264, 294
inducer, 260, 262, 267, 269
induction, 71, 82, 96, 97, 160, 186, 202, 258, 260, 261, 263, 269, 270, 271, 272
infant mortality, 2, 50
infants, x, xii, 2, 3, 4, 7, 39, 40, 41, 54, 55, 62, 63, 96, 102, 176, 187, 235, 236, 238, 239, 240, 241, 243, 244, 245, 246, 250, 251, 252, 253, 254, 255, 293, 294, 295, 297, 299, 300, 302
infarction, 92, 107, 108, 195, 196
infection, 2, 163, 170, 286, 297, 301
inflammation, 69, 80, 85, 88, 206, 280, 282, 285, 286
infliximab, 284
infusion model, 157
ingestion, 173
inheritance, 88
inherited disorder, x, 184
inhibition, xii, 141, 145, 150, 277, 282, 284, 290
inhibitor, 139, 163, 208, 284, 285
initiation, 285, 288
injuries, 154, 281, 290, 291
injury, iv, vii, viii, xi, 70, 153, 154, 165, 171, 177, 178, 236, 238, 277, 278, 279, 280, 281, 282, 283, 284, 285, 286, 287, 288, 289, 290, 291
inner ear, 239, 253
insertion, 74, 125, 161, 187, 190, 192, 203
institutions, xii, 41, 293, 295
insulin, 71, 97, 285
insulin resistance, 71, 97
integration, 124, 125
integrity, x, 54, 179, 235, 236, 237, 245, 250
intellect, 238
intercellular adhesion molecule, 70, 185
interface, 220
interference, 130, 228, 282
intervention, 54, 82, 86, 106, 169, 270, 278
intestinal malabsorption, viii, 153
intestine, 138, 168, 169, 288
intima, 79, 104, 194
intoxication, 2, 170
intracranial aneurysm, 109
intravenously, 158
intron, 206, 268
inversion, 25
iodine, 284
ionization, 129, 130
ions, x, 120, 121, 131, 132, 133, 135, 136, 137, 146, 152, 211, 212, 213, 214, 216, 218, 219
iron, 68, 70, 96, 103, 113, 116, 184, 197, 207

irradiation, 7, 31, 32, 33, 36, 37, 38, 61, 64, 70, 71, 216
ischemia, 166, 167, 169, 177, 179, 252
islands, 295
isolation, 61, 115
isomerization, 3, 20, 29, 37, 56, 58, 59, 61
isomers, 3, 12, 29, 34, 35, 37, 39, 55, 99
isoniazid, 283, 284
isotope, 145, 159
Israel, 206
issues, xii, 244, 294, 297, 299
Italy, 188, 192

J

Japan, 43, 104, 188, 197, 211, 221, 230, 257
Japanese women, 82
jejunum, 275

K

K^+, 136, 137, 146, 161, 162
Kawasaki disease, 87, 88, 106
kidney, viii, 148, 153, 154, 155, 157, 158, 164, 169, 174, 175, 177, 195
kidneys, 139
kinase activity, 97
kinetic studies, 205, 227
kinetics, 6, 8, 9, 10, 13, 23, 31, 37, 39, 40, 49, 58, 100, 145, 151

L

labeling, 192
laboratory tests, 190, 281, 288
lactate dehydrogenase, 6, 29, 40, 57
lactic acid, 7
languages, 140
laparotomy, 164
large intestine, 211
laser radiation, 7, 28, 31, 57, 61
lasers, 7, 31
latency, 240, 241, 242, 245, 246, 247, 249
LDL, 69, 70, 78, 83, 89, 97, 185
lead, 6, 16, 27, 34, 37, 42, 49, 50, 52, 69, 142, 192, 193, 200, 215, 239, 264, 275, 278, 285, 288, 294
leakage, 193
leaks, 212
learning, 238
learning disabilities, 238
LED, vii, 1, 45, 50, 51, 52, 53, 64
lens, 150
lesions, 238, 239, 263, 273
lethargy, 238
leucine, 97
leucocyte, 291

Lewis acids, 131
lifetime, 8, 9, 10, 13, 16, 17, 18, 19, 23, 26, 29, 32, 219
ligand, 70, 96, 133, 194, 216, 218, 226, 227, 261, 266
light, vii, 1, 2, 3, 4, 5, 6, 19, 28, 29, 30, 34, 38, 39, 40, 41, 42, 43, 46, 47, 48, 49, 50, 51, 52, 53, 54, 55, 56, 57, 58, 59, 61, 63, 64, 114, 138, 212, 231
light conditions, 6
light emitting diode, 50, 64
light transmission, 30
light-emitting diodes, vii, 1, 50, 51
lignin, 224
lipid oxidation, x, 184
lipid peroxidation, 69, 195, 224
lipids, 69, 185
lipoproteins, 29, 69, 185, 202
liposomes, 61
liquid chromatography, 231, 296, 300
liquids, 135
liver cells, 114, 169, 182, 193, 237
liver cirrhosis, 158
liver damage, 158, 280, 281, 282, 288
liver disease, 89, 154, 156, 171, 195, 197, 258, 263, 274, 278, 280, 281, 285, 286, 289, 290, 299, 302
liver enzymes, 79, 193, 284
liver failure, 169, 170, 171, 172, 174, 180, 181, 182, 286, 288
liver function tests, 193, 279
liver transplant, 171, 181
liver transplantation, 171, 181
localization, 216
loci, ix, 97, 107, 109, 183
locus, viii, 67, 73, 94, 95, 99, 108, 109, 187, 192, 202, 204, 205, 258, 272
loss of appetite, 286
low temperatures, 60, 141
luciferase, 71, 72, 82, 260, 261, 262, 267, 269, 270
lumen, 162, 165
luminescence, 4, 6, 7, 8, 11, 12, 18, 23, 34, 57, 60, 228
lung disease, 252, 253
Luo, 101, 203, 301
lymphocytosis, 286
lymphoma, ix, 183
lysine, 151

M

macrolide antibiotics, 280, 284
macromolecules, 38, 39
macrophages, 68
magnesium, 163, 176
magnitude, 22, 33, 38, 216, 242, 244

majority, 170, 171, 187, 189, 192, 195, 196, 197, 240, 264, 278, 280, 282, 287
malaise, 286
malignancy, 163
malignant melanoma, 48
mammals, 56, 150, 175, 184, 194
man, 142, 174, 204, 274, 301
management, 54, 62, 169, 173, 255, 271, 288, 289, 291
manganese, 214
mannitol, 167, 177
materials, 181, 224, 294, 295
matrix, 16, 17, 18, 21, 23, 28, 53
matter, iv, 120, 139, 288
MB, 99, 108, 177, 202, 207
measurement, 7, 140, 151, 159, 161, 228, 231
measurements, 5, 6, 119, 121, 125, 128, 129, 136, 138, 227, 229
meat, 207
media, 79, 104, 194, 226, 230
median, 83, 86, 87, 90
medical, xi, xii, 3, 41, 47, 96, 97, 98, 105, 140, 147, 171, 190, 224, 258, 259, 270, 293, 294, 295
medical care, 295
medical science, 105
medication, 83, 278, 279, 280, 283
medicine, 2, 104, 106, 251, 283, 289
melanin, 52
melanoma, 48, 98, 197, 207
mellitus, 81
membranes, viii, 111, 130, 139, 140, 141, 169, 181
menopause, 78
mental retardation, 238
mercury, vii, 1, 30, 44, 46, 48, 49, 53
meta-analysis, 77, 82, 99, 103, 171, 181, 189, 194, 196, 204, 206, 274
metabolic acidosis, viii, 153
metabolic disorder, 236
metabolic disorders, 236
metabolic syndrome, 104, 194
metabolism, vii, viii, 48, 67, 68, 70, 72, 74, 77, 89, 94, 98, 101, 112, 133, 150, 154, 174, 175, 176, 189, 190, 199, 200, 202, 203, 205, 207, 208, 210, 252, 255, 260, 273, 275
metabolites, 68, 154, 282, 285
metabolized, 260
metabolizing, 258, 259, 270, 278
metals, 116
metastasis, 280
meter, 7
metformin, 284
methanol, 149
methodology, 145, 146

methyl group, 130, 131
methyl groups, 130, 131
mice, 104, 166, 177, 178, 272
microcirculation, 165, 178
microemulsion, 218, 226
microenvironments, 16, 24
micronutrients, 201
microorganism, 212, 225
microscopy, 222
microsomes, 150, 203, 208
Middle East, 193
migration, 27
miniature, 232
Ministry of Education, 67
mission, 115
mitochondria, 136
MNDO, 15
MNDO method, 15
model system, 40, 141
modelling, 150, 253
models, 56, 60, 112, 121, 122, 126, 146, 156, 157, 160, 165, 167, 175, 177, 236, 243, 244
modifications, 70, 224
molecular biology, xii, 107, 293, 296
molecular dynamics, 141
molecular mass, 219, 221
molecular medicine, 97
molecular oxygen, 34
molecular structure, 29, 147
molecular weight, 39, 170
molecules, 5, 6, 8, 10, 16, 17, 18, 19, 24, 25, 28, 29, 30, 36, 37, 39, 40, 49, 59, 60, 68, 69, 96, 118, 125, 133, 136, 142, 145, 149, 184, 202, 208, 215, 220
monolayer, 133, 161, 282
morbidity, 79, 166, 173, 198, 273
Morocco, 188
mortality, ix, xii, 83, 87, 88, 106, 109, 154, 164, 165, 166, 167, 168, 170, 173, 180, 194, 197, 201, 273, 293, 295
mortality rate, ix, xii, 154, 164, 165, 166, 293, 295
mortality risk, 106, 201
Moscow, 225
motif, xi, 114, 258, 261, 263
motivation, 119
motor control, 238
Mozambique, 188
MR, 100, 175, 179, 204, 206, 258
mRNA, 104, 260, 262
mucosa, 179
mucous membrane, 138
mucous membranes, 138
mucus, 154
multi-ethnic, 73
multivariate calibration, 226
muscle contraction, 176
mutagenesis, 226, 263
mutant, 204, 205, 218, 219, 263, 270, 274, 300
mutation, xi, xii, 98, 102, 103, 190, 193, 204, 205, 206, 218, 219, 224, 229, 258, 262, 263, 264, 266, 267, 268, 270, 274, 283, 293, 294, 296, 297, 298, 299, 302
mutational analysis, 301
mutations, xii, 187, 189, 190, 192, 193, 204, 215, 218, 227, 260, 261, 264, 265, 266, 268, 273, 274, 275, 277, 283, 297, 299, 300
mycelium, 225
myelosuppression, 199
myocardial infarction, 78, 87, 92, 102, 109, 194, 195, 196, 207
myocyte, 159, 160
myopathy, 89, 107, 198, 199, 202

N

Na^+, viii, 132, 136, 137, 153, 158, 159, 161, 162, 290
NaCl, 162, 229
NAD, 113, 133, 146, 232
NADH, 32, 33, 133, 146
nanotube, 232
naphthalene, xii, 293, 295
National Academy of Sciences, 1, 41, 253
National Health and Nutrition Examination Survey, 79, 104
National Health and Nutrition Examination Survey (NHANES), 79, 104
National Research Council, 41
nausea, 285, 286
Nd, 75
necrosis, 185, 280, 286
negative effects, 48
neonates, x, xii, 55, 56, 62, 63, 73, 99, 100, 186, 203, 235, 237, 238, 240, 241, 242, 243, 244, 245, 246, 247, 248, 249, 250, 251, 252, 253, 254, 293, 294, 295, 297, 298, 299
nephrosis, 164
nephrotoxic drugs, 168, 172
nerve, 239, 240, 253, 254
nervous system, 254
neural function, 236
neural network, 150
neurohormonal, 159, 160
neurons, 237, 239, 242, 254
neuropathy, 238, 239, 241, 254
neurotoxicity, x, 149, 235, 236, 237, 238, 239, 240, 241, 242, 245, 246, 250, 251, 254

neutral, 6, 61, 126, 129, 133, 147, 171, 212, 229
neutropenia, 199, 202, 209
nevus, 63
NHANES, 79, 104, 197
nicotinamide, 32, 69, 133, 146, 184
nicotinic acid, 114, 126, 127, 132, 133, 150
Nigeria, 188
NIR, 214
nitric oxide, ix, 70, 78, 154, 165, 179, 207
nitric oxide synthase, 165, 179, 207
nitrite, 214
nitrogen, 6, 13, 127, 129, 132
NMR, 119, 128, 129, 145, 148, 149
nonsmokers, 79
non-smokers, 80
norepinephrine, 161
North America, 188
Nrf2, xi, 207, 257, 263, 272
NSAIDs, 284
nuclear magnetic resonance, 252
nuclear receptors, xi, 257, 258, 260, 272
nuclei, 236, 238, 239
nucleotide sequence, 214
nucleotides, 72, 133, 146, 263
nucleus, 238, 239
null, xii, 78, 277, 283, 291
nursing, 54
nursing care, 54
nutrition, 103

O

objective tests, 239
obstruction, 154, 155, 157, 158, 159, 160, 161, 163, 167, 173, 178
oculomotor, 238
OH, 115, 121, 216
oil, 218, 226
operations, 2
opisthotonus, 238
optical activity, 26
optical density, 5, 7, 31, 32, 34, 35, 36, 38, 51
optical power density, 46
optimization, 7, 15, 130, 143
organ, 154, 239, 278
organelles, 29
organic solvents, 11, 12, 23
organism, vii, 1, 2, 3, 20
organize, 133
organs, 70, 184
osmium, 232
osmolality, ix, 154
otoacoustic emissions, 239, 254
ototoxicity, 239, 240, 243, 244

overlap, 49, 51, 218, 281
overlay, 124
overproduction, 113, 190
oxidation, x, 3, 29, 69, 78, 97, 99, 184, 185, 186, 202, 211, 213, 218, 219, 220, 226, 230
oxidative damage, 70
oxidative stress, 71, 82, 83, 96, 97, 186, 194, 207, 263
oxygen, 3, 6, 7, 19, 29, 30, 32, 36, 38, 39, 56, 116, 127, 129, 132, 181, 196, 212, 218, 219, 221, 225, 227, 231, 232
oxygen consumption, 181
ozone, 231

P

pain, 285, 286
parallel, 141, 159, 160, 166, 250, 251
parallel simulation, 141
parenchyma, 286
parents, xii, 293, 295
participants, 77, 83, 90, 100, 195, 196
partition, 149
pathogenesis, 169, 179, 285
pathology, 82, 101, 239, 251
pathophysiological, 147, 281, 282
pathophysiology, ix, 154, 156, 157, 163, 165, 166, 173, 252, 282, 288, 289, 291
pathways, 9, 12, 14, 207, 238, 239, 281
PCM, 124, 151
PCR, 296
pedigree, 71
penetrance, 195
penicillin, 284, 285
peptidase, 214
peptide, 159, 160, 164, 175
peptides, 175
perfusion, 166, 167, 176, 181, 270
perinatal, 240, 244, 245, 252, 253
peritonitis, 168
permeability, 29, 151
permission, iv, 140
permittivity, 26
peroxide, 219, 221, 227, 228
pH, 5, 6, 16, 20, 23, 24, 30, 58, 60, 114, 119, 120, 121, 132, 149, 204, 216, 217, 221, 222, 226, 229, 231
pharmaceuticals, 198
pharmacogenetics, 198, 199, 206
pharmacogenomics, 273
pharmacokinetics, 73, 100, 209
pharmacology, 100, 106, 133
pharmacotherapy, 287, 288
phenol, 217

phenolic compounds, 217
phenothiazines, 280
phenotype, 90, 99, 189, 274
phenotypes, x, 184, 273
phenylalanine, 30, 204
phenytoin, 284
Philadelphia, 151, 272, 299
phosphate, xii, 5, 69, 103, 137, 163, 184, 186, 187, 203, 217, 293, 294, 299, 300, 301
phosphatidylcholine, 282
phospholipids, 56, 112, 113, 114, 132
phosphorescence, 58
phosphorylation, 242, 252
photodegradation, 3, 4, 29, 30, 31, 33, 37, 38, 48, 49
photoirradiation, 61
photolysis, 33, 37, 49
photonics, 7, 12
photons, 17, 25, 31, 34, 38, 40, 49
photooxidation, 12, 34, 35, 36, 49, 61, 62
photopolymerization, 37
photosensitivity, 5
photosensitizers, 37
photosynthesis, 223, 229
phototoxicity, 61
phycobilin, 212
phycocyanin, 224
Physiological, 150
physiology, 100, 147, 159
pigmentation, 48, 138, 193
pigs, 70
pioglitazone, 284
plants, xi, 212, 223, 258
plaque, 70, 79, 104, 185
plasma levels, 63, 168, 175
plasma membrane, 114, 132, 136, 148
plasma proteins, 139
plasmapheresis, 170, 181, 190
plasmid, 261, 266, 267, 269
platelet aggregation, 70
platinum, 232
PM, 108, 260, 291
PM3, 15
PMMA, 43
point mutation, xii, 143, 145, 190, 192, 194, 293, 296, 297, 300
polar, 17, 60, 130, 132
polar groups, 132
polarity, 219
polarization, 4, 5, 8, 9, 10, 11, 12, 13, 17, 18, 19, 20, 21, 22, 23, 24, 25, 56, 126, 145
pollutants, 260
polymer, 226, 229
polymerization, 232

polymers, 232
polymethylmethacrylate, 30, 43, 46
polymorphism, viii, ix, xi, xii, 67, 71, 72, 74, 75, 76, 77, 80, 81, 82, 83, 88, 90, 93, 94, 95, 98, 99, 100, 101, 102, 105, 106, 107, 183, 186, 195, 197, 201, 202, 203, 206, 207, 209, 210, 258, 259, 264, 267, 268, 270, 273, 275, 277, 283, 285, 300, 301
polymorphisms, vii, x, 67, 68, 74, 77, 80, 81, 82, 83, 84, 86, 87, 88, 89, 91, 92, 93, 94, 95, 98, 100, 102, 103, 106, 107, 109, 184, 186, 192, 201, 202, 203, 207, 208, 209, 210, 260, 268, 274, 275, 283, 290
polypeptide, 16, 24, 68, 100, 101, 154, 282, 290, 298
polypeptides, 282
polyvinylalcohol, 218
population, 72, 73, 74, 76, 77, 79, 81, 82, 83, 88, 89, 93, 98, 99, 101, 102, 105, 106, 107, 167, 186, 187, 189, 196, 197, 198, 201, 203, 205, 207, 210, 242, 251, 283, 285, 290
population control, 285
porphyrins, 63
portal vein, 161
potassium, 131, 132, 137, 146, 150, 162, 163
potential benefits, 168
precipitation, 285
pregnancy, 193, 290
premature infant, 2, 41, 62, 251
prematurity, 41, 55
preparation, iv, 12, 218
preterm infants, 54, 55, 57, 63, 64, 252, 253
prevention, 54, 68, 70, 104, 172, 271, 278, 289, 300
primary biliary cirrhosis, 171
primary teeth, 238
probability, 18, 26, 27, 28, 33, 53, 187, 295
progesterone, 78, 104
prognosis, ix, 154, 190, 242, 284, 286, 288
pro-inflammatory, 290
proliferation, 70
promoter, ix, xi, xiii, 71, 72, 74, 76, 82, 98, 101, 106, 183, 186, 187, 189, 190, 192, 194, 197, 201, 202, 203, 205, 206, 207, 209, 210, 226, 257, 258, 261, 263, 266, 267, 268, 272, 273, 274, 275, 294, 298, 301
prophylaxis, 166
prostaglandins, 178, 179
protection, xi, 70, 82, 96, 202, 232, 258, 270
protective factors, 102, 198, 207
protein kinase C, 107
protein sequence, 114
protein structure, 22
protein synthesis, 154

proteins, 28, 69, 112, 132, 137, 139, 141, 147, 149, 184, 185, 186, 198, 199, 215, 221, 225, 226, 227, 230
protons, 120, 121, 125, 136, 146, 219, 220, 227
prototype, 64
pruritus, 284, 285, 286, 288
puberty, 190
public health, 200
pulmonary edema, 170
pumps, 136
pure water, 145
purification, 5, 23, 226
purity, 12, 52
P-value, 73, 76
pyrolytic graphite, 222

Q

quality assurance, 295, 299
quality control, 295
quantization, 146
quantum chemistry, 125
quantum yields, 12, 57, 59
quartz, 45, 46, 64
quercetin, 269, 270, 275
quinone, 222

R

race, 294
radiation, vii, 1, 4, 5, 6, 7, 8, 12, 13, 16, 18, 19, 22, 24, 27, 28, 29, 30, 31, 32, 33, 34, 35, 36, 37, 38, 39, 40, 41, 42, 44, 46, 47, 48, 49, 51, 53, 54, 57, 62, 64
Radiation, 30, 40, 41, 51, 53
radical mechanism, 6, 30
radicals, 69
radius, 26, 133, 135, 146
Raman spectra, 216
RE, 99, 109, 253
reactant, 143
reaction mechanism, x, 145, 148, 211, 212, 219, 224
reaction rate, 139
reactions, 3, 6, 12, 29, 30, 35, 48, 49, 53, 61, 70, 97, 116, 151, 185, 222, 225, 227, 229, 278, 279, 291
reactive oxygen, 69
reactivity, 139, 142
reading, 214
reality, 115
receptors, xi, 133, 260, 272
recognition, 241
recommendations, iv, 41, 171, 192
reconstruction, 27
recovery, 171, 243, 285, 288
recycling, 154

red blood cells, vii, 68, 70, 112, 186, 187, 212, 301
regeneration, 180
regression, 77, 196
regression analysis, 77
regression model, 196
rehydration, 166, 172
reintroduction, 168, 169
rejection, 82, 87
relaxation, 12, 16, 17, 18, 57
relaxation process, 18, 57
relaxation rate, 16
relevance, 147, 230
reliability, 141, 287, 295
renal artery stenosis, 160
renal dysfunction, 155, 156, 164, 166, 173, 179
renal failure, 155, 173, 174, 178, 179, 181
renin, 160, 164, 179
renormalization, 124
repair, 196
replication, 108
repressor, 260
repulsion, 132
requirements, 258, 275
researchers, 47, 115, 142, 190, 194, 196
resection, 156
residues, 30, 36, 37, 38, 40, 114, 116, 143, 144, 215, 218
resins, 288
resistance, 95, 135, 188, 206, 290, 291
resolution, 224, 225, 288
response, viii, x, xi, 54, 61, 70, 80, 81, 85, 101, 105, 112, 131, 137, 153, 159, 160, 161, 166, 174, 176, 178, 235, 236, 239, 251, 252, 253, 254, 255, 257, 258, 260, 261, 263, 270, 271, 272, 274
responsiveness, 161
restenosis, 80, 81, 82, 88, 105, 106, 195, 206
restoration, 172
restriction enzyme, 296
retardation, 47, 48, 241
reticulum, 274
RH, 100, 104, 207, 209
riboflavin, 48, 49, 63
rings, 12, 14, 22, 113, 118, 131, 132, 138, 212
risk factors, x, xi, 78, 80, 101, 105, 106, 155, 184, 194, 195, 201, 206, 244, 277, 281, 298
risks, 90, 93, 283
risperidone, 284
rodents, 165
room temperature, 8, 10, 12, 13, 16, 18, 22, 23, 132
rotational mobility, 12, 13, 22, 24
rotations, 118
Russia, 7

S

safety, 170, 172, 279, 285
salt concentration, 154
salts, 112, 139, 149, 154, 165, 166, 168, 169, 173, 176, 177, 178, 179, 285, 289
sampling error, 287
saturation, 7, 42
scaling, 124, 143
schizophrenia, 69
science, viii, 96, 106, 111
scope, ix, 126, 154
secretin, 285
secretion, 63, 112, 138, 155, 163, 205, 282, 285
segregation, 187, 200
selective serotonin reuptake inhibitor, 63
selectivity, 19, 20, 28, 58
senses, 212
sensing, 222, 231
sensitivity, 5, 78, 129, 221, 245, 287
sensitization, 40
sensorineural hearing loss, 239
sensors, 219, 223, 224
sepsis, 164, 166, 172, 177, 245
sequencing, x, 184, 296
serine, 97, 114, 147
serum albumin, 5, 18, 19, 29, 40, 42, 52, 56, 57, 58, 59, 60, 61, 62, 114, 125, 132, 148, 158, 228, 237
serum iron level, 196
services, iv
sex, 104, 196
sex steroid, 104
sham, 157
shape, 17
shock, 161, 175
short supply, 171
showing, 74, 133, 137, 241
sickle cell, 98
side chain, 112, 215, 218
side effects, vii, xi, 1, 29, 39, 40, 46, 48, 49, 50, 53, 198, 199, 258, 270
signal peptide, 274
signal transduction, 133
signaling pathway, 272
signalling, 97
signals, 129, 136, 218, 236, 260
significance level, 74, 76
signs, 13, 205
silica, 222
silver, 284
simulation, vii, viii, 111, 112, 115, 120, 121, 124, 132, 137, 141, 142, 143, 145, 146, 147, 149, 151
Singapore, 299

single-nucleotide polymorphism, 100
skeleton, 24
skin, 2, 42, 48, 52, 116, 138
SLAC, 219
sleep deprivation, 288
small intestine, 64, 112, 211
smoking, 78, 79, 81, 201
smooth muscle, 70
SNP, 71, 72, 73, 74, 75, 76, 77, 82, 88, 89, 90, 92, 93, 95, 107, 108, 186, 187, 196, 199, 285
society, 61
sodium, viii, 32, 114, 128, 131, 136, 137, 146, 150, 153, 154, 155, 156, 158, 159, 160, 162, 166, 168, 169, 174, 175, 282, 284
sol-gel, 222, 229, 232
solid tumors, 198
solubility, 5, 62, 68, 114, 132, 149, 151, 184, 211, 212
solution, 5, 6, 7, 9, 10, 12, 13, 16, 17, 18, 20, 22, 23, 30, 31, 33, 34, 35, 36, 37, 38, 39, 40, 49, 51, 56, 60, 62, 112, 113, 119, 120, 121, 122, 125, 134, 135, 139, 147, 149, 152, 167, 170, 221, 228, 229, 232, 284
solvation, 112, 115, 121, 122, 124, 125, 126, 128, 146, 147
solvent molecules, 124, 129
solvents, 5, 17, 130
South Africa, 188
South America, 188
Southeast Asia, 299
spasticity, 238
specialists, 4
species, 60, 69, 114, 122, 125, 132, 133, 134, 135, 136, 139, 149, 155, 156, 212, 219, 221, 258
Specord M, 5
spectroscopy, 7, 57, 60, 128, 129, 149, 150, 252
speech, 241
sphincter, 285
spin, 149
spleen, 70, 116, 211
squamous cell, 201
squamous cell carcinoma, 201
SS, 98, 100, 101, 201, 203, 206, 207, 300, 301, 302
St. Petersburg, 7
stability, 6, 9, 114, 136, 141, 218, 232
standard deviation, 240
stasis, 285
state, 6, 7, 8, 10, 12, 13, 15, 17, 18, 19, 22, 26, 27, 29, 35, 36, 37, 57, 58, 59, 129, 132, 140, 143, 145, 216, 220, 226, 274, 279
states, 18, 26, 59, 60, 112, 113, 114, 119, 121, 125, 127, 130, 135, 141, 146, 147, 216, 228
statin, 89, 107, 198, 199, 202, 208

steatorrhea, 163
stenosis, 80
stent, 105
steroids, 147
Stevens-Johnson syndrome, 282, 290
stimulation, 70, 71, 72, 161, 162, 236, 242
stimulus, 250
stock, 140
storage, 194
stratification, 89
stress, 116
stretching, 159, 160
stroke, 78, 83, 86, 87, 90, 91, 92, 93, 102, 104, 106, 107, 108, 109, 194
structural characteristics, 142
structural defects, 192
structure, 14, 16, 19, 20, 22, 23, 24, 28, 39, 52, 53, 56, 59, 60, 101, 112, 114, 115, 118, 119, 129, 130, 133, 137, 142, 148, 149, 203, 215, 216, 218, 219, 224, 225, 259, 268, 274
subgroups, 80, 81, 83, 214
substitution, 190, 205, 218, 263, 268, 296, 298, 300
substrate, x, 97, 114, 132, 133, 143, 146, 150, 189, 198, 211, 212, 213, 215, 217, 218, 219, 220, 224, 226, 229, 260, 296
Sudan, 188
sulfate, 260
sulfonamides, 237
sulfur, 216
Sun, 98
supplementation, 176, 270, 289
suppression, 176
surface area, 46
surface component, 165
surface modification, 225
surfactant, 154
surgical intervention, 3
surgical technique, 156
survival, ix, 154, 170, 171, 172, 190, 196, 205
susceptibility, 82, 98, 99, 106, 107, 108, 166, 194, 202, 206, 207, 281, 283, 285, 291
suspensions, 226
Swan-Ganz catheter, 161
swelling, 167
Switzerland, 43, 102, 104, 108
symptoms, 138, 238, 278, 280, 286, 288
synaptic transmission, 239, 250, 254
syndrome, vii, viii, xi, xii, 1, 2, 28, 31, 39, 48, 63, 70, 95, 96, 101, 153, 155, 162, 163, 173, 176, 185, 187, 190, 191, 192, 193, 194, 200, 202, 203, 205, 206, 209, 210, 257, 258, 263, 266, 268, 273, 274, 277, 280, 282, 284, 285, 286, 288, 298, 301
synergistic effect, 298

synthesis, viii, 70, 111, 230, 236
systolic blood pressure, 78, 79, 160

T

tactics, 54
Taiwan, vi, xii, 83, 106, 197, 293, 295, 296, 299, 300
tamoxifen, 284
Tanzania, 188
target, viii, 2, 6, 30, 111, 116, 278
technical assistance, 54
techniques, 114, 119, 120, 135, 141, 169, 170, 171, 237, 245
technologies, 2, 4
technology, vii, x, 1, 13, 19, 57, 184
teeth, 238
telomere, 97, 203
temperature, 6, 9, 10, 12, 13, 18, 22, 46, 57, 126, 133, 141, 146
testing, 6, 29, 99, 106, 151, 198, 202, 236, 241, 250
tetracyclines, 284
tetrad, 238
textbook, 125, 299
Thailand, 153, 188
thalassemia, 103, 204, 298, 301, 302
theoretical approaches, 120
therapeutic agents, 288
therapeutic effects, xi, 258, 270
therapeutic interventions, 2
therapeutics, 97, 177
therapy, ix, 2, 41, 42, 47, 50, 54, 89, 102, 104, 168, 170, 171, 173, 180, 181, 184, 208, 280, 284, 288, 289, 290, 297
thermodynamic parameters, 145
thermodynamic properties, 142
thermodynamics, 61, 141
thermoregulation, 48
threonine, 97
thrombocytopenia, 170
thymine, 71
thyroid, 88
time resolution, 23
tissue, 40, 42, 49, 63, 70, 96, 169, 170, 178, 208, 221, 228, 237, 272
TNF, 70, 185, 282
TNF-alpha, 282
TNF-α, 70, 185
Togo, 105, 206
top-down, 231
total cholesterol, 83
toxic effect, 161, 241, 242, 243, 244, 250
toxic substances, 154
toxic waste, 68, 69

toxicity, viii, xi, 95, 153, 154, 165, 166, 198, 199, 209, 238, 243, 245, 250, 254, 258, 269, 271, 273, 278, 279, 286
toxicology, 101
toxin, 97, 170, 171, 291
Toyota, 59
tracks, 72
traits, 100
transcription, 71, 72, 96, 97, 187, 194, 260, 261, 262, 264, 268
transcription factors, 96, 260
transduction, 133, 239
transfection, 261
transformation, 38, 145, 229
transfusion, 2, 241
transient ischemic attack, 92
translocation, 166, 179
transmembrane region, 115
transmission, 30, 39, 43, 239
transplant, 82, 87, 195
transplantation, 86, 106, 171, 190, 207, 288
transport, vii, viii, 8, 12, 19, 25, 28, 29, 73, 95, 111, 112, 113, 114, 115, 127, 128, 132, 133, 134, 135, 136, 137, 146, 147, 150, 282, 290, 291
transport processes, 137
transportation, 114
treatment, vii, x, xi, 1, 2, 3, 29, 41, 42, 46, 47, 50, 51, 55, 57, 58, 62, 63, 64, 105, 126, 138, 141, 142, 166, 168, 169, 170, 172, 180, 181, 190, 197, 198, 208, 209, 228, 229, 235, 236, 241, 243, 250, 253, 257, 259, 260, 261, 270, 271, 278, 280, 285, 289, 291
trial, xi, 55, 64, 65, 167, 170, 172, 177, 179, 180, 182, 199, 208, 258, 275
triggers, 158
tryptophan, 30, 36, 48, 63
tumor, 70, 185, 201, 207
tumor invasion, 201, 207
tumor necrosis factor, 70
tumors, 171, 285
tunneling, 233
Turkey, 188
Turks, 72
turnover, 99, 116, 219, 220
tyrosine, 30, 97

U

Ukraine, 7
ultrastructure, 178
umbilical cord, 295
unconjugated bilirubin, x, 69, 73, 96, 113, 114, 119, 132, 138, 139, 147, 148, 149, 155, 186, 211, 212, 219, 230, 235, 236, 237, 238, 294

uniform, 53, 172, 259
United Kingdom, 235
United States, 185, 290
unstable compounds, 34
uric acid, ix, 104, 153, 163, 201
urine, ix, 56, 62, 139, 140, 154, 157, 158, 163, 166, 184, 187, 199, 211, 212, 221, 286
USA, 5, 41, 43, 51, 57, 60, 96, 97, 101, 202, 210, 252, 273
UV, 5, 30, 34, 36, 39, 42, 46, 48, 49, 57, 61, 70, 231
UV radiation, 46

V

vacuole, 133
vacuum, 3
valence, 139
vapor, 54
variables, 120, 196, 242, 243, 244, 246, 250
variations, 74, 190, 209, 298
vascular cell adhesion molecule, 70, 185
vascular system, 154
vascular wall, 78
vasculature, 70
vasculitis, 88
vasoactive intestinal peptide, 285
vasoconstriction, 165
vasodilation, 104, 161, 166
vasodilator, 70
vasomotor, 70
VCAM, 70, 185
vector, 5, 26, 262, 267, 270
vegetables, 270
velocity, 134, 135, 142
vesicle, 242, 252
vessels, 165
viscosity, 12, 13, 16, 58
visual system, 241
visualization, 145
vitamin B2, 49
vitamin D, 176
vitamins, 163, 289
vomiting, 286

W

Washington, 253
waste, 70, 113
water, ix, xiii, 5, 13, 33, 37, 63, 64, 69, 112, 113, 114, 115, 117, 118, 119, 120, 123, 128, 132, 133, 138, 148, 154, 155, 157, 158, 159, 160, 161, 166, 167, 171, 175, 185, 211, 212, 215, 216, 218, 219, 220, 226, 231, 232, 237, 294, 297
wavelengths, 5, 6, 10, 16, 18, 21, 23, 25, 26, 27, 29, 31, 34, 36, 40, 46, 49, 52, 53

weight loss, 157, 163
Western Europe, 192
wild type, 123, 166
withdrawal, xii, 277, 278, 288
workers, 112, 117, 122, 124, 125, 128, 140, 145
worldwide, 294

X

X-ray analysis, 225

Y

yeast, 226

Yemen, 188
yield, vii, viii, 1, 8, 9, 10, 12, 13, 15, 16, 18, 22, 27, 28, 34, 35, 36, 37, 39, 51, 53, 56, 58, 59, 111, 219, 260
young adults, 80, 104, 105

Z

Zimbabwe, 188
zinc, 123, 141